Self-Assembly, Pattern Formation and Growth Phenomena in Nano-Systems

NATO Science Series

A Series presenting the results of scientific meetings supported under the NATO Science Programme.

The Series is published by IOS Press, Amsterdam, and Springer (formerly Kluwer Academic Publishers) in conjunction with the NATO Public Diplomacy Division

Sub-Series

I. **Life and Behavioural Sciences**	IOS Press
II. **Mathematics, Physics and Chemistry**	Springer (formerly Kluwer Academic Publishers)
III. **Computer and Systems Science**	IOS Press
IV. **Earth and Environmental Sciences**	Springer (formerly Kluwer Academic Publishers)

The NATO Science Series continues the series of books published formerly as the NATO ASI Series.

The NATO Science Programme offers support for collaboration in civil science between scientists of countries of the Euro-Atlantic Partnership Council. The types of scientific meeting generally supported are "Advanced Study Institutes" and "Advanced Research Workshops", and the NATO Science Series collects together the results of these meetings. The meetings are co-organized by scientists from NATO countries and scientists from NATO's Partner countries – countries of the CIS and Central and Eastern Europe.

Advanced Study Institutes are high-level tutorial courses offering in-depth study of latest advances in a field.
Advanced Research Workshops are expert meetings aimed at critical assessment of a field, and identification of directions for future action.

As a consequence of the restructuring of the NATO Science Programme in 1999, the NATO Science Series was re-organized to the four sub-series noted above. Please consult the following web sites for information on previous volumes published in the Series.

http://www.nato.int/science
http://www.springeronline.com
http://www.iospress.nl

Series II: Mathematics, Physics and Chemistry – Vol. 216

Self-Assembly, Pattern Formation and Growth Phenomena in Nano-Systems

edited by

Alexander A. Golovin

Northwestern University,
Evanston, IL, U.S.A.

and

Alexander A. Nepomnyashchy

Israel Institute of Technology,
Haifa, Israel

 Springer

Published in cooperation with NATO Public Diplomacy Division

Proceedings of the NATO Advanced Study Institute on
Self-Assembly, Pattern Formation and Growth
Phenomena in Nano-Systems
St. Etienne de Tinee, France
August 28–September 11, 2004

A C.I.P. Catalogue record for this book is available from the Library of Congress.

ISBN-10 1-4020-4353-8 (PB)
ISBN-13 978-1-4020-4353-6 (PB)
ISBN-10 1-4020-4354-6 (HB)
ISBN-13 978-1-4020-4354-3 (HB)
ISBN-10 1-4020-4355-4 (e-book)
ISBN-13 978-1-4020-4355-0 (e-book)

Published by Springer,
P.O. Box 17, 3300 AA Dordrecht, The Netherlands.

www.springer.com

Printed on acid-free paper

Contents

Contents

ix

Contributing Authors

Dr. Igor S. Aranson, Argonne National Laboratory, 9700 South Cass Avenue, Argonne, Illinois 60439, USA

Prof. Alvin Bayliss, Department of Engineering Sciences and Applied Mathematics, Northwestern University, 2145 Sheridan Rd, Evanston, IL 60208, USA

Prof. Agnes Buka, Research Institute for Solid State Physics and Optics of the Hungarian Academy of Sciences, Konkoly-Thege Miklos u. 29-33, H-1121 Budapest, Hungary

Prof. Stephen H. Davis, Department of Engineering Sciences and Applied Mathematics, Northwestern University, 2145 Sheridan Rd, Evanston, IL 60208, USA

Dr. Gabor Demeter, Research Institute for Particle and Nuclear Physics of the Hungarian Academy of Sciences, Konkoly-Thege Miklos ut 29-33, H-1121 Budapest, Hungary

Prof. Nandor Eber, Research Institute for Solid State Physics and Optics of the Hungarian Academy of Sciences, Konkoly-Thege Miklos u. 29-33, H-1121 Budapest, Hungary

Prof. Maxim D. Frank-Kamenetskii, Department of Biomedical Engineering, Boston University, 36 Cummington St., Boston, MA 02215, USA

Prof. Alexander A. Golovin, Department of Engineering Sciences and Applied Mathematics, Northwestern University, 2145 Sheridan Rd, Evanston, IL 60208, USA

Prof. Lorenz Kramer (deceased).

Dr. Dmitry Krimer, Physikalisches Institut der Universitaet Bayreuth, D-95440 Bayreuth, Germany

Prof. Bernard J. Matkowsky, Department of Engineering Sciences and Applied Mathematics, Northwestern University, 2145 Sheridan Rd, Evanston, IL 60208, USA

Prof. Alexander A. Nepomnyashchy, Department of Mathematics, Technion – Israel Institute of Technology, Haifa 32000, Israel

Prof. Werner Pesch, Institute of Physics, University of Bayreuth, D-95440 Bayreuth, Germany

Prof. Leonid M. Pismen, Department of Chemical Engineering, Technion – Israel Institute of Technology, Haifa 32000, Israel

Prof. Yves Pomeau, Laboratoire de Physique Statistique de l'Ecole normale superieure, 24 Rue Lhomond, 75231 Paris Cedex 05, France

Dr. Lev S. Tsimring, Institute for Nonlinear Science, University of California, San Diego, La Jolla, CA 92093-0402, USA

Prof. Vladimir A. Volpert, Department of Engineering Sciences and Applied Mathematics, Northwestern University, 2145 Sheridan Rd, Evanston, IL 60208, USA

Prof. Peter W. Voorhees, Department of Materials Science and Engineering, Northwestern University, 2145 Sheridan Rd, Evanston, IL 60208, USA

Foreword

Lorenz Kramer, who was the main driving force behind the PHYSBIO program, died suddenly on 5 April 2005. This was a shock to his numerous friends who were not aware that this strong, vigorous and buoyant man struggled for many years with a deeply rooted decease which all of a sudden went out of control.

Lorenz has always been an innovator working with enthusiasm and persuasion on the leading edge of scientific exploration. The three main subjects of his work: superconductivity from 1968 to 1981; pattern formation out of equilibrium starting from 1982 ; biophysics from 1975 to 1980 and again from the start of this century – were intertwined and supplied each other with ideas and techniques. One example will suffice here: a direct interpretation of superflow solutions as saddle points in transitions between different patterns – thereby connecting two opposites: conservative and gradient systems. He was one of the principal players in the field of nonlinear science during its acme in 1990s, and one who directed this field, when it came of age, to new applications, in particular, in physics of liquid crystals and biophysics. Some of the most brilliant nonlinear scientists, both theorists and experimentalists, now in their forties, are his former students and junior colleagues.

Lorenz's e-mail signature once read: "Basic schedule: 0-24 - but variety is the spice of life". And he didn't lack variety: running alone to the highest peak in the Pyrenees after a busy week of lectures, testing the sturdy design of his Mac laptop in a backpack while biking to his office. It is thanks to him that the PHYSBIO workshops were invariably carried out at high altitude as well as on the highest scientific level. Variety was his birthmark: his father German, mother Italian, both biologists; his children branched off to industry, music and architecture; and physics, music, poetry and mountaineering merged in his personality. There will be no one like him.

Leonid M. Pismen
Technion, Israel

Igor S. Aranson
Argonne National Laboratory, USA

Preface

Non-linear dynamics and pattern formation in non-equilibrium system have been attracting a great deal of attention for several decades due to their tremendous importance in many physical, chemical and biological processes. However, only recently was it realized that similar phenomena play a crucial role in the vast majority of processes that occur on nanoscales. While on the macro- and micro-scales one has the advantage of controlling the processes by special instruments and devices, on nanoscales, such instruments are absent, or their use is prohibitively expensive. Therefore, spontaneous pattern formation, self-organization and self-assembly promise a unique route to the control of these processes. The investigation of self-organization on nanoscales requires substantial revision of the available scientific knowledge in pattern formation and nonlinear dynamics due to essentially new mechanisms and phenomena, that can be ignored in macro- and microworld, but play a decisive role in the world where typical distances are measured in nanometers. The understanding of the basic physical principles and mechanisms of self-assembly and pattern formation on nanoscales can lead to a real breakthrough in nanotechnology and to the creation of a new generation of electronic devices, sensors, detectors, as well as "labs-on-a-chip".

The need for intensive investigation of basic mechanisms of self-assembly and self-organization on nanoscales, as well as the need to draw the attention of the broad scientific research community specializing in nonlinear dynamics and pattern formation in nonequilibrium systems to the fascinating area of self-assembly and self-organization on nanoscales inspired the organization of the NATO Advanced Study Institute "Self-Assembly, Pattern Formation and Growth Phenomena in Nano-Systems" that took place in St. Etienne de Tinee in France, August 28 - September 11, 2004. Fifteen lecturers from France, Germany, Hungary, Israel and the USA gave series of lectures to an audience of graduate students and postdocs from Belgium, France, Germany, Israel, Italy, Romania, Russia, Spain, the USA and Uzbekistan. The lectures were devoted to various aspects of self-assembly, pattern formation and nonlinear dynamics in nano-scale physical, chemical and biological systems, or systems in which nanoscale processes play a crucial role and determine the macroscopic behavior.

The present book consists of ten articles containing lecture notes written by the lecturers of the NATO ASI. The first article discusses general aspects of pattern formation and universal features of self-organization in non-equilibrium systems. The next two articles are devoted to pattern formation and nonlinear phenomena in liquid crystals – a most remarkable system in which nano-scale anisotropic structure determines quite unusual and complex macroscopic behavior. The fourth article describes the self-assembly of quantum dots – spatially-regular nano-scale structures – from thin semiconductor films. These structures have been attracting a great deal of attention as a promising route to creating a new generation of electronic devices. The fifth and sixth articles discuss a remarkable example of the failure of a traditional approach to an "every day" macroscopic hydrodynamic phenomenon – a moving contact line. It is shown that it is only by introducing new, mesoscopic physics based on the liquid structure at nano-scales, that this phenomenon can be explained and understood. The seventh and eighth articles are devoted to self-organization phenomena in systems where chemical processes that occur at nano-scales lead, due to nonlinear coupling with thermal and diffusion processes, to macroscopic non-stationary structures which, in turn, as a result of instabilities, produce microscopic texture in initially homogeneous media. Namely, these articles discuss the propagation and instability of combustion fronts in self-propagating high-temperature synthesis of solid materials, and the propagation and instabilities of polymerization fronts in frontal polymerization processes. The last two articles deal with micro- and nano-scale self-organization phenomena in biological systems. The ninth article considers the recently discovered, very interesting phenomenon of self-organization of biological micro-tubules and motors. Finally, it would not be an exaggeration to say that the last, tenth article, is devoted to the most remarkable and the most important example of nano-scale self-assembly – the self-organization and behavior of DNA molecules. This article presents a comprehensive, contemporary review of the physics of DNA.

To summarize, this book attempts to give examples of self-organization phenomena on micro- and nano-scale as well as examples of the interplay between phenomena on nano- and macro-scales leading to complex behavior in various physical, chemical and biological systems. It is not accidental therefore that this NATO ASI was organized in conjunction with the European School PHYSBIO-04. Moreover, it was mainly due to the inspiration of the organizers of PHYSBIO-04 – Prof. Agnes Buka, Prof. Pierre Coullet, Prof. Lorenz Kramer and Prof. Yves Pomeau – that the organization of the NATO ASI became possible. We are very grateful to them for their enthusiasm and support. Tragically, one of the organizers of PHYSBIO-04, Professor Lorenz Kramer, who was one of the world leading experts in pattern formation and nonlinear phenomena in non-equilibrium systems, suddenly passed away in April 2005, while this book was in preparation. This was a great loss to all of us as well as many others. This book is dedicated to his memory.

ALEXANDER A. GOLOVIN

ALEXANDER A. NEPOMNYASHCHY

Acknowledgments

It took the efforts of many people to make the NATO ASI a very useful, successful, and pleasant event. It would have been impossible without the tremendous organizational work of Dr. Jean-Luc Beaumont of the Nonlinear Institute in Nice, France. It would also have been impossible without the encouragement and help of Dr. Igor Aronson of the Argonne National Laboratory in the USA. We are very grateful to the lovely people of St Etienne de Tinee, a small, beautiful mountain village near Nice in France, and to the helpful efforts of St. Etienne's major, Mr. Pierre Brun. Their hospitality created a warm atmosphere that made the stay of the NATO ASI participants in St Etienne de Tinee most pleasant and made them want to return to this wonderful place again. Also, our special thanks go to Dr. Denis Melnikov and Ms Jessica Conway for their invaluable help in translation between French, English and Russian; this was a bridge that helped connect together people from so many different countries. One of the editors and co-director of the NATO ASI (A.A.G.) is also very grateful to his son, Peter Golovin, for constant help in many organizational matters. Finally, we would like to thank Prof. Bernard Matkowsky and Prof. Vladimir Volpert of Northwestern University as well as the graduate students: Jessica Conway, Lael Fisher, Margo Levine, Christine Sample, Gogi Singh and Liam Stanton, for enormous help with the preparation of this book. Without their help this book would not have appeared.

This NATO Advanced Study Institute was financially supported by the NATO ASI grant #980684. This support is gratefully acknowledged. Travel of some of the student participants from the US to the NATO ASI was partially supported by the US National Science Foundation.

Lorenz Kramer lecturing at a conference "Pattern Formation at the Turn of the Millenium", organized in La Foux d'Allos, France, in June 2002.

Lorenz Kramer at the summit of Mount Aneto (3400m) in the Pyrenees, Spain, during the European School PHYSBIO-2003.

GENERAL ASPECTS OF PATTERN FORMATION

Alexander A. Nepomnyashchy[a] and Alexander A. Golovin[b]

[a] *Department of Mathematics*
Technion – Israel Institute of Technology
32000 Haifa, Israel;

nepom@math.technion.ac.il

[b] *Department of Engineering Sciences and Applied Mathematics*
Northwestern University
Evanston, IL 60208, USA;

a-golovin@northwestern.edu

Abstract Pattern formation is a widespread phenomenon observed in different physical, chemical and biological systems on various spatial scales, including the nanometer scale. In this chapter discussed are the universal features of pattern formation: pattern selection, modulational instabilities, structure and dynamics of domain walls, fronts and defects, as well as non-potential effects and wavy patterns. Principal mathematical models used for the description of patterns (Swift-Hohenberg equation, Newell-Whitehead-Segel equation, Cross-Newell equation, complex Ginzburg-Landau equation) are introduced and some asymptotic methods of their analysis are presented.

Keywords: pattern formation, block copolymers, thermal convection, Swift-Hohenberg equation, pattern selection, Newell-Whitehead-Segel equation, modulational instabilities, dislocations, domain walls, Cross-Newell equation, disclinations, complex Ginzburg-Landau equation, spiral waves

1. Introduction

The spontaneous development of spatial or spatio-temporal nonuniformities under homogeneous external conditions is a characteristic feature of nonequilibrium systems. This phenomenon is called *pattern formation* [1]-[3].

The most well-known example of pattern formation is Rayleigh-Benard convection which appears when a fluid layer is uniformly heated from below [4], [5]. When the heating is sufficiently intensive, convective motion of the fluid is developed spontaneously: the hot fluid moves upward, and the cold fluid moves downward. It is remarkable that near the convection onset the regions of upward flow and downward flow form an ordered pattern. There are

A. A. Golovin and A. A. Nepomnyashchy (eds.),
Self-Assembly, Pattern Formation and Growth Phenomena in Nano-Systems, 1–54.
© 2006 *Springer. Printed in the Netherlands.*

two kinds of patterns that are observed especially often. The first one is the *roll pattern* (or stripe pattern) in which the fluid streamlines form cylinders. These cylinders may be bent, and they may form spirals or target-like patterns [6], [7]. Another typical pattern is the hexagonal one in which the liquid flow is divided into honeycomb cells. For some fluids, the motion is downward in the center of each cell and upward on the border between the cells; for other fluids, the motion is in the opposite direction [8].

The same patterns, stripes and hexagons, appear in completely different physical systems and on different spatial scales. For instance, stripe patterns are observed in human fingerprints, on zebra's skin and in the visual cortex [9]. Hexagonal patterns result from the propagation of laser beams through a nonlinear medium [10] and in systems with chemically reacting and diffusing species [11]. In many systems the typical scale of periodic spatial structures is very small, from micro- to nanometers. Examples include hexagonal Abrikosov vortex lattices in superconductors [12], magnetic stripe phases in ferromagnetic garnet films [13], spatially-periodic phases of diblock-copolymers [14], spatially-regular surface structures in epitaxial solid films [15], hexagonal arrays of nano-pores in aluminum oxide produced by anodization [16], etc. Self-assembly of spatially regular nano-scale structures is especially important in several areas of nano-technology [17, 18].

The development of patterns is not necessarily a manifestation of a non-equilibrium process. A spatially non-uniform state can correspond to the minimum of the free-energy functional of a system in thermodynamic equilibrium, as Abrikosov vortex lattices, stripe ferromagnetic phases and periodic diblock-copolymer phases mentioned above. In the latter, a linear chain molecule of a diblock-copolymer consists of two blocks, say, A and B. Above the critical temperature T_c, there is a mixture of both types of blocks. Below T_c, the copolymer melt undergoes phase separation that leads to the formation of A-rich and B-rich microdomains. In the bulk, these microdomains typically have the shape of lamellae, hexagonally ordered cylinders or body-centered cubic (bcc) ordered spheres. On the surface, one again observes stripes or hexagonally ordered spots.

We will explain why the above-mentioned two kinds of patterns are so widespread, and the conditions of their appearance will be formulated. It should be noted, however, that other kinds of patterns, e.g. square patterns [19] and even quasiperiodic patterns [20] can be also observed. Moreover, there exist non-stationary, wavy, patterns in the form of traveling [21] and standing waves [22], rolls with alternating directions [23], traveling squares [24] etc. We will formulate principles of pattern selection that can be applied to any system.

2. Basic models for domain coarsening and pattern formation

Patterns usually appear due to the instability of a uniform state. However, such an instability does not necessarily lead to pattern formation. Let us consider, e.g., phase separation of a van-der-Waals fluid near the critical point T_c. For $T > T_c$, there exists only one phase, while for $T < T_c$, there exist two stable phases, corresponding to gas and liquid, and an unstable phase whose density is intermediate between those of the gas and the liquid. When an initially uniform fluid is cooled below T_c, the unstable phase is destroyed, and in the beginning one observes a mixture of stable-phase domains, i.e. liquid droplets and gas bubbles, which can be considered as a disordered pattern. However, the domain size of each phase grows with time (this phenomenon is called Ostwald ripening or coarsening). Finally, one observes a full separation of phases: a liquid layer is formed in the bottom part of the cavity, and a gas layer at the top. Thus, the instability of a certain uniform state is not sufficient for getting stable patterns. Below we formulate some mathematical models that describe both phenomena, domain coarsening and pattern formation.

Phase separation in binary alloys: Cahn-Hilliard equation

To analyse the phenomenon of domain size growth in a quantitative way, let us consider a simpler physical system, a metallic alloy. There are two kinds of atoms, A and B, with volume fractions ϕ_A and ϕ_B, respectively. For the sake of simplicity, assume that the averaged volume fractions $\langle\phi_A\rangle$ and $\langle\phi_B\rangle$ are equal. There exists a temperature T_c such that for $T > T_c$ the fractions are mixed, i.e. *the order parameter* $\phi = \phi_A - \phi_B$ vanishes anywhere, while for $T < T_c$ they are separated, i.e there exist two thermodynamically stable phases, one with $\phi > 0$ ("A-rich phase") and the other with $\phi < 0$ ("B-rich phase"). A mathematical model of this phenomenon has been suggested by Cahn and Hilliard [25]. From the point of view of thermodynamics, phase separation can be described by means of the *Ginzburg-Landau free energy functional*

$$F\{\phi(\mathbf{r}),\, T\} = \int d\mathbf{r} \left[\frac{1}{2}(\nabla\phi)^2 + W(\phi)\right], \tag{1}$$

where

$$W(\phi) = -\frac{1}{2}a(T)\phi^2 + \frac{1}{4}\phi^4. \tag{2}$$

Above the critical temperature $(T > T_c)$, $a(T) < 0$, and the function $W(\phi)$ has a unique minimum at $\phi = 0$. Below the critical temperature $(T < T_c)$, $a(T) > 0$, so that the function $W(\phi)$ has the maximum at $\phi = 0$, which corresponds to an unstable phase, and two minima $\phi = \pm\phi_e(T)$, $\phi_e(T) = \sqrt{a(T)}$, which correspond to stable, A-rich and B-rich, phases; see Fig.1.

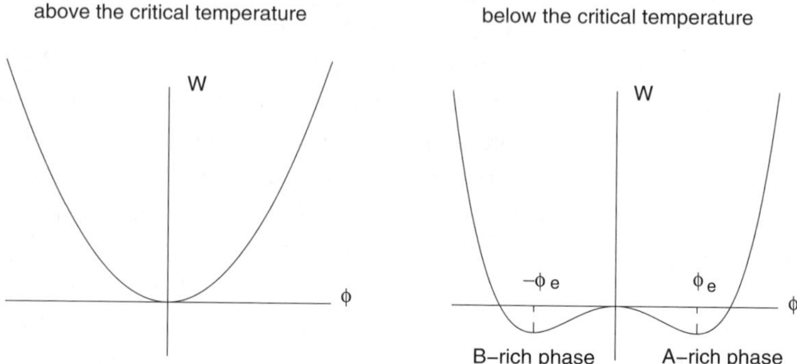

Figure 1.   Free energy density of a binary system as a function of the order parameter above and below the critical temperature.

Because of the conservation of the total number of atoms of each phase, the equation that describes the evolution of the order parameter $\phi(\mathbf{r}, t)$, has the form of the conservation law,

$$\frac{\partial \phi}{\partial t} + \nabla \cdot \mathbf{j} = 0, \tag{3}$$

where $\mathbf{j}$ is the flux of the order parameter. The Cahn-Hilliard model is based on the assumption

$$\mathbf{j} = -M(\phi)\nabla \frac{\delta F}{\delta \phi}, \tag{4}$$

where $M(\phi)$ is a positive function, and

$$\frac{\delta F}{\delta \phi} = -\nabla^2 \phi - a(T)\phi + \phi^3 \tag{5}$$

is the variational derivative of the functional (1). For the sake of simplicity, let us take $M(\phi) = 1$, $a = 1$; then the kinetics of the phase separation is described by the Cahn-Hilliard equation

$$\frac{\partial \phi}{\partial t} = \nabla^2(-\phi + \phi^3 - \nabla^2 \phi). \tag{6}$$

Consider first the one-dimensional version of equation (6),

$$\phi_t = (-\phi + \phi^3 - \phi_{xx})_{xx} \tag{7}$$

in an infinite domain $-\infty < x < \infty$. It is assumed that ϕ is bounded, and the mean value $\langle \phi \rangle = 0$.

Equation (7) has a set of solutions that do not depend on time. These solutions satisfy the ordinary differential equation (ODE)

$$(-\phi + \phi^3 - \phi_{xx})_{xx} = 0. \tag{8}$$

Integrating (8) twice, we find

$$-\phi + \phi^3 - \phi_{xx} = Ax + B,$$

where A and B are constants. The conditions of boundness and vanishing mean value prescribe the choice $A = B = 0$. Note that the equation

$$-\phi + \phi^3 - \phi_{xx} = 0$$

is equivalent to

$$\frac{\delta F}{\delta \phi} = 0,$$

so all the solutions of that equation correspond to extrema (but not necessarily minima) of the free energy functional F. One more integration leads to the relation

$$\frac{\phi_x^2}{2} + \frac{\phi^2}{2} - \frac{\phi^4}{4} = E, \tag{9}$$

where E is a constant. Equation (9) is formally equivalent to the energy conservation law of a particle with unit mass in the potential $U(\phi) = \phi^2/2 - \phi^4/4$. The phase portrait corresponding to such a particle is shown in Fig.2.

The bounded solutions can be separated into three groups:

1. *Constant solutions.* Solutions $\phi = 0$ $(E = 0)$ and $\phi = \pm 1$ $(E = 1/4)$ correspond to uniform phases, unstable and stable, respectively.

2. *Periodic solutions.* For any $0 < E < 1/4$ there exist periodic solutions which correspond to a layered pattern with the alternating sign of ϕ. The solutions can be described analytically by means of Jacobi elliptic sine-functions.

3. *Separatrices.* For $E = 1/4$, in addition to the solutions $\phi = \pm 1$, which correspond to saddle points in the phase plane (ϕ, ϕ_x), there exist two families of solutions

$$\phi(x) = \pm \tanh \frac{x - x_0}{\sqrt{2}} \tag{10}$$

which correspond to *domain walls* between the phases $\phi = \pm 1$.

In the course of evolution governed by the Cahn-Hilliard equation (7), the system tends to an *attracting* stationary state which corresponds to a *minimum* of the Ginzburg-Landau functional. In order to distinguish between this stationary state and all other stationary solutions, it is necessary to investigate the stability of all the solutions listed in the previous paragraph.

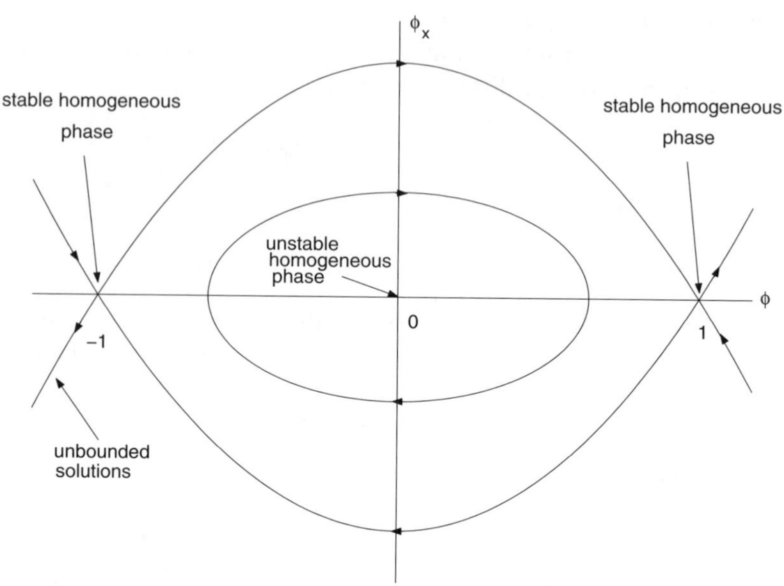

Figure 2.   Phase portrait of a system described by eq.(9).

It is easy to check the linear stability of solutions corresponding to uniform phases. Linearizing equation (7) around the solution $\phi = 0$, i.e. taking $\phi = \tilde{\phi}$, $|\tilde{\phi}| \ll 1$, we get

$$\tilde{\phi}_t = -\tilde{\phi}_{xx} - \tilde{\phi}_{xxxx}.$$

For the normal modes,

$$\tilde{\phi} = \exp[ikx + \sigma(k)t],$$

we find

$$\sigma(k) = k^2 - k^4. \tag{11}$$

This dispersion relation is shown in Fig.3.

Thus, the uniform phase $\phi = 0$ is unstable ($\sigma(k) > 0$) with respect to sufficiently long-wave disturbances, $|k| < 1$. The disturbance with $k_m = 1/\sqrt{2}$ has the maximum growth rate, $\sigma_m = \sigma(k_m) = 1/4$.

Linearizing equation (7) around the solutions $\phi = \pm 1$, i.e. taking $\phi = \pm 1 + \tilde{\phi}$, we get

$$\tilde{\phi}_t = 2\tilde{\phi}_{xx} - \tilde{\phi}_{xxxx}$$

and hence $\sigma(k) = -2k^2 - k^4 < 0$ for any $k \neq 0$ (the disturbance with $k = 0$ violates the condition $\langle \tilde{\phi} \rangle = 0$). Hence, the uniform phases are linearly stable. However, the mean value of ϕ is different from 0 for those phases. Therefore,

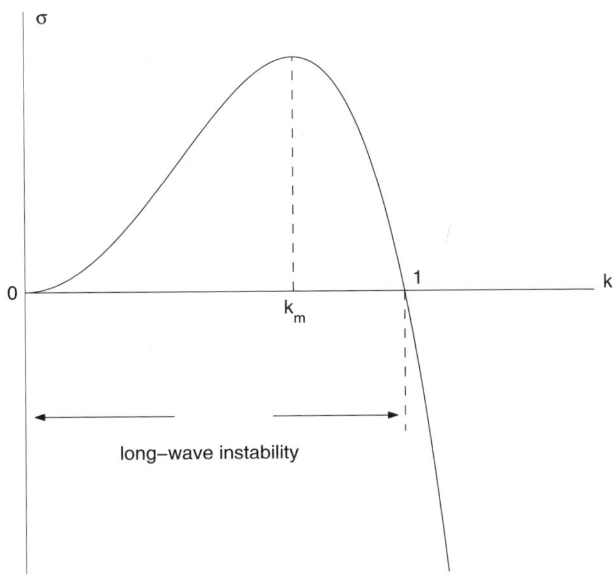

Figure 3. Dispersion relation (11).

the decomposition of the unstable phase $\phi = 0$ cannot lead to the development of each of the uniform states in the whole space.

Similarly, it can be shown that the domain wall solutions are (neutrally) stable, while all the spatially periodic solutions are unstable [26]. The unstable mode corresponds to a collective motion of pairs of domain walls towards each other [27]. This motion leads finally to the collision and annihilation of two domain walls, which leads to the doubling of the structure period, see Fig.4. Kawahara and Ohta [27] have also derived the equations of motion for arbitrary (not necessarily periodic) systems of interacting domain walls that are valid when the distances between the walls are sufficiently large (see also [28]). The time evolution of the structure leads eventually to a state with one domain wall, i.e. to complete phase separation. Note that the rate of the domain coarsening is very slow: for large distances between domain walls, their velocities are exponentially small, and the average distance $L(t)$ between adjacent domain walls grows as $\ln t$.

For two- and three-dimensional Cahn-Hilliard equations, there are no stable spatially periodic solutions as well, and the development of stable spatially nonuniform patterns is impossible. The evolution leads to the complete separation of phases. The growth law of the domain size L is $t^{1/3}$ [29], [30]. This

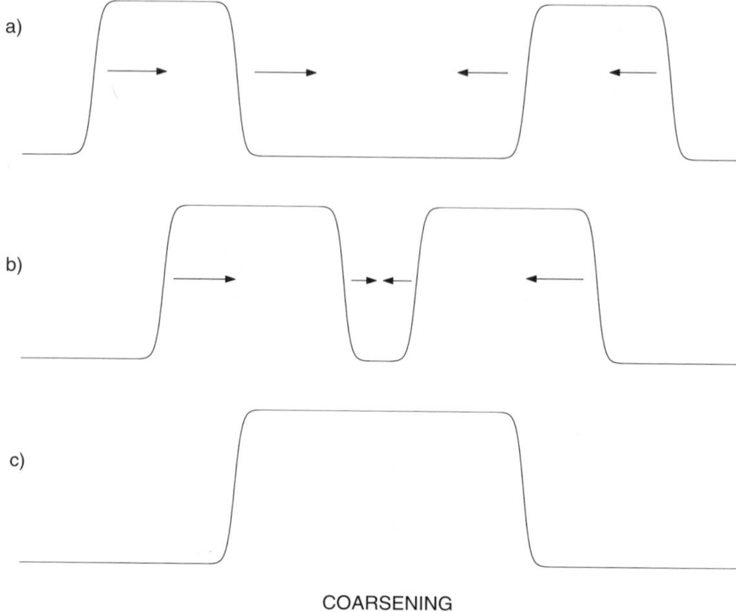

Figure 4.   Motion of domain walls during the coarsening process described by the Cahn-Hilliard equation.

power law is a manifestation of the *self-similarity* of the coarsening process (for more details, see [31], [32]).

Phase separation in diblock copolymer melts: modified Cahn-Hilliard equation

Let us apply the idea of the Cahn-Hilliard approach to a diblock copolymer, where ϕ_A and ϕ_B are now the reduced local densities of *monomers* A and B which are chemically bonded in the diblock-copolymer linear chain molecule. As before, we shall assume that $\langle \phi_A \rangle = \langle \phi_B \rangle$, and use $\phi = \phi_A - \phi_B$ ($\langle \phi \rangle = 0$) as the order parameter. It has been shown [33]-[35] that the long-range interaction of monomers in a copolymer chain can be described by an additional nonlocal term in the Ginzburg-Landau free energy functional:

$$F\{\phi\} = \int d\mathbf{r} \left[\frac{1}{2}(\nabla \phi)^2 - \frac{\phi^2}{2} + \frac{\phi^4}{4} \right] + \frac{\Gamma}{2} \int d\mathbf{r} d\mathbf{r}' \phi(\mathbf{r}) G(\mathbf{r}, \mathbf{r}') \phi(\mathbf{r}'), \quad (12)$$

where Γ is a constant and G is the Green's function for the Laplace equation,

$$\nabla^2 G(\mathbf{r}, \mathbf{r}') = -\delta(\mathbf{r} - \mathbf{r}'), \quad (13)$$

with appropriate boundary conditions. Substituting (12), (13) into (3), (4), one obtains the following modification of the Cahn-Hilliard equation:

$$\frac{\partial \phi}{\partial t} = \nabla^2(-\phi + \phi^3 - \nabla^2\phi) - \Gamma\phi, \quad \langle\phi\rangle = 0. \qquad (14)$$

For $\Gamma = 0$, the standard Cahn-Hilliard equation is recovered.

If $\Gamma \neq 0$, the only stationary uniform solution is $\phi = 0$. The stability of this solution is determined by the linearized equation for the disturbance $\tilde{\phi}$,

$$\tilde{\phi}_t = -\nabla^2\tilde{\phi} - \nabla^4\tilde{\phi} - \Gamma\tilde{\phi}.$$

For the normal mode

$$\tilde{\phi} = \exp[i\mathbf{k} \cdot \mathbf{r} + \sigma(k)t],$$

one obtains

$$\sigma(k) = k^2 - k^4 - \Gamma. \qquad (15)$$

Note that because of the rotational invariance of the problem (14) the growth rate depends only on the modulus of the wavevector $k = |\mathbf{k}|$ but not on its orientation.

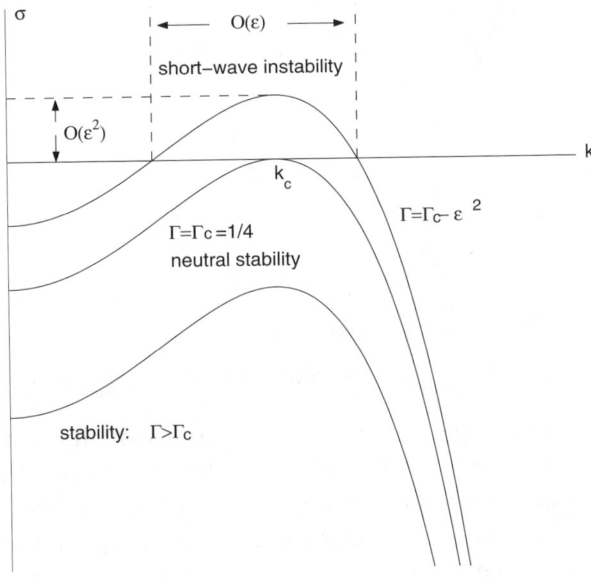

Figure 5.   Dispersion relation (15) for different values of Γ.

One can see that for $\Gamma > \Gamma_c = 1/4$, $\sigma(k) < 0$ for any k, therefore the uniform phase $\phi = 0$ is stable. The instability appears when Γ becomes less than Γ_c. If $\Gamma = \Gamma_c - \epsilon^2$, $0 < \epsilon \ll 1$, the growth rate $\sigma(k) > 0$ in a narrow interval of k, $k_- < k < k_+$ around the critical value $k_c = 1/\sqrt{2}$, see Fig.5. Thus, there appears a *short-wave* instability. The length of this interval is $\Delta k = k_+ - k_- = O(\epsilon)$, while the maximum growth rate $\sigma(k_c) = O(\epsilon^2)$. In the plane of the wavevector's components (k_x, k_y) the instability region is the ring $k_-^2 < k_x^2 + k_y^2 < k_+^2$; see Fig.6. Note that equation (14) can also be written in the form

$$\frac{\partial \phi}{\partial t} = \epsilon^2 \phi - \left(\frac{1}{2} + \nabla^2\right)^2 \phi + \nabla^2 \phi^3. \tag{16}$$

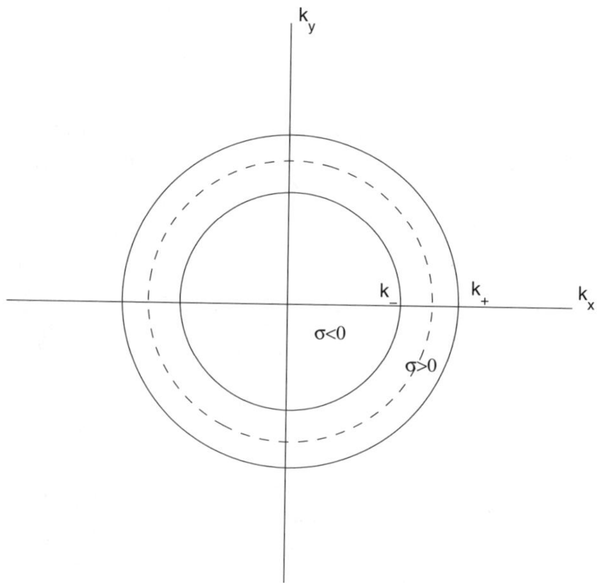

Figure 6.   Ring of unstable wavenumbers in the Fourier plane.

Let us emphasize that there are no other spatially uniform stationary solutions except $\phi = 0$. Thus, when the latter solution is unstable, the system tends to a non-uniform state, i.e. pattern formation takes place. A direct simulation shows that stripe patterns are formed [36], with the stripes wavelength near $2\pi/k_c$. Note that because of the rotational invariance of problem (14) the orientation of the stripes is arbitrary. Initially, a disordered system of stripes is developed from random initial conditions, and then some large domains are developed with a definite orientation of stripes inside each domain. The mean domain size grows with time, i.e. domain coarsening takes place for differently oriented stripe patterns rather than for different uniform phases.

Heat convection patterns: Swift-Hohenberg equation

We shall use one more model similar to (16) but with a simpler nonlinear term:

$$\frac{\partial \phi}{\partial t} = \epsilon^2 \phi - \left(1 + \nabla^2\right)^2 \phi - \phi^3. \tag{17}$$

Equation (17) has been suggested by Swift and Hohenberg [37] as a model for the description of convective pattern in a fluid layer heated from below. Here ϕ is the order parameter proportional to the vertical fluid velocity, and ϵ^2 is proportional to the difference between the actual temperature drop across the layer and its critical value corresponding to the instability threshold. Numerical simulations show that pattern formation in the framework of equation (17) is fully similar to that described by equation (16) [36]. The growth rate of the disturbance with the wavevector $\mathbf{k} = (k_x,\ k_y)$ is determined by

$$\sigma(k_x,\ k_y) = \epsilon^2 - (1 - k^2)^2, \tag{18}$$

hence it is positive inside the ring $k_-^2 < k^2 < k_+^2$, where $k_\pm^2 = 1 \pm \epsilon$. Because of its relative simplicity, we shall use equation (17) as the basic model for the description of pattern selection.

3.   Pattern selection

As noticed above, in the case of a rotationally invariant problem, linear stability theory predicts the growth of disturbances with arbitrary directions of wavevectors. One could expect that the generation of disturbances with different orientations would produce a spatially disordered state (*weak turbulence* [38] or *turbulent crystal* [39]). We shall see however that the strong nonlinear interaction between disturbances typically leads to the selection of spatially ordered patterns.

Selection of roll patterns

Let us consider the Swift-Hohenberg (SH) equation (17) in the case $0 < \epsilon \ll 1$. It is natural to expect that the nontrivial solutions of this equation will be proportional to the square root of the governing parameter ϵ^2, i.e. $\phi = O(\epsilon)$. Also, because the maximum value of the linear growth rate $\sigma_m = \epsilon^2$, we can expect that, at least at the linear stage of growth, the characteristic time scale of growing disturbances is $T = \epsilon^2 t$. Let us look for bounded solutions of equation (17) in the infinite region $\mathbf{r} \in R^2$ in the form:

$$\phi(\mathbf{r}, t) = \epsilon(\phi_1(\mathbf{r}, T) + \epsilon^2 \phi_3(\mathbf{r}, T)) + \dots. \tag{19}$$

We substitute (19) into (17) and take into account that $\partial/\partial t = \epsilon^2 \partial/\partial T$.

To leading order, $O(\epsilon)$, we find:

$$-(1 + \nabla^2)^2 \phi_1 = 0. \tag{20}$$

A bounded solution can be described as a sum (integral) of a finite (infinite) number of plane waves with wavevectors $\mathbf{k}_n$, $|\mathbf{k}_n| = 1$. Below, we shall assume that the number of plane waves is finite. Because the order parameter ϕ is real, we get

$$\phi_1(\mathbf{r}, T) = \sum_{n=1}^{N} \left[A_n(T)e^{i\mathbf{k}_n \cdot \mathbf{r}} + A_n^*(T)e^{-i\mathbf{k}_n \cdot \mathbf{r}} \right], \qquad (21)$$

where * denotes complex conjugate. The particular cases $N = 1$, $N = 2$ and $N = 3$ correspond to roll, square and hexagonal patterns, respectively; see Fig.7.

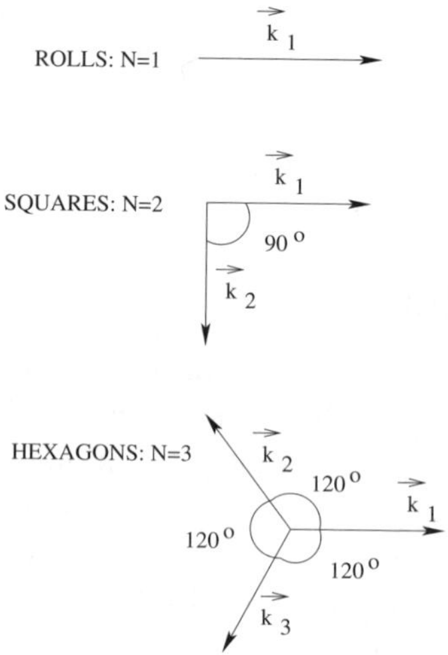

Figure 7.   Wavevector systems corresponding to roll, square and hexagonal patterns.

At order $O(\epsilon^3)$, the following equation is obtained:

$$-(1 + \nabla^2)^2 \phi_3 = \frac{\partial \phi_1}{\partial T} - \phi_1 + \phi_1^3. \qquad (22)$$

Equation (22) has a bounded solution only if the function on the right-hand side, $L + NL$, where

$$L = \sum_{n=1}^{N} \left[\left(\frac{dA_n(T)}{dT} - A_n(T) \right) e^{i\mathbf{k}_n \cdot \mathbf{r}} + \left(\frac{dA_n^*(T)}{dT} - A_n^*(T) \right) e^{-i\mathbf{k}_n \cdot \mathbf{r}} \right],$$

and

$$NL = \sum_{l=1}^{N}\sum_{p=1}^{N}\sum_{q=1}^{N} \left[A_l(T)e^{i\mathbf{k}_l \cdot \mathbf{r}} + A_l^*(T)e^{-i\mathbf{k}_l \cdot \mathbf{r}} \right]$$

$$\times \left[A_p(T)e^{i\mathbf{k}_p \cdot \mathbf{r}} + A_p^*(T)e^{-i\mathbf{k}_p \cdot \mathbf{r}} \right] \left[A_q(T)e^{i\mathbf{k}_q \cdot \mathbf{r}} + A_q^*(T)e^{-i\mathbf{k}_p \cdot \mathbf{r}} \right],$$

has no Fourier components corresponding to the wavevectors $\mathbf{k}$ on the circle $|\mathbf{k}| = 1$. Let us collect all the terms that contain $\exp(i\mathbf{k}_n \cdot \mathbf{r})$ for a definite n. The contribution of the linear expression L is obvious. The nonlinear expression NL contains three terms of the form $A_n^2 A_n^* \exp(i\mathbf{k}_n \cdot \mathbf{r})$, and $6(N-1)$ terms of the form $A_n A_m A_m^* \exp(i\mathbf{k}_n \cdot \mathbf{r})$, $m \neq n$; see Fig.8. The sum of all these terms for each n has to vanish, thus we get a set of *amplitude equations*

$$\frac{dA_n}{dT} - A_n + 3A_n^2 A_n^* + 6 \sum_{m \neq n} A_n A_m A_m^* = 0, \ n = 1, \ldots, N,$$

or

$$\frac{dA_n}{dT} = (1 - 3|A_n|^2 - 6 \sum_{m \neq n} |A_m|^2)A_n. \tag{23}$$

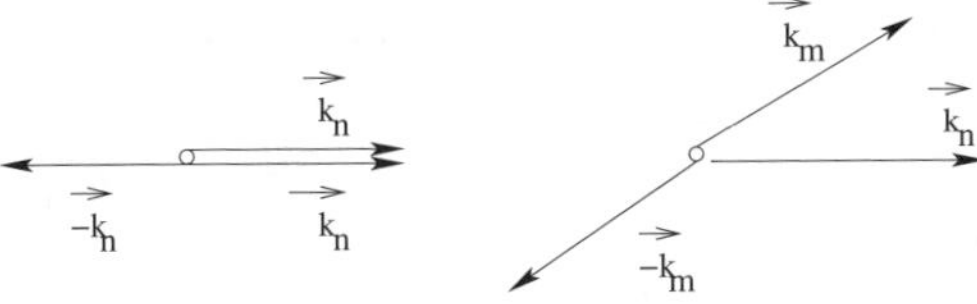

Figure 8. Wavevector arrangements corresponding to self-interaction (left) and cross-interaction (right) terms in the amplitude equation (23).

If we present the complex amplitudes in the form $A_n = R_n \exp(i\theta_n)$, $n = 1, \ldots, N$, we find that the phases do not change with time,

$$\frac{d\theta_n}{dT} = 0, \tag{24}$$

while the time evolution of real amplitudes is governed by the following system of equations:

$$\frac{dR_n}{dT} = (1 - 3R_n^2 - 6 \sum_{m \neq n} R_m^2)R_n, \ n = 1, \ldots, N. \tag{25}$$

The latter equation can be written in the form

$$\frac{dR_n}{dT} = -\frac{dU(R_1, \ldots, R_N)}{dR_n}, \tag{26}$$

where the *Lyapunov function*

$$U(R_1,\ldots,R_n) = \sum_{n=1}^{N}\left(-\frac{R_n^2}{2} + \frac{3R_n^4}{4} + \sum_{m\neq n}\frac{3R_n^2 R_m^2}{2}\right) \tag{27}$$

decreases monotonically in time for any $(R_1,\ldots,R_n)$ except for the stationary points of the dynamical system (25):

$$\frac{dU(R_1,\ldots,R_n)}{dt} = \sum_{n=1}^{N}\frac{dU(R_1,\ldots,R_N)}{dR_n}\frac{dR_n}{dT} = -\sum_{n=1}^{N}\left(\frac{dR_n}{dT}\right)^2 < 0$$

except for the points where all the derivatives dR_n/dT vanish. Thus, the time evolution leads generally to a (local) minimum of the function $U(R_1,\ldots,R_n)$ which corresponds to a stable stationary solution of the system (25). The stationary solutions that correspond to maxima or saddle points of U, are unstable.

As an example, let us consider the particular case $N = 2$, $\mathbf{k}_2 \perp \mathbf{k}_1$. Without loss of generality, we can assume $\theta_1 = \theta_2 = 0$, because the values of θ_n can be changed arbitrarilly by shifting the origin in the plane of $\mathbf{r}$. The system (25) has the following stationary points:

(i) $R_1 = R_2 = 0$, i.e. $\phi_1 = 0$ (no convection);

(ii) $R_1 = 1/\sqrt{3}$, $R_2 = 0$, i.e. $\phi_1 = (2/\sqrt{3})\cos(\mathbf{k}_1 \cdot \mathbf{r})$ (rolls with axes perpendicular to $\mathbf{k}_1$);

(iii) $R_1 = 0$, $R_2 = 1/\sqrt{3}$, i.e. $\phi_1 = (2/\sqrt{3})\cos(\mathbf{k}_2 \cdot \mathbf{r})$ (rolls with axes perpendicular to $\mathbf{k}_2$);

(iv) $R_1 = R_2 = 1/3$, i.e. $\phi_1 = (2/3)[\cos(\mathbf{k}_1 \cdot \mathbf{r}) + \cos(\mathbf{k}_2 \cdot \mathbf{r})] = (4/3)\cos[(\mathbf{k}_1 + \mathbf{k}_2)/2]\cos[(\mathbf{k}_1 - \mathbf{k}_2)/2]$ (square patterns).

It can be shown that solution (i) corresponds to a maximum of $U(R_1, R_2)$, and solution (iv) corresponds to a saddle point, and these solutions are unstable. Roll solutions (ii) and (iii) correspond to local minima of $U(R_1, R_2)$, and these solutions are stable. The corresponding phase portrait is shown in Fig.9a.

A similar analysis can be carried out in the general case. One can show that solutions $r_n = (1/\sqrt{3})\delta_{nm}$ $n = 1,\ldots,N$ for any m, $m = 1,\ldots,N$, which correspond to rolls of different orientation, are stable, while all other stationary solutions are unstable. Generally, the same result on the stability of roll patterns and instability of any other patterns is obtained for the more general system,

$$\frac{dR_n}{dT} = (1 - M_{nn}R_n^2 - M_{nm}\sum_{m\neq n}R_m^2)R_n, \ n = 1,\ldots,N. \tag{28}$$

where M_{nm} are elements of a symmetric matrix satisfying the conditions

$$M_{11} = M_{22} = \ldots = M_{nn} > 0; \ M_{nm} > M_{nn}, \ m \neq n \tag{29}$$

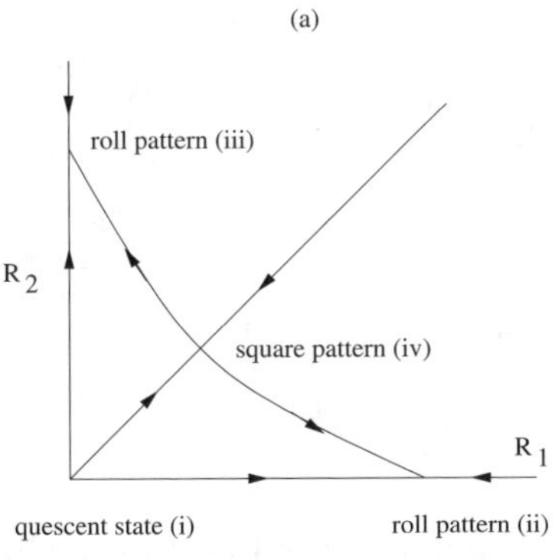

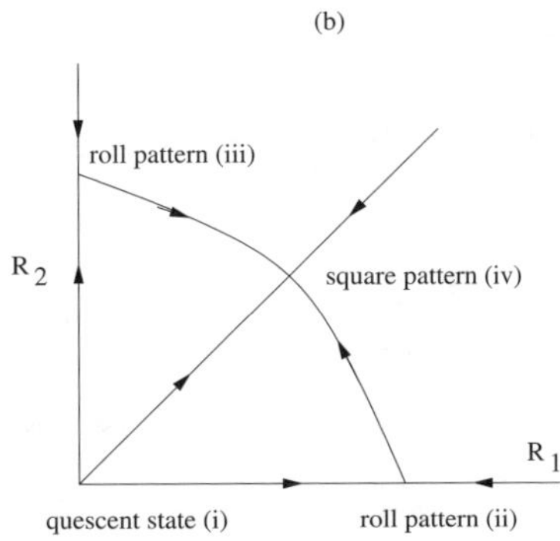

Figure 9.   Phase portrait of the system (25) for N=2 corresponding to the selection of (a) roll patterns; (b) square patterns.

(see [40]). Condition (29) means that the growth of each amplitude R_n is more strongly suppressed by the action of other amplitudes R_m, $m \neq n$ than by a self-action. In other words, the problem (28), which can be also written as

$$\frac{dI_n}{dT} = \frac{1}{2}(1 - M_{nn}I_n - M_{nm} \sum_{m \neq n} I_m)I_n, \ n = 1, \ldots, N, \qquad (30)$$

where $I_n = R_n^2 \geq 0$, belongs to the class of problems of *competition of species* [9]. Because of the competitive nonlinear interaction, only one species, i.e. a particular roll pattern, survives at large T.

Selection of square patterns

If condition (29) is violated, more complicated patterns may appear. As an example, let us consider *the Gertsberg-Sivashinsky equation* that was derived in the problem of Rayleigh-Benard convection in a layer between weakly conducting boundaries [41],

$$\frac{\partial \phi}{\partial t} = \epsilon^2 \phi - \left(1 + \nabla^2\right)^2 \phi + \nabla \cdot (|\nabla \phi|^2 \nabla \phi). \qquad (31)$$

Here, the order parameter ϕ has the meaning of the mean temperature across the layer. Let us repeat the analysis done in the previous subsection for the SH equation. For the sake of simplicity, let us consider the interaction of two orthogonal roll systems taking $N = 2$ in (21) and assuming $\mathbf{k}_1 \perp \mathbf{k}_2$:

$$\phi_1 = \left(A_1 e^{i\mathbf{k}_1 \cdot \mathbf{r}} + A_1^* e^{-i\mathbf{k}_1 \cdot \mathbf{r}}\right) + \left(A_2 e^{i\mathbf{k}_2 \cdot \mathbf{r}} + A_2^* e^{-i\mathbf{k}_2 \cdot \mathbf{r}}\right),$$

$$k_1^2 = k_2^2 = 1, \ \mathbf{k}_1 \cdot \mathbf{k}_2 = 0.$$

Substituting

$$\nabla\phi_1 = i\mathbf{k}_1\left(A_1 e^{i\mathbf{k}_1\cdot\mathbf{r}} - A_1^* e^{-i\mathbf{k}_1\cdot\mathbf{r}}\right) + i\mathbf{k}_2\left(A_1 e^{i\mathbf{k}_2\cdot\mathbf{r}} - A_2^* e^{-i\mathbf{k}_2\cdot\mathbf{r}}\right),$$

$$|\nabla\phi_1|^2 = -\left(A_1^2 e^{2i\mathbf{k}_1\cdot\mathbf{r}} - 2|A_1|^2 + A_1^{*2} e^{-2i\mathbf{k}_1\cdot\mathbf{r}}\right)$$
$$-\left(A_2^2 e^{2i\mathbf{k}_2\cdot\mathbf{r}} - 2|A_2|^2 + A_2^{*2} e^{-2i\mathbf{k}_2\cdot\mathbf{r}}\right)$$

into the right-hand side of the equation for ϕ_3, we find the following solvability conditions:

$$\frac{dA_1}{dT} = (1 - 3|A_1|^2 - 2|A_2|^2)A_1; \quad \frac{dA_2}{dT} = (1 - 3|A_2|^2 - 2|A_1|^2)A_2, \quad (32)$$

or

$$\frac{dR_1}{dT} = (1 - 3R_1^2 - 2R_2^2)R_1; \quad \frac{dR_2}{dT} = (1 - 3R_2^2 - 2R_1^2)R_2, \quad (33)$$

where we use the polar form of the complex amplitudes, $A_n = R_n \exp(i\theta_n)$, $n = 1, 2$. Note, that now the non-diagonal elements $M_{12} = M_{21} = 2$ of the matrix M defined by (28) are smaller than the diagonal elements $M_{11} = M_{22} = 3$, hence condition (29) is violated. As in the case of the SH equation, we obtain 4 stationary solutions: (i) $R_1 = R_2 = 0$ (quiescent state); (ii) $R_1 = 1/\sqrt{3}$, $R_2 = 0$ (rolls); (iii) $R_1 = 0$, $R_2 = 1/\sqrt{3}$ (rolls); (iv) $R_1 = R_2 = 1/\sqrt{5}$ (squares). Now, however, solutions (ii) and (iii) are saddle points, while the solution (iv) is a stable node. The relation $M_{12} < M_{11}$ corresponds to the case of *symbiosis of species* [9]. Hence, the stable stationary state is characterized by a "symbiosis" of two rolls systems, i.e. it is a square pattern. The corresponding phase portrait is shown in Fig.9b.

Selection of hexagonal patterns

Generally, one can expect the selection of hexagonal patterns due to the "symbiotic" mechanism described above in the case where the nonlinear interaction coefficient M_{mn} is smaller than M_{nn} for wavevectors $\mathbf{k}_m$ and $\mathbf{k}_n$ with a 60° angle between them. However, the ubiquity of hexagonal patterns has another explanation. In order to describe it, let us consider some modifications of the models described in Section 2.

Diblock copolymers with different lengths of components chains. If the chain lengths of the A and B monomers are different, $\langle\phi_A\rangle \neq \langle\phi_B\rangle$, then $\langle\phi\rangle = \langle\phi_A - \phi_B\rangle = \beta \neq 0$. Ohta and Kawasaki [34] showed that the expression for the free energy functional (12) has to be modified in the following way:

$$F\{\phi\} = \int d\mathbf{r} \left[\frac{1}{2}(\nabla\phi)^2 - \frac{\phi^2}{2} + \frac{\phi^4}{4} \right]$$

$$+ \frac{\Gamma}{2} \int d\mathbf{r} d\mathbf{r}' [\phi(\mathbf{r}) - \beta] G(\mathbf{r}, \mathbf{r}')[\phi(\mathbf{r}') - \beta], \tag{34}$$

hence

$$\frac{\partial\phi}{\partial t} = \nabla^2(-\phi + \phi^3 - \nabla^2\phi) - \Gamma(\phi - \beta), \quad \langle\phi\rangle = \beta. \tag{35}$$

Define $\psi = \phi - \beta$ and rewrite equation (35) in the form

$$\frac{\partial\psi}{\partial t} = \nabla^2[-\nabla^2\psi + (3\beta^2 - 1)\psi + 3\beta\psi^2 + \psi^3] - \Gamma\psi. \tag{36}$$

The crucial difference between equation (36) and equation (14) is the appearance of a *quadratic* nonlinear term violating the symmetry between ψ and $-\psi$. Below we shall see that this term is the origin of the generation of hexagonal patterns.

Non-Boussinesq convection.   The Swift-Hohenberg model (17) is appropriate for the description of the so-called Boussinesq convection [42], when the dependence of thermophysical fluid parameters on temperature is disregarded. If this dependence is taken into account, a quadratic nonlinearity appears in the amplitude equations [43]. We shall use the following phenomenological modification of the SH equation for non-Boussinesq convection:

$$\frac{\partial\phi}{\partial t} = \gamma\phi - \left(1 + \nabla^2\right)^2 \phi + \alpha\phi^2 - \phi^3. \tag{37}$$

The coefficient α characterizing the non-Boussinesq properties of the fluid can have either sign. Also, we shall consider the system both above the linear instability threshold ($\gamma > 0$) and below that threshold ($\gamma < 0$).

Amplitude equations for hexagonal patterns.   Let us consider model (37) with $|\alpha| \ll 1$ and $\gamma = \Gamma\alpha^2$, $\Gamma = O(1)$. We assume that the characteristic time scale is $T = \alpha^2 t$, hence $\partial/\partial t = \alpha^2\partial/\partial T$, and construct the solution in the form $\phi = \alpha\phi_1 + \alpha^3\phi_3 + \dots$.

Again, to leading order $O(\alpha)$ we obtain equation (20),

$$-(1 + \nabla^2)^2\phi_1 = 0.$$

We are interested in a specific solution of that equation,

$$\phi_1(\mathbf{r}, T) = \sum_{n=1}^{3} \left[A_n(T)e^{i\mathbf{k}_n \cdot \mathbf{r}} + A_n^*(T)e^{-\mathbf{k}_n \cdot \mathbf{r}} \right], \tag{38}$$

where the angle between the unit vectors $\mathbf{k}_n$, $n = 1, 2, 3$, is $120°$, so that

$$\mathbf{k}_1 + \mathbf{k}_2 + \mathbf{k}_3 = \mathbf{0}. \tag{39}$$

At order $O(\alpha^3)$ we obtain the following equation:

$$-(1 + \nabla^2)^2 \phi_3 = \frac{\partial \phi_1}{\partial T} - \Gamma \phi_1 - \phi_1^2 + \phi_1^3. \tag{40}$$

Because of the relations

$$(-\mathbf{k}_2) + (-\mathbf{k}_3) = \mathbf{k}_1, \ \ (-\mathbf{k}_3) + (-\mathbf{k}_1) = \mathbf{k}_2, \ \ (-\mathbf{k}_1) + (-\mathbf{k}_2) = \mathbf{k}_3,$$

the quadratic nonlinear term on the right-hand side of (40) generates additional quadratic terms in the solvability conditions. For instance, the condition of vanishing of the Fourier component of the right-hand side with the wavevector $\mathbf{k}_1$ leads to the following amplitude equation:

$$\frac{dA_1}{dT} = \Gamma A_1 + 2A_2^* A_3^* - 3|A_1|^2 A_1 - 6(|A_2|^2 + |A_3|^2)A_1. \tag{41}$$

Two additional amplitude equations are obtained from (41) by the cyclic permutation of the subscripts 1, 2 and 3:

$$\frac{dA_2}{dT} = \Gamma A_2 + 2A_3^* A_1^* - 3|A_2|^2 A_2 - 6(|A_3|^2 + |A_1|^2)A_2, \tag{42}$$

$$\frac{dA_3}{dT} = \Gamma A_3 + 2A_1^* A_2^* - 3|A_3|^2 A_3 - 6(|A_1|^2 + |A_2|^2)A_3. \tag{43}$$

The system (41)-(43) can be written as

$$\frac{dA_n}{dT} = -\frac{dU}{dA_n^*}, \ \ n = 1, 2, 3, \tag{44}$$

where the Lyapunov function

$$U(A_1, A_1^*, A_2, A_2^*, A_3, A_3^*) = \sum_{n=1}^{3}\left(-\Gamma|A_n|^2 + \frac{3}{2}|A_n|^4\right) \tag{45}$$

$$-2(A_1 A_2 A_3 + A_1^* A_2^* A_3^*) + 3(|A_1|^2|A_2|^2 + |A_1|^2|A_3|^2 + |A_2|^2|A_3|^2)$$

(A_n and A_n^* are considered as independent variables). It can be shown that $dU/dT \leq 0$, and $dU/dT = 0$ only for a stationary solution.

Stationary solutions.   Let us consider the stationary solutions of the system of amplitude equations (41)-(43), and their stability.

Quiescent state. The solution $A_1 = A_2 = A_3 = 0$ corresponds to the quiescent state (no convection). The linearized equations for disturbances are:

$$\frac{d\tilde{A}_1}{dT} = \Gamma\tilde{A}_1, \ \frac{d\tilde{A}_2}{dT} = \Gamma\tilde{A}_2, \ \frac{d\tilde{A}_3}{dT} = \Gamma\tilde{A}_3. \tag{46}$$

For normal modes,

$$\tilde{A}_1, \tilde{A}_2, \tilde{A}_3 \sim e^{\sigma T},$$

the eigenvalue $\sigma = \Gamma$, hence the quiescent state is stable for $\Gamma < 0$ and unstable for $\Gamma > 0$.

Rolls. Consider the solution $A_1 = \sqrt{\Gamma/3}\exp i\theta_1$, $A_2 = A_3 = 0$. This solution exists only for $\Gamma > 0$. Linearizing the system (41)-(43) around this solution, we find that the system splits into two sub-systems: a separate equation for $\tilde{A}_1$, and a coupled system of equations for $\tilde{A}_2$ and $\tilde{A}_3$. The equation for $\tilde{A}_1$ is

$$\frac{d\tilde{A}_1}{dT} = \Gamma\tilde{A}_1 - 6|A_1|^2\tilde{A}_1 - 3A_1^2\tilde{A}_1^*. \tag{47}$$

Substituting the expression for A_1 and define $\tilde{A}_1 = a_1\exp i\theta_1$, we get

$$\frac{da_1}{dT} = -\Gamma(a_1 + a_1^*). \tag{48}$$

Thus, the real (amplitude) disturbance decays with the rate $\sigma = -2\Gamma$, while the imaginary (phase) disturbance is neutral: $\sigma = 0$. The neutral disturbance corresponds to an infinitesimal change of the phase θ_1, i.e. a spatial shift of the rolls system as a whole. The system of equations for $\tilde{A}_2$ and $\tilde{A}_3$ reads:

$$\frac{d\tilde{A}_2}{dT} = \Gamma\tilde{A}_2 + 2A_1^*\tilde{A}_3^* - 6|A_1|^2\tilde{A}_2, \tag{49}$$

$$\frac{d\tilde{A}_3}{dT} = \Gamma\tilde{A}_3 + 2A_1^*\tilde{A}_2^* - 6|A_1|^2\tilde{A}_3. \tag{50}$$

Taking the complex conjugate of (50) and assuming

$$\tilde{A}_2, \tilde{A}_3^* \sim e^{\sigma T},$$

we find that the resulting algebraic system has a nontrivial solution if

$$(\Gamma - 6|A_1|^2 - \sigma)^2 - 4|A_1|^2 = 0,$$

hence

$$\sigma = -\Gamma \pm 2\sqrt{\Gamma/3}. \tag{51}$$

Thus, the rolls are unstable in the interval $0 < \Gamma < \Gamma_1$ and become stable for $\Gamma > \Gamma_1$, where $\Gamma_1 = 4/3$.

Obviously, the same result is obtained for the two other rolls solutions, (i) $A_2 = \sqrt{\Gamma/3}\exp i\theta_2$, $A_3 = A_1 = 0$ and (ii) $A_3 = \sqrt{\Gamma/3}\exp i\theta_3$, $A_1 = A_2 = 0$.

So, in the interval $0 < \Gamma < \Gamma_1$ neither quiescent state nor roll patterns are stable. One has to investigate other critical points of the Lyapunov function (45).

Hexagons. Let us now assume that all the amplitudes A_n, $n = 1, 2, 3$ are not equal to zero, and present them in the form $A_n(T) = R_n(T)\exp[i\theta_n(T)]$. Equation (41) gives rise to the following equations for the real functions:

$$\frac{dR_1}{dT} = (\Gamma - 3R_1^2 - 6R_2^2 - 6R_3^2)R_1 + 2R_2R_3\cos(\theta_1 + \theta_2 + \theta_3), \qquad (52)$$

$$R_1\frac{d\theta_1}{dT} = -2R_2R_3\sin(\theta_1 + \theta_2 + \theta_3). \qquad (53)$$

Four additional equations are obtained by the cyclic permutation of the subscripts 1, 2, and 3.

Note that due to the quadratic terms in the amplitude equations, the phases θ_n, $n = 1, 2, 3$ are not constant. Adding the equations for θ_n and denoting $\Theta = \theta_1 + \theta_2 + \theta_3$, we obtain the following equation for the time evolution of Θ which describes the phase synchronization of the roll subsystems:

$$\frac{d\Theta}{dT} = -Q\sin\Theta, \qquad (54)$$

where

$$Q = \frac{2R_2R_3}{R_1} + \frac{2R_3R_1}{R_2} + \frac{2R_1R_2}{R_3} > 0. \qquad (55)$$

There are two different stationary solutions of equation (54), $\Theta = 0$ and $\Theta = \pi$ (adding $2\pi n$ with integer n to Θ does not change the planform of ϕ_1). Linearizing the equation for a disturbance $\tilde{\Theta}$, we find that

$$\frac{d\tilde{\Theta}}{dT} = -Q\cos\Theta \cdot \tilde{\Theta}. \qquad (56)$$

Hence, the invariant manifold $\Theta = 0$ is attracting, while the manifold $\Theta = \pi$ is repelling. Below, we shall consider the dynamics on the manifold $\Theta = 0$:

$$\frac{dR_1}{dT} = (\Gamma - 3R_1^2 - 6R_2^2 - 6R_3^2)R_1 + 2R_2R_3; \qquad (57)$$

two additional equations are obtained by the cyclic permutation of the subscripts 1, 2, and 3.

There is a stationary solution which corresponds to a hexagonal pattern: $R_1 = R_2 = R_3 \equiv R$, where R satisfies the equation

$$15R^2 - 2R - \Gamma = 0. \tag{58}$$

Taking into account that $R > 0$ by definition, we find that there are two branches of solutions:

$$R_+ = \frac{1 + \sqrt{1 + 15\Gamma}}{15}, \quad \Gamma \geq \Gamma_2 = -\frac{1}{15} \tag{59}$$

and

$$R_- = \frac{1 - \sqrt{1 + 15\Gamma}}{15}, \quad \Gamma_2 \leq \Gamma < 0. \tag{60}$$

At the point $\Gamma = \Gamma_2$, both branches merge, $R_+ = R_- = 1/15$.

Let us now consider the stability of the hexagons on the manifold $\Theta = 0$. Linearizing (57) and the other two dynamic equations, we obtain the following system for the evolution of disturbances:

$$\frac{d\tilde{R}_1}{dT} = a + b(\tilde{R}_2 + \tilde{R}_3), \quad \frac{d\tilde{R}_2}{dT} = a + b(\tilde{R}_3 + \tilde{R}_1), \quad \frac{d\tilde{R}_3}{dT} = a + b(\tilde{R}_1 + \tilde{R}_2), \tag{61}$$

where

$$a = -2R - 6R^2, \quad b = 2R - 12R^2. \tag{62}$$

For the normal modes, $R_n \sim \exp(\sigma T)$, we obtain the relation:

$$\sigma^3 - 3a\sigma^2 + 3(a^2 - b^2)\sigma - (a - b)^2(a + 2b) = 0. \tag{63}$$

According to the Descartes rule, all the roots of (63) are negative, so that the hexagons are stable, if the following conditions are satisfied: (i) $-3a > 0$; (ii) $3(a^2 - b^2) > 0$; (iii) $-(a - b)^2(a + 2b) > 0$.

Condition (i) is satisfied for any $R > 0$. Substituting (62) into condition (ii), we find $R < 2/3$; the latter inequality is satisfied for the upper branch (59), if $\Gamma < \Gamma_3 = 16/3$, as well as for the entire lower branch (60). The condition (iii) gives $R > 1/15$, hence the lower branch is unstable. Finally, we obtain that the upper branch is stable in the interval $\Gamma_2 < \Gamma < \Gamma_3$, where $\Gamma_2 = -1/15$, $\Gamma_3 = 16/3$. The bifurcation diagram showing stable and unstable branches is shown in Fig.10.

Because $\Gamma_2 < 0$, both the quiescent state and the upper branch of hexagons are stable, i.e. provide a local minimum of the Lyapunov function U, in the interval $\Gamma_2 < \Gamma < 0$. Thus, the transition between the quiescent state and the hexagonal pattern in the presence of a cubic term in the Lyapunov function is similar to a *first order phase transition* which takes place in the presence of a cubic term in the free energy Ginzburg-Landau functional [44], in contradistinction to the transition between the quiescent state and the roll pattern in the

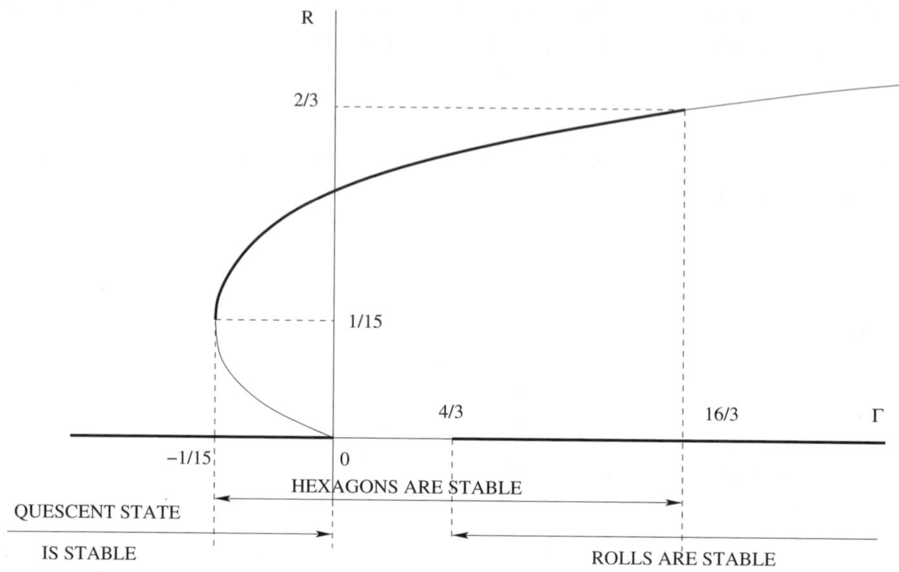

Figure 10.   Bifurcation diagram of an equilateral hexagonal pattern with the amplitude described by eq.(58).

absence of a cubic term in the Lyapunov function, which is similar to a *second order phase transition*. The lower branch of hexagons corresponds to a saddle point. Its stable manifold separates the attraction basins of two stable nodes corresponding to the quiescent state and to the upper branch of hexagons.

Recall that the roll pattern becomes stable for $\Gamma > \Gamma_1 = 4/3$. Hence, in the interval $\Gamma_1 < \Gamma < \Gamma_3$ the Lyapunov function has 4 local minima, three of them correspond to three types of roll patterns, and one of them corresponds to hexagons. The basins of attractions between them are separated by stable manifolds of some additional saddle-point stationary solutions, corresponding to squares (e.g. $R_1 = R_2 \neq 0$, $R_3 = 0$) and "skewed hexagons" (e.g. $R_1 = R_2 \neq R_3 \neq 0$). Finding the latter solutions is suggested to the readers as an exercise.

4.   Modulated patterns

In the previous section, we considered perfectly periodic patterns with a discrete set of basic wavevectors satisfying the condition $|\mathbf{k}_n| = 1$ for any n. Recall that the set of wavevectors corresponding to unstable modes is actually a ring which contains a continuum of wavevectors. In the present section, we shall discuss a wider class of solutions corresponding to large-scale modulations of periodic patterns.

Newell-Whitehead-Segel equation

Let us return to the standard Swift-Hohenberg equation (17),

$$\frac{\partial \phi}{\partial t} = \epsilon^2 \phi - \left(1 + \nabla^2\right)^2 \phi - \phi^3,$$

and consider a roll pattern with the basic wavevector $\mathbf{k} = (1, 0)$. As we know, the roll pattern is stable with respect to disturbances with other wavevectors $\mathbf{k}_n$, $|\mathbf{k}_n| = 1$. The analysis done in the previous section used implicitly the assumption that the difference between the wavevectors $\mathbf{k}$ and $\mathbf{k}_n$ is $O(1)$, i.e. the angle between them is $O(1)$. Now we shall take into account the possibility of large-scale distortion of patterns by means of disturbances with wavevectors close to each other. For this goal, we shall apply multiscale analysis. The ring of unstable modes, $1 - \epsilon < k_x^2 + k_y^2 < 1 + \epsilon$, around the point $k_x = 1$, $k_y = 0$ has width $O(\epsilon)$ in the x-direction and width $O(\epsilon^{1/2})$ in the y-direction, as shown in Fig.11. Therefore, it is natural to assume that the function ϕ depends on the following scaled variables:

$$x_0 = x, \ x_1 = \epsilon x, \ y_{1/2} = \epsilon^{1/2} y. \tag{64}$$

The Fourier transform of such a function will be concentrated around the instability ring. As in the previous section, we shall use the scaled time variable $T = \epsilon^2 t$.

Following the idea of multiscale expansions (see, e.g., [45]), we substitute

$$\frac{\partial}{\partial x} = \frac{\partial}{\partial x_0} + \epsilon \frac{\partial}{\partial x_1}, \ \frac{\partial}{\partial y} = \epsilon^{1/2} \frac{\partial}{\partial y_{1/2}}, \ \frac{\partial}{\partial t} = \epsilon^2 \frac{\partial}{\partial T}$$

into (17) and obtain:

$$\epsilon^2 \frac{\partial \phi}{\partial T} = \epsilon^2 \phi - \left[1 + \frac{\partial^2}{\partial x_0^2} + \epsilon \left(2 \frac{\partial^2}{\partial x_0 \partial x_1} + \frac{\partial^2}{\partial y_{1/2}^2}\right) + \epsilon^2 \frac{\partial^2}{\partial x_1^2}\right]^2 \phi - \phi^3. \tag{65}$$

Now we substitute the solution in the form

$$\phi = \epsilon \phi_1 + \epsilon^2 \phi_2 + \epsilon^3 \phi_3 + \ldots, \tag{66}$$

and demand boundness with respect to all the variables at each order.

At order ϵ, we obtain:

$$-\left(1 + \frac{\partial^2}{\partial x_0^2}\right)^2 \phi_1 = 0. \tag{67}$$

The most general bounded solution of this equation,

$$\phi_1 = A(T, \ x_1, \ y_{1/2})e^{ix_0} + A^*(T, \ x_1, \ y_{1/2})e^{-ix_0}, \tag{68}$$

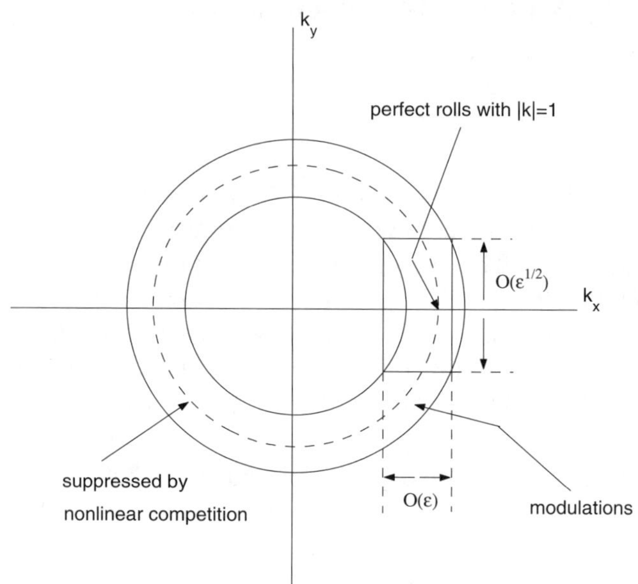

Figure 11.   Ring of unstable modes in the Fourier plane and regions corresponding to spatial modulations in different directions on different scales.

describes a large-scale modulation of a roll pattern.

In the order ϵ^2, we get:

$$-\left(1 + \frac{\partial^2}{\partial x_0^2}\right)^2 \phi_2 = 2\left(2\frac{\partial^2}{\partial x_0 \partial x_1} + \frac{\partial^2}{\partial y_{1/2}^2}\right)\left(1 + \frac{\partial^2}{\partial x_0^2}\right)\phi_1. \qquad (69)$$

Because

$$\left(1 + \frac{\partial^2}{\partial x_0^2}\right)\phi_1 = 0,$$

the right-hand side of (69) vanishes, so that the equation is always solvable, and its solution is similar to (68):

$$\phi_2 = B(T,\ x_1,\ y_{1/2})e^{ix_0} + B^*(T,\ x_1,\ y_{1/2})e^{-ix_0}.$$

Finally, at order ϵ^3, taking into account that

$$\left(1 + \frac{\partial^2}{\partial x_0^2}\right)\phi_1 = \left(1 + \frac{\partial^2}{\partial x_0^2}\right)\phi_2 = 0,$$

we obtain:

$$-\left(1 + \frac{\partial^2}{\partial x_0^2}\right)^2 \phi_3 = \frac{\partial \phi_1}{\partial T} - \phi_1 + \phi_1^3 + \left(2\frac{\partial^2}{\partial x_0 \partial x_1} + \frac{\partial^2}{\partial y_{1/2}^2}\right)^2 \phi_1. \quad (70)$$

The term proportional to $\exp i x_0$ on the right-hand side of (70) must vanish, otherwise the solution of (70) will not be bounded as $x_0 \to \pm\infty$. That gives us the following evolution equation for the envelope function $A(T, x_1, y_{1/2})$, which is called the *Newell-Whitehead-Segel (NWS) equation* [46, 47]:

$$\frac{\partial A}{\partial T} = A - 3|A|^2 A - \left(2i\frac{\partial}{\partial x_1} + \frac{\partial^2}{\partial y_{1/2}^2} \right)^2 A. \tag{71}$$

Rescaling the variables,

$$a = A\sqrt{3}, \ X = x_1/2, \ Y = y_{1/2}/\sqrt{2}, \tag{72}$$

we transform the NWS equation to its standard form:

$$\frac{\partial a}{\partial T} = a - |a|^2 a + \left(\frac{\partial}{\partial X} - \frac{i}{2}\frac{\partial^2}{\partial Y^2} \right)^2 a. \tag{73}$$

Equation (73) is not specific for patterns described by the Swift-Hohenberg equations, but is generic for any patterns generated by a short-wave monotonic instability in a rotationally invariant system. Therefore, all the results obtained below by means of that equation, are generic.

The NWS equation can be written in the form

$$\frac{\partial a}{\partial T} = -\frac{\delta F}{\delta a^*}, \tag{74}$$

where the *Lyapunov functional* is defined as

$$F = \int d\mathbf{r} \left[\frac{1}{2}(|a|^2 - 1)^2 + \left| \left(\frac{\partial}{\partial X} - \frac{i}{2}\frac{\partial^2}{\partial Y^2} \right) a \right|^2 \right]. \tag{75}$$

Note that

$$\frac{dF}{dT} = \int d\mathbf{r} \left(\frac{\delta F}{\delta a}\frac{\partial a}{\partial T} + \frac{\delta F}{\delta a^*}\frac{\partial a^*}{\partial T} \right) = -2 \int d\mathbf{r} \left| \frac{\partial a}{\partial T} \right|^2 \leq 0. \tag{76}$$

Therefore, the system tends to a stationary solution and has no time-periodic or chaotic solutions.

Modulational instabilities of rolls

Equation (74) has a family of stationary solutions which do not depend on Y:

$$a = a_K(X) = \sqrt{1 - K^2}e^{iK(X - X_0)}, \ -1 < K < 1, \tag{77}$$

where X_0 is an arbitrary constant. For the sake of simplicity, choose $X_0 = 0$. Taking into account the relations (72), we find that the corresponding order-parameter field is

$$\phi_1 = Ae^{ix_0} + A^* e^{-ix_0} = 2\sqrt{3}\sqrt{1 - K^2}\cos\left(x_0 + \frac{1}{2}Kx_1\right)$$
$$= 2\sqrt{3(1 - K^2)}\cos\left[\left(1 + \frac{1}{2}K\epsilon\right)x\right].$$

Thus, these solutions correspond to roll solutions with wavenumbers

$$k = 1 + K\epsilon/2, \quad -1 < K < 1 \tag{78}$$

inside the instability interval, generally different from 1.

Let us investigate the stability of roll solutions in the framework of the NWS equation. Linearizing equation (73) around the solution (77), we obtain the following equation:

$$\frac{d\tilde{a}}{dT} = -(1 - 2K^2)\tilde{a} - (1 - K^2)\tilde{a}^* e^{2iKX} + \left(\frac{\partial}{\partial X} - \frac{i}{2}\frac{\partial^2}{\partial Y^2}\right)^2 \tilde{a}. \tag{79}$$

The dependence of the coefficient on X can be eliminated by the transformation

$$\tilde{a}(X, T) = b(X, T)e^{iKX};$$

we obtain

$$\frac{db}{dT} = -(1 - 2K^2)b - (1 - K^2)b^* + \left(\frac{\partial}{\partial X} - \frac{i}{2}\frac{\partial^2}{\partial Y^2} + iK\right)^2 b. \tag{80}$$

Now we can find the normal modes in the form

$$b = b_1 e^{i(\tilde{K}_X X + \tilde{K}_Y Y) + \sigma T} + b_2 e^{-i(\tilde{K}_X X + \tilde{K}_Y Y) + \sigma^* T} \tag{81}$$

(actually the eigenvalues σ are real, because the equation is variational). The condition of existence of nontrivial solutions for the coupled algebraic system for b_1 and b_2^* gives the following expression for two branches of eigenvalues:

$$\sigma_{\pm}(\tilde{K}_X, \tilde{K}_Y; K) = -1 + K^2 - \tilde{K}_X^2 - \frac{1}{2}K\tilde{K}_Y^2 - \frac{1}{16}\tilde{K}_Y^4$$
$$\pm \sqrt{(1 - K^2)^2 + 4\tilde{K}_X^2\left(K + \frac{1}{2}\tilde{K}_Y^2\right)^2}. \tag{82}$$

Recall that $\tilde{K}_X, \tilde{K}_Y$ are the components of the wavevector of the disturbance, while K is a parameter of the basic roll solution related to its wavenumber k according to (78).

To reveal the instability modes, it is sufficient to consider the branch with the higher value of σ in the limit of longwave disturbances, i.e. for small $\tilde{K}_X, \tilde{K}_Y$:

$$\sigma_+ = -\frac{1-3K^2}{1-K^2}\tilde{K}_X^2 - \frac{1}{2}K\tilde{K}_Y^2 + o(\tilde{K}_X^2, \tilde{K}_Y^2). \tag{83}$$

One can see that the roll solutions within the interval $1/3 < K^2 < 1$ are unstable with respect to disturbances with $\tilde{K}_X^2 \neq 0$, $\tilde{K}_Y^2 = 0$, i.e. to *longitudinal modulations*. This kind of instability in nonlinear dissipative systems was discovered by Eckhaus [48] and is called the *Eckhaus instability*.

If the wavenumber of the roll solution satisfies the condition $K < 0$ (i.e. $k < 1$), a *transverse* modulational instability takes place with respect to disturbances with $\tilde{K}_X^2 = 0$, $\tilde{K}_Y^2 \neq 0$. It is also called the *zigzag instability*.

We come to the conclusion that only the roll patterns inside the *stability interval* $0 < K < 1/\sqrt{3}$ are stable. The stability interval is also called the *Busse balloon*, for it was first discovered by Busse *et al.* in the context of the Rayleigh-Benard convection patterns [40]. See the diagram in Fig.12.

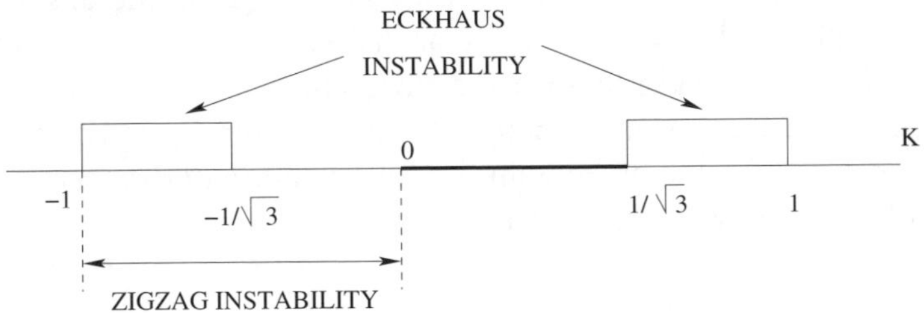

Figure 12. Intervals of the modulation wavenumber K corresponding to Eckhaus and zigzag instabilities.

Nonlinear phase diffusion equation

Derivation of the nonlinear phase diffusion equation. The longwave nature of the two basic instabilities of roll patterns described above shows that longwave distortions of rolls are of major interest. Let us consider longwave solutions of the NWS equation (73)

$$a = R(\xi,\,\eta,\,\tau)\exp[i\theta(\xi,\,\eta,\,\tau)], \tag{84}$$

where

$$\xi = \delta X,\ \eta = \delta^{1/2}Y,\ \tau = \delta^2 T;\ 0 < \delta \ll 1.$$

Needless to say that the small parameter δ has nothing to do with the small parameter ϵ used in the derivation of the NWS equation: the smallness of ϵ is due to the smallness of the governing parameter in the SH equation, while δ characterizes the scale of a specific class of solutions of the NWS equation.

Substituting

$$\frac{\partial}{\partial X} = \delta \frac{\partial}{\partial \xi}, \quad \frac{\partial}{\partial Y} = \delta^{1/2} \frac{\partial}{\partial \eta}, \quad \frac{\partial}{\partial T} = \delta^2 \frac{\partial}{\partial \tau}$$

into (73) and using the representation (84), we obtain:

$$\delta^2 \frac{\partial}{\partial \tau} \left(Re^{i\theta} \right) = (R - R^3)e^{i\theta} + \delta^2 \left(\frac{\partial}{\partial \xi} - \frac{i}{2} \frac{\partial^2}{\partial \eta^2} \right)^2 \left(Re^{i\theta} \right). \tag{85}$$

Let us consider solutions in the form

$$R = R_0 + \delta^2 R_2 + \ldots, \quad \theta = \theta_0 + \delta^2 \theta_2 + \ldots. \tag{86}$$

After substituting (86) into (85), we find to leading order:

$$R_0 - R_0^3 = 0.$$

Because we are interested in distorted rolls rather than in the unstable trivial solution, we choose $R_0 = 1$. Note, that the leading-order roll amplitude is constant under the action of longwave distortions.

At the next order, a system of coupled equations for R_2 and θ_0 is obtained. After eliminating R_2, one can obtain the following *nonlinear phase diffusion equation* [49] (below we drop the subscript $_0$):

$$\theta_\tau = \theta_{\xi\xi} - \frac{1}{4}\theta_{\eta\eta\eta\eta} + 2\theta_\eta\theta_{\xi\eta} + \theta_\xi\theta_{\eta\eta} + \frac{3}{2}\theta_\eta^2\theta_{\eta\eta}, \tag{87}$$

which governs the longwave phase distortions of roll patterns.

We shall apply this equation for the consideration of *defects* in roll patterns.

Dislocation.   According to relation (84), the phase θ is defined modulo 2π at points where $R \neq 0$, and undefined at points where $R = 0$. The roll pattern can contain a *point defect* of the following structure. The amplitude $R = 0$ at a certain point, say the point $X = Y = 0$. Except for this point, the phase is smooth, but going around the point $X = Y = 0$ along a closed circle $X^2 + Y^2 = const$ leads to a phase increment $2n\pi$, where $n = \pm 1$. Such a defect in the roll pattern is called a *dislocation*. We shall apply the nonlinear phase equation for the description of the dislocations in the *far field*, i.e. at large distances from the center (i.e. for $\xi, \eta = O(1)$).

One can easily show [50] that the stationary equation (87) is satisfied by any solution of the *Burgers equation*

$$\theta_\xi = \pm \frac{1}{2}(\theta_{\eta\eta} - \theta_\eta^2). \tag{88}$$

A positive dislocation ($n = 1$) is described by the solution of equation (88), with the + sign on the right-hand side, which satisfies the boundary conditions on the cut $x = 0$, $y > 0$:

$$\theta(0^{\pm}, y) = \pm\pi, \; y > 0. \tag{89}$$

By means of the Hopf-Cole transformation

$$\theta = \mp \ln(f) \tag{90}$$

equation (88) is transformed to

$$f_{\xi} = \pm\frac{1}{2} f_{\eta\eta}. \tag{91}$$

First, we shall find the solution which corresponds to a positive dislocation, in the region $\xi > 0$ in the form of a *self-similar solution*: $f = f(\zeta)$, $\zeta = \eta/\sqrt{2\xi}$. The corresponding boundary value problem,

$$f'' + 2\zeta f' = 0, \; -\infty < \zeta < \infty; \; f(-\infty) = 1, \; f(\infty) = \exp(-\pi), \tag{92}$$

has the following solution:

$$f = \frac{1 + \exp(-\pi)}{2} - \frac{1 - \exp(-\pi)}{2}\mathrm{erf}(\zeta).$$

The solution in the region $\xi < 0$ is calculated in a similar way. Finally, we obtain the following expression for $\theta(\xi, \eta)$:

$$\theta(\xi, \eta) = -\mathrm{sign}(\xi) \ln\left[\frac{1 + \exp(-\pi)}{2} - \frac{1 - \exp(-\pi)}{2}\mathrm{erf}\left(\frac{\eta}{\sqrt{2|\xi|}}\right)\right]. \tag{93}$$

The solution for a negative dislocation is obtained similarly. Contour lines of the function defined by eq.(93) showing the dislocation structure are shown in Fig.13.

Nonlinear theory of the zigzag instability. In the previous subsection, we found that a roll pattern with $K < 0$ is subject to a transverse (zigzag) instability with $\tilde{K}_X = 0$, $\tilde{K}_Y \neq 0$. In order to investigate the temporal evolution of a zigzag disturbance on the background of a roll pattern, substitute $\theta = K\xi + \Phi(\eta, \tau)$ into the nonlinear phase equation (87). We obtain:

$$\Phi_{\tau} = -\frac{1}{4}\Phi_{\eta\eta\eta\eta} + K\Phi_{\eta\eta} + \frac{3}{2}\Phi_{\eta}^2\Phi_{\eta\eta}. \tag{94}$$

For the η-component of the wavevector, $Q = \Phi_{\eta}$, we obtain a *Cahn-Hilliard equation*,

$$Q_{\tau} = -\frac{1}{4}Q_{\eta\eta\eta\eta} - (-K)Q_{\eta\eta} + \frac{1}{2}\left(Q^3\right)_{\eta\eta}. \tag{95}$$

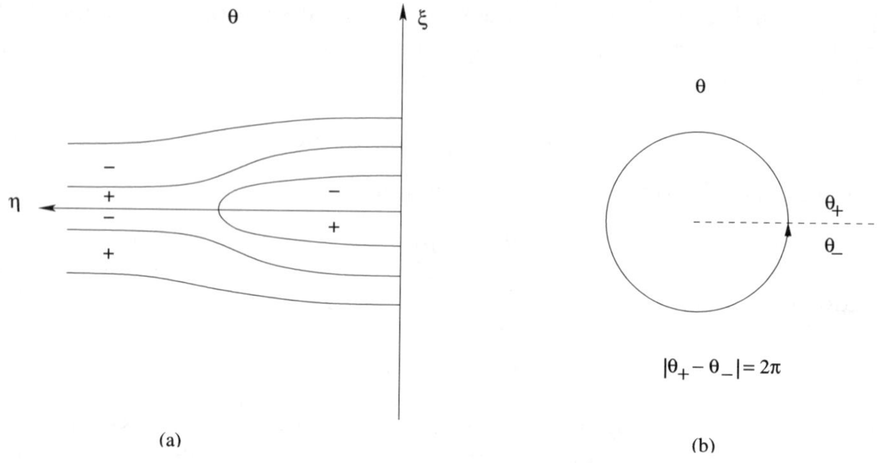

Figure 13.   (a) Contour lines of the function (93) showing the structure of a dislocation. (b) Change of phase around the dislocation.

As we know, the spatially-periodic solutions of equation (95) are unstable. The coarsening process leads to the kink solution

$$Q = \sqrt{2(-K)} \tanh \sqrt{2(-K)}\eta,$$

which corresponds to a smooth *domain wall*

$$\theta = K\xi + \ln \cosh \sqrt{2(-K)}\eta \tag{96}$$

between two sets of inclined roll patterns. The contour lines of the function defined by eq.(96) showing the domain wall structure is presented in Fig.14.

Domain walls and fronts

As we found in Sec.3.1, a roll pattern is selected when the off-diagonal elements of the nonlinear interaction matrix are larger than the diagonal elements (see (29)), because all other patterns are unstable. Specifically, it is true for systems governed by the Swift-Hohenberg equation. The stability arguments cannot determine, however, the direction of the pattern's wave vector: due to the isotropy of the problem, roll patterns with different orientation of the wave vector are equally stable. In reality, rolls with different orientations can develop in different parts of the system forming *domains* of ordered structures separated by *domain walls*. In order to describe this situation, we shall consider a wider class of solutions to the Swift-Hohenberg equation (14),

$$\frac{\partial \phi}{\partial t} = \epsilon^2 \phi - \left(1 + \nabla^2\right)^2 \phi - \phi^3,$$

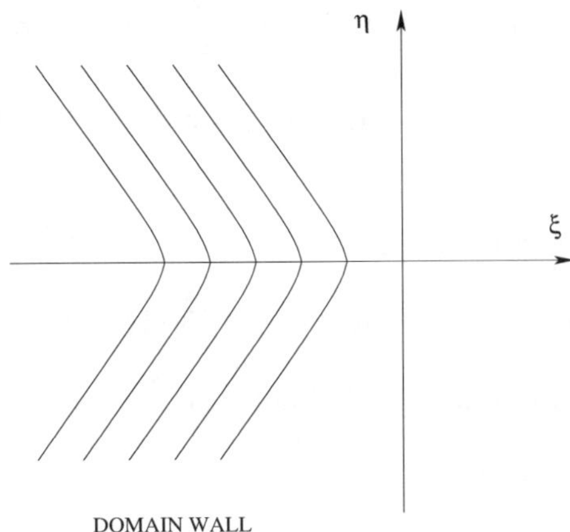

Figure 14. Contour line of the function (96) showing the structure of a domain wall.

than those studied in Sec. 4.1.

Coupled Newell-Whitehead-Segel equations. Because we are not going to consider the zigzag instability, we shall use a simpler assumption on the scaling of the solution:

$$\phi = \phi(T, \mathbf{r}_0, \mathbf{r}_1),$$

where $T = \epsilon^2 t$, $\mathbf{r}_0 = \mathbf{r}$, $\mathbf{r}_1 = \epsilon \mathbf{r}$. The rescaled Swift-Hohenberg equation reads:

$$\epsilon^2 \frac{\partial \phi}{\partial T} = \epsilon^2 \phi - \phi^3 - [(1 + \nabla_0^2) + 2\epsilon(\nabla_0 \cdot \nabla_1) + \epsilon^2 \nabla_1^2]^2 \phi. \tag{97}$$

As in (19), we assume

$$\phi = \epsilon \phi_1 + \epsilon^3 \phi_3 + \dots$$

At leading order, $O(\epsilon)$, we find:

$$-(1 + \nabla_0^2)^2 \phi_1 = 0. \tag{98}$$

In order to consider rolls of different orientations, we take

$$\phi_1 = \sum_{n=1}^{N} \left[A_n(T, \mathbf{r}_1) e^{i\mathbf{k}_n \cdot \mathbf{r}_0} + A_n^*(T, \mathbf{r}_1) e^{-i\mathbf{k}_n \cdot \mathbf{r}_0} \right], \tag{99}$$

where $|\mathbf{k}_n| = 1$.

The ansatz (99) resembles (21), but there is an essential difference: now the functions A_n depend on the slow coordinate $\mathbf{r}_1$. Therefore we can consider different roll patterns localized in different regions rather than uniformly superposed.

At order $O(\epsilon^3)$, we obtain the following generalization of equations (23):

$$\frac{\partial A_n}{\partial T} = (1 - 3|A_n|^2 - 6 \sum_{m \neq n} |A_m|^2)A_n + 4(\mathbf{k}_n \cdot \nabla_1)^2 A_n, \ n = 1, \ldots, N.$$

$$(100)$$

By means of the scale transformation,

$$a_n = A_n\sqrt{3}, \ n = 1, \ldots, N; \mathbf{R} = \mathbf{r}_1/2, \tag{101}$$

the obtained system of coupled NWS equations is transformed to its standard form,

$$\frac{\partial a_n}{\partial T} = (1 - |a_n|^2 - 2 \sum_{m \neq n} |a_m|^2)a_n + (\mathbf{k}_n \cdot \nabla_{\mathbf{R}})^2 a_n, \ n = 1, \ldots, N. \tag{102}$$

Generally, the system of coupled NWS equations can be written in the form

$$\frac{\partial a_n}{\partial T} = -\frac{\delta F}{\delta a_n^*}. \tag{103}$$

where the Lyapunov functional has the form

$$F = \int d\mathbf{R} L[\{a\}, \{a^*\}, \{\mathbf{k} \cdot \nabla_{\mathbf{R}} a\}, \{\mathbf{k} \cdot \nabla_{\mathbf{R}} a^*\}], \tag{104}$$

so that

$$\frac{dF}{dT} \leq 0. \tag{105}$$

Here $\{(\cdot)\}$ means the corresponding set of $(\cdot)_{1,\ldots,N}$. If the Lyapunov functional density L does not contain cubic terms, the amplitude equations look like

$$\frac{\partial a_n}{\partial T} = \left(1 - |a_n|^2 - \sum_{m \neq n} g_{mn}|a_m|^2\right) a_n + (\mathbf{k}_n \cdot \nabla_{\mathbf{R}})^2 a_n, \ n = 1, \ldots, N.$$

$$(106)$$

Otherwise, they can be written as

$$\frac{\partial a_n}{\partial T} = \left(\Gamma - |a_n|^2 - \sum_{m \neq n} g_{mn}|a_m|^2\right) a_n$$
$$+ a_{n'}^* a_{n''}^* + (\mathbf{k}_n \cdot \nabla_{\mathbf{R}})^2 a_n, \ n = 1, \ldots, N, \tag{107}$$

where

$$\mathbf{k}_n + \mathbf{k}_{n'} + \mathbf{k}_{n''} = 0.$$

Stationary domain walls.   For the sake of simplicity, let us consider a plane domain wall perpendicular to the axis X, which separates two semi-infinite roll systems with wave vectors $\mathbf{k}_1$ and $\mathbf{k}_2$ [51], as shown in Fig.15a. Because the Lyapunov functional densities of both roll patterns are equal, there is no reason for a motion of the domain wall, hence it is motionless [52]. The problem is governed by the following system of ordinary differential equations:

$$D_1 a_1'' + a_1 - |a_1|^2 a_1 - g|a_2|^2 a_1 \;=\; 0,$$
$$D_2 a_2'' + a_2 - |a_2|^2 a_2 - g|a_1|^2 a_2 \;=\; 0, \qquad (108)$$
$$-\infty < X < \infty,$$

where $D_n = (k_n)_X$, and $'$ denotes differentiation with respect to X. In the case of the Swift-Hohenberg equation, $g = 2$. Generally, roll patterns are selected, if $g > 1$. We assume that at large distances from the domain wall there are perfect roll patterns. This leads to the boundary conditions:

$$X \to -\infty, \; a_2 \to 0; \; X \to \infty, \; a_1 \to 0. \qquad (109)$$

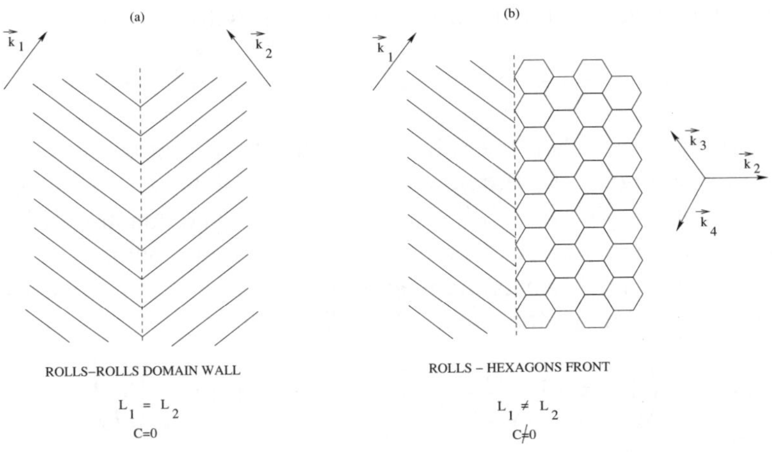

Figure 15.   (a) Stationary domain wall between two systems of rolls with different wavevectors; (b) moving front between rolls and hexagons.

Introduce $a_n = R_n \exp(i\theta_n)$, $i = 1, 2$. The imaginary parts of the equations give the relations

$$r_n \theta_n'' + 2 r_n' \theta_n' = 0, \; n = 1, 2,$$

or

$$(r_n^2 \theta_n')' = 0, \; n = 1, 2,$$

hence

$$r_n^2 \theta_n' = M_n = const.$$

Taking into account the boundary conditions (109), we conclude that $M_1 = M_2 = 0$, so the phases θ_1 and θ_2 are constant. Taking into account the meaning of the phase modulations, we conclude that a steady domain wall can exist only between patterns with the critical values of wavenumbers $|\mathbf{k}_n|$ ($|\mathbf{k}_n| = 1$ in the case of the SH equation). So, the existence of a domain wall acts as a factor which selects the wavenumber in a much more definite way than the stability arguments presented in Sec.4.2.

The real parts of the equations read (we take into account that $\theta_1' = \theta_2' = 0$):

$$D_1 r_1'' + r_1(1 - r_1^2 - gr_2^2) = 0, \quad D_2 r_2'' + r_2(1 - r_2^2 - gr_1^2) = 0, \tag{110}$$

$$X \to -\infty : \; r_2 \to 0, \; r_1 \to 1; X \to \infty : \; r_1 \to 0, \; r_2 \to 1. \tag{111}$$

The system of equations (110) describes the two-dimensional motion of a particle with the Lagrange function (equal to the Lyapunov functional density) $L = K(r_1', r_2') - U(r_1, r_2)$, where

$$K(r_1', r_2') = \frac{1}{2} D_1 (r_1')^2 + \frac{1}{2} D_2 (r_2')^2$$

(note that the "mass" of the fictitious particle is anisotropic) and

$$U(r_1, r_2) = \frac{1}{2}(r_1^2 + r_2^2) - \frac{1}{4}(r_1^4 + 2gr_1^2 r_2^2 + r_2^4). \tag{112}$$

In the case $g > 1$, when the rolls are stable, the potential $U(r_1, r_2)$ has a minimum in the point $(0, 0)$, two maxima in the points $(1, 0)$ and $(0, 1)$, and a saddle point $(1/\sqrt{1 + g}, 1/\sqrt{1 + g})$, see Fig.16. The domain wall solution corresponds to the trajectory that starts at the maximum point $(1, 0)$ at $X \to -\infty$ and tends to another maximum point $(0, 1)$ as $X \to \infty$. An exact solution of (110), (111) is known for $g = 3$, $D_1 = D_2 = D$:

$$r_1 = \frac{1}{2}\left(1 - \tanh\frac{X}{\sqrt{2D}}\right), \quad r_2 = \frac{1}{2}\left(1 + \tanh\frac{X}{\sqrt{2D}}\right). \tag{113}$$

Moving fronts. Fronts between patterns of different types can be considered in a similar way. Assuming that phase modulations of patterns are absent, so that $a_n = r_n$ is real for any $n = 1, \ldots, N$, and the front is flat, we can rewrite the system of equations (106) in the form

$$\frac{\partial r_n}{\partial T} = D_n \frac{\partial^2 r_n}{\partial X^2} + \frac{\partial U(r_1, \ldots, r_N)}{\partial r_n}, \quad n = 1, \ldots, N, \tag{114}$$

where

$$U(r_1, \ldots, r_N) = \frac{1}{2}\sum_{n=1}^{N} r_n^2 - \frac{1}{4}\sum_{n=1}^{N} r_n^4 - \frac{g}{2}\sum_{n=1}^{N}\sum_{m=1}^{n-1} r_n^2 r_m^2. \tag{115}$$

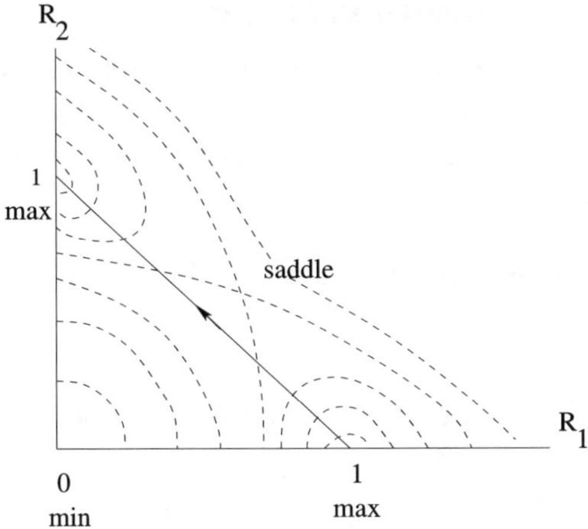

Figure 16. Contour lines of the potential energy (112). The line with an arrow corresponds to the domain wall solution.

A front propagating with a constant velocity c is described by the solution

$$r_n = r_n(\xi), \ n = 1, \ldots, N; \ \xi = X - cT \tag{116}$$

of the equation

$$D_n \frac{d^2 r_n}{d\xi^2} + c \frac{dr_n}{d\xi} + \frac{\partial U}{\partial r_n} = 0, \ n = 1, \ldots, N, \tag{117}$$

satisfying the boundary conditions

$$r_n(-\infty) = r_n^-, \ r_n(\infty) = r_n^+, \tag{118}$$

where $\{r_n^-\}$ and $\{r_n^+\}$ correspond to steady patterns on both sides of the front. Multiplying the nth equation of the set (117) by $dr_n/d\xi$, integrating with respect to ξ from $-\infty$ to ∞ and taking the sum over n from 1 to N, we find that

$$c = -\frac{U(\infty) - U(-\infty)}{\sum_{n=1}^{N} \int_{-\infty}^{\infty} d\xi (dr_n/d\xi)^2} = \frac{L^+ - L^-}{\sum_{n=1}^{N} \int_{-\infty}^{\infty} d\xi (dr_n/d\xi)^2}, \tag{119}$$

where $L^{\pm} = L(r_1^{\pm}, r_2^{\pm}, \ldots, r_N^{\pm})$ are the densities of the Lyapunov functionals for the uniform patterns situated at $X \to \pm\infty$.

Note that $c > 0$ if $L^- < L^+$, and $c < 0$ if $L^- > L^+$, i.e. the pattern with the lower Lyapunov functional density ousts the pattern with the higher Lyapunov functional density.

Cross-Newell phase diffusion equation

The Newell-Whitehead-Segel equations are valid only near the instability threshold, $\epsilon^2 \ll 1$. However, even far from the instability threshold the patterns may be subject to large-scale modulations. For their description, another approach can be applied [53].

Derivation of the Cross-Newell equation.   Let us consider the Swift-Hohenberg equation,

$$\phi_t = \gamma\phi - \left(1 + \nabla^2\right)^2 \phi - \phi^3, \tag{120}$$

with $\gamma = O(1)$. Equation (120) has a class of periodic stationary solutions corresponding to roll patterns:

$$\phi(\mathbf{r}) = \sum_{n=1}^{\infty} A_n \cos n\mathbf{k} \cdot \mathbf{r} = f(\theta), \tag{121}$$

where $\theta = \mathbf{k} \cdot \mathbf{x}$, $\mathbf{k}$ is a constant wavevector, and $f(\theta)$ is a 2π-periodic function (one can see that only Fourier components with odd n are present in the expansion (121) for $f(\theta)$).

As we know, the wavevector $\mathbf{k}$ of the roll pattern is not unique. Therefore, we can imagine a situation when the local wavevector of a roll pattern is a slow function of the coordinate $\mathbf{r}$, and it can slowly change in time:

$$\mathbf{k} = \mathbf{k}(\mathbf{R}, \ \mathbf{T}), \tag{122}$$

where $\mathbf{R} = \delta\mathbf{r}$, $\delta \ll 1$; the appropriate time scale is $T = \delta^2 t$. Note that the wavevector for the real function ϕ is defined up to its sign (thus it is similar to the order parameter of a nematic crystal, the "director"). The wavevector can be considered as a gradient of a certain phase function θ,

$$\mathbf{k} = \nabla\theta.$$

The scaling of θ is

$$\theta = \frac{1}{\delta}\Theta(\mathbf{R}, \ \mathbf{T}).$$

Let us find the solution of the Swift-Hohenberg equation (120), corresponding to a slowly distorted roll pattern with the wavevector field (122),

$$\phi = \phi(\theta, \mathbf{R}, T),$$

where ϕ is 2π-periodic in θ. Using the transformation formulas for derivatives

$$\begin{aligned}
\nabla_{\mathbf{r}}\phi &= (\mathbf{k} \cdot \partial_\theta + \delta\nabla_{\mathbf{R}})\,\phi, \\
\nabla_{\mathbf{r}}^2\phi &= (\mathbf{k} \cdot \partial_\theta + \delta\nabla_{\mathbf{R}})(\mathbf{k} \cdot \partial_\theta + \delta\nabla_{\mathbf{R}})\,\phi \\
&= \left[k^2\partial_\theta^2 + \delta\,(2\mathbf{k} \cdot \nabla_{\mathbf{R}} + \nabla_{\mathbf{R}} \cdot \mathbf{k})\,\partial_\theta + \delta^2\nabla_{\mathbf{R}}^2\right]\phi, \\
\partial_t\phi &= \partial_t\theta\partial_\theta\phi + \delta^2\partial_T\phi = \delta\partial_T\Theta\partial_\theta\phi + \delta^2\partial_T\phi,
\end{aligned}$$

we get:

$$\delta\partial_T\Theta\partial_\theta\phi + \delta^2\partial_T\phi = \gamma\phi - \phi^3 - \left[\left(1 + k^2\partial_\theta^2\right)\right.$$
$$\left. + \delta\left(2\mathbf{k}\cdot\nabla_\mathbf{R} + \nabla_\mathbf{R}\cdot\mathbf{k}\right)\partial_\theta + \delta^2\nabla_\mathbf{R}^2\right]^2\phi. \quad (123)$$

It is natural to expect that the solution can be expanded in powers of δ as

$$\phi = \phi_0 + \delta\phi_1 + \dots. \quad (124)$$

At order δ^0, we obtain the nonlinear equation,

$$\gamma\phi_0 - \phi_0^3 - \left(1 + k^2\partial_\theta^2\right)^2\phi_0. \quad (125)$$

The solution ϕ_0 coincides with the 2π-periodic function $f(\theta)$ described above, which corresponds to an undistorted roll pattern with the wavenumber k.

At order δ^1, we obtain an inhomogeneous linear equation,

$$\gamma\phi_1 - 3\phi_0^2\phi_1 - \left(1 + k^2\partial_\theta^2\right)^2\phi_1 = \partial_T\Theta\partial_\theta f \quad (126)$$

$$+ \left[\left(1 + k^2\partial_\theta^2\right)\left(2\mathbf{k}\cdot\nabla_\mathbf{R} + \nabla_\mathbf{R}\cdot\mathbf{k}\right) + \left(2\mathbf{k}\cdot\nabla_\mathbf{R} + \nabla_\mathbf{R}\cdot\mathbf{k}\right)\left(1 + k^2\partial_\theta^2\right)\right]\partial_\theta f.$$

Equation (126) is solvable on the class of 2π-periodic functions only if its right-hand side is orthogonal to the eigenfunction of the homogeneous equation, $\partial_\theta f$. Using the notation

$$\langle g\rangle \equiv \frac{1}{2\pi}\int_0^{2\pi} g d\theta,$$

we find the following solvability condition:

$$\tau(k)\partial_T\Theta + \nabla_\mathbf{R}\cdot[B(k)\mathbf{k}] = 0, \quad (127)$$

where

$$\tau(k) = \langle(\partial_\theta f)^2\rangle, \quad (128)$$

$$B(k) = 2\left[\langle(\partial_\theta f)^2\rangle - k^2\langle(\partial_\theta^2 f)^2\rangle\right]. \quad (129)$$

The relation between $\mathbf{k}$ and Θ is $\mathbf{k} = \nabla_\mathbf{R}\Theta$, hence

$$\nabla_\mathbf{R}\times\mathbf{k} = 0. \quad (130)$$

Equation (127) is called *Cross-Newell equation*. This equation is universal and can be derived for any rotationally isotropic system which produces roll patterns due to a short-wave monotonic instability. Each particular problem is characterized by specific $\tau(k)$ and $B(k)$.

Monoharmonic approximation. To calculate the functions $\tau(k)$, $B(k)$, one needs the solution of equation (125), $f(\theta)$, which cannot be found analytically. However, in a rather wide interval of γ, the higher harmonics A_n in the expansion (121) are small with respect to the basic harmonics A_1. Let us truncate the expansion (121) using only one term,

$$f \approx A\cos\theta, \tag{131}$$

and apply the Galerkin approximation with one Galerkin function, $\cos\theta$. The projection of the equation (125) on that basic function gives

$$A^2(k^2) = \frac{4}{3}\left[\gamma - (1 - k^2)^2)\right], \ 1 - \sqrt{\gamma} < k^2 < 1 + \sqrt{\gamma}. \tag{132}$$

Using the approximation (131), (132), we find that

$$\langle(\partial_\theta f)^2\rangle = \langle(\partial_\theta^2 f)^2\rangle = \frac{A^2}{2},$$

hence

$$\tau(k) = \frac{A^2}{2} = \frac{2}{3}\left[\gamma - (1 - k^2)^2)\right], \tag{133}$$

$$B(k) = (1 - k^2)A^2 = \frac{4}{3}(1 - k^2)\left[\gamma - (1 - k^2)^2)\right]. \tag{134}$$

The functions $\tau(k)$ and $B(k)$ are defined in the interval $k_L < k < k_R$, where $k_L^2 = 1 - \sqrt{\gamma}$, $k_R^2 = 1 + \sqrt{\gamma}$. The function $\tau(k)$ is positive in the whole interval $k_L < k < k_R$, while the function $B(k)$ is positive in the interval $k_L < k < k_B$ and negative in the interval $k_B < k < k_R$, where $k_B = 1$.

Equation (127) can be used for studying the stability of rolls with respect to large-scale modulations far from threshold (i.e. for $\gamma = O(1)$) rather than $\gamma \ll 1$ [53]. It turns out that the rolls are subject to a zigzag (transverse) instability if the wavenumber k of the roll pattern has $B(k) > 0$, i.e. in the interval $k_L < k < k_B$. The Eckhaus (longitudinal) modulational instability appears as $d(kB(k))/dk > 0$. The function $kB(k)$ (see formula (134) and Fig.17) has a maximum at a certain point $k = k_{EL}$ in the interval (k_L, k_B) and a minimum at a point $k = k_{ER}$ in the interval (k_B, k_R). Hence, the rolls are subject to the Eckhaus instability if their wavenumber is either in the interval (k_L, k_{EL}) or in the interval (k_{ER}, k_R). Finally, we come to the conclusion that the stability interval ("Busse balloon") is (k_B, k_{ER}), i.e. it is situated between the point k_B where $B(k)$ changes its sign, and the point k_{ER} where the function $kB(k)$ has its minimum.

Disclinations. The Cross-Newell equation can be used for studying special type of defects in roll patterns, *disclinations* [3], [54]. When going around the

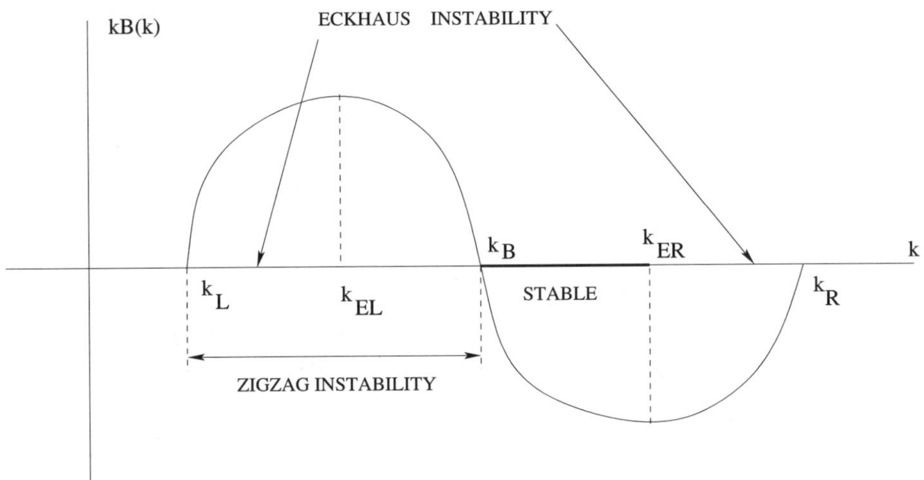

Figure 17.   Function $kB(k)$ defined by (134) and intervals of k corresponding to a stable pattern ($k_B < k < k_{ER}$), Eckhaus instability ($k_L < k < k_{EL}$) and zigzag instability ($k_L < k < k_B$).

core of the defect of this type, one observes the pattern vector rotating. As noticed above, the local wavevector **k** of a roll pattern is defined up to the sign, hence **k** is a *director* rather than a vector. Thus, the roll pattern "comes back to itself" after making a full circle around the core if the rotation angle is an integer number n multiplied by π (rather than 2π). Typical examples of disclinations observed in experiments are: (i) focus disclinations, generating a target pattern ($n = 2$), (ii) convex disclinations ($n = 1$), (iii) concave disclinations ($n = -1$), and (iv) saddle disclinations ($n = -2$). Patterns corresponding to these four types of disclinations are schematically shown in Fig.18. In the case of a motionless, steady dislocation, the corresponding field $\mathbf{k}(\mathbf{R})$ of the wavevector is governed by the equations

$$\nabla_{\mathbf{R}} \cdot [\mathbf{k}B(k)] = 0, \ \nabla_{\mathbf{R}} \times \mathbf{k} = 0 \qquad (135)$$

(except at the central point where the wavevector field $\mathbf{k}(\mathbf{R})$ has a singularity, and the wavevector is not defined).

Integration of the nonlinear equation (135) can be significantly simplified by the Legendre transformation [54]

$$\Theta(X, Y) + \hat{\Theta}(k_X, k_Y) = \mathbf{k} \cdot \mathbf{R}, \qquad (136)$$

which is equivalent to the inversion of the dependence $\mathbf{k}(\mathbf{R})$,

$$\mathbf{R} = \mathbf{R}(\mathbf{k}),$$

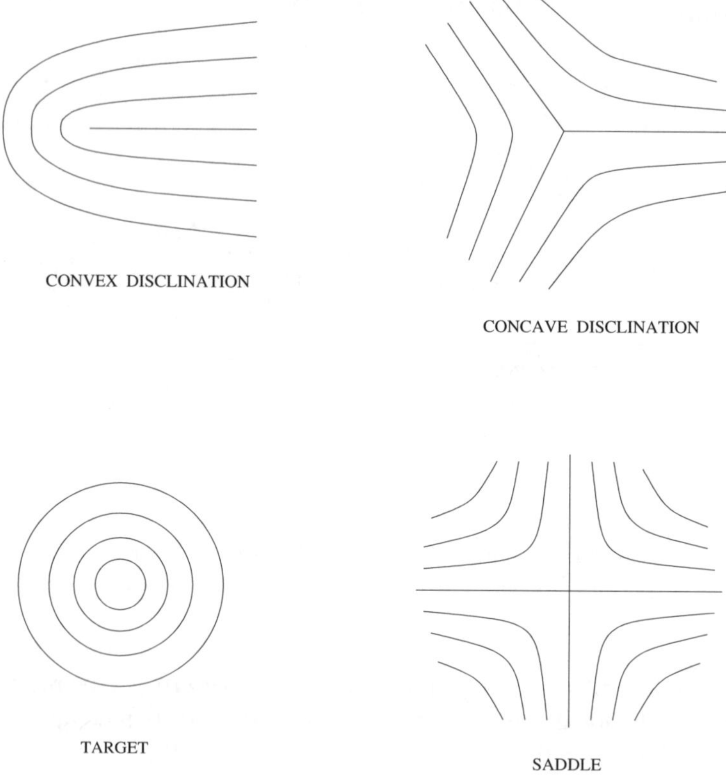

Figure 18.   Various types of disclination patterns.

and the introduction of the dual function $\hat{\Theta}(\mathbf{k})$,

$$\mathbf{R} = \nabla_{\mathbf{k}}\hat{\Theta}(\mathbf{k}).$$

For $\hat{\Theta}(\mathbf{k})$, one obtains a *linear* equation, which can be written in polar coordinates (k, φ) as

$$\frac{\partial}{\partial k}(kB)\frac{\partial \hat{\Theta}}{\partial k} + \frac{1}{k}\frac{\partial}{\partial k}(kB)\frac{\partial^2 \hat{\Theta}}{\partial^2 \varphi} = 0. \tag{137}$$

The latter equation has a set of solutions in the form

$$\hat{\Theta}(k, \varphi) = F_m(k)\cos(m\varphi) \tag{138}$$

with an integer m. One can show [54] that the solution

$$\hat{\Theta} \sim \int \frac{dk}{kB(k)}$$

corresponds to a focus disclination with a target pattern, the solution

$$\hat{\Theta} \sim k \int \frac{dk}{k^3 B(k)} \cos\varphi$$

describes a convex disclination, the solution

$$\hat{\Theta} \sim \ln(k_B - k) \cos 3\varphi$$

describes the asymptotics of a concave disclination, etc.

5.   Beyond the Swift-Hohenberg model

As shown above, the Swift-Hohenberg model (120) and its modification (37) are sufficient for the explanation of many features of pattern formation. However, there are several phenomena which need an extension of the Swift-Hohenberg model for their description.

Non-potential effects

First of all, the Swift-Hohenberg equation (120) is a *potential* equation which can be written as

$$\frac{\partial \phi}{\partial t} = -\frac{\delta F(\phi)}{\delta \phi}, \tag{139}$$

where the Lyapunov functional

$$F = \int d\mathbf{r} \left[\frac{1}{2}(\nabla^2\phi)^2 - (\nabla\phi)^2 + \frac{1 - \epsilon^2}{2}\phi^2 + \frac{1}{4}\phi^4 \right]. \tag{140}$$

Hence, the system's dynamics is fully relaxational, i.e. the evolution of the system is characterized by a monotonic decrease of the Lyapunov functional and the approach to a final stationary state, which excludes the possibility of oscillatory instabilities, spatio-temporal chaos etc.

However, dissipative physical systems usually have no Lyapunov functionals. Hence, some features of pattern formation can be overlooked by using the model (17).

In order to extend considerations to non-potential systems and to improve the coincidence of the model predictions with the results of observations and direct numerical simulations, some additional terms can be added to the right-hand side of (120), e.g. (see [55])

$$\partial_t \phi = \left[\gamma - \left(1 + \nabla^2\right)^2 \right] \phi - a\phi^3 \tag{141}$$

$$-b\phi(\nabla\phi)^2 + c\phi^2\nabla^2\phi + d\nabla^2\phi(\nabla\phi)^2 + e(\partial_i\phi)(\partial_j\phi)\partial_i\partial_j\phi.$$

Tuning of the coefficients allows us to reproduce details of the numerically obtained stability diagrams. Specifically, in addition to the Eckhaus instability (the disturbance wavevector is parallel to that of the roll) and zigzag instability (the disturbance wavevector is orthogonal to that of the roll), we can predict a *skewed-varicose* instability characterized by a disturbance wavevector inclined with respect to the wavevector of the roll.

Mean-flow effects

Spiral-defect chaos in Rayleigh-Benard convection.   The most remarkable phenomenon that needs an extension of the Swift-Hohenberg model for its explanation, is the development of *spiral-defect chaos* in Rayleigh-Benard convection [56], [6], which involves rotating spirals, target patterns, dislocations etc. The origin of this complicated behavior is the creation of a two-dimensional *mean flow*

$$\mathbf{U} = (\partial_y \zeta, -\partial_x \zeta), \tag{142}$$

with vertical vorticity $-\nabla^2 \zeta$, in the case when the rolls are curved so that

$$\left(\nabla \left(\nabla^2 \phi\right) \times \nabla \phi\right) \cdot \mathbf{e}_z = \partial_x \nabla^2 \phi \cdot \partial_y \phi - \partial_x \phi \cdot \partial_y \nabla^2 \phi \neq 0$$

[57].

In the presence of a mean flow, the generalized Swift-Hohenberg equation includes the advection of the convective rolls. It can be written, e.g., in the form

$$(\partial_t + g_m \mathbf{U} \cdot \nabla) \phi = \left[\gamma - \left(1 + \nabla^2\right)^2\right] \phi - a\phi^3 + d\nabla^2 \phi (\nabla \phi)^2 \tag{143}$$

[58], [59]. The generation of the mean flow by the curved rolls is described by a phenomenological equation [58], [60]

$$\left[\tau_\zeta \partial_t - P \left(\eta \nabla^2 - c^2\right)\right] \nabla^2 \zeta = \left[\nabla \left(\nabla^2 \phi\right) \times \nabla \phi\right] \cdot \mathbf{e}_z, \tag{144}$$

where P is the Prandtl number, τ_ζ, η and c^2 are positive constants. The system (142)-(144) describes the transient spiral-defect chaos reasonably well [58], [59], though the long-time dynamics may be questionable [60].

Hydrodynamic effects in diblock copolymer films.   Similarly, the model (14) can be extended by adding a flow. The phenomenological system of equations looks as follows [61]:

$$\frac{\partial \phi}{\partial t} + (\mathbf{v} \cdot \nabla)\phi = \nabla^2 \frac{\delta F\{\phi\}}{\delta \phi}, \tag{145}$$

$$\rho \left(\frac{\partial}{\partial t} - \nu \nabla^2\right) \mathbf{v} = -\mathbf{T}\left[\phi \nabla \frac{\delta F\{\phi\}}{\delta \phi}\right], \tag{146}$$

where $F\{\phi\}$ is the free energy functional (12), the operator $\mathbf{T}$ selects the transverse component of the vector field it is applied to, so that the incompressibility condition $\nabla \cdot \mathbf{v}$ is satisfied, ρ and ν are the density and kinematic viscosity. Taking the curl of equation (146), one obtains

$$\rho \left(\frac{\partial}{\partial t} - \nu \nabla^2 \right) \mathbf{\Omega} = \nabla \frac{\delta F\{\phi\}}{\delta \phi} \times \nabla \phi, \qquad (147)$$

where $\mathbf{\Omega} = \nabla \times \mathbf{v}$ is the vorticity. In thin films, one can average equation (147) across the film, take its z-component, and disregard the horizontal derivatives of $\mathbf{\Omega}$ compared to the vertical one. Finally, one obtains the following system of equations:

$$\frac{\partial \phi}{\partial t} + (\mathbf{v} \cdot \nabla)\phi = \nabla^2(-\phi + \phi^3 - \nabla^2 \phi) - \Gamma \phi, \qquad (148)$$

$$\nabla^2 \zeta = g\mathbf{e}_z \cdot \left[\nabla \left(\nabla^2 + \Gamma \nabla^{-2} \right) \phi \times \nabla \phi \right], \qquad (149)$$

where $\nabla^{-2}\phi(\mathbf{r})$ denotes $- \int d\mathbf{r}' G (\mathbf{r} - \mathbf{r}') \phi(\mathbf{r}')$.

The stability analysis of roll patterns performed in the framework of the system (148), (149), reveals a skewed-varicose instability. Numerical simulations predict the development of labyrinthine, spiral and target patterns.

6. Wavy patterns

Until now, we considered patterns which are developed due to a primary monotonic instability. Let us now consider the case when this instability is oscillatory.

Oscillatory instability

As the simplest example, let us take a one-dimensional two-component *reaction-diffusion system*

$$u_t = D_u u_{xx} + f(u, v), \quad v_t = D_v v_{xx} + g(u, v) \qquad (150)$$

(see, e.g., [9]). A uniform steady state (u_0, v_0) is the solution of the system

$$f(u_0, v_0) = g(u_0, v_0) = 0.$$

Assuming

$$u = u_0 + \tilde{u}e^{ikx + \sigma t}, \quad v = v_0 + \tilde{v}e^{ikx + \sigma t}$$

and linearizing around the steady state, we obtain the following equation for the growth rate σ:

$$\begin{vmatrix} f_u - D_u k^2 - \sigma & f_v \\ g_u & g_v - D_v k^2 - \sigma \end{vmatrix},$$

or

$$\sigma^2 - (\text{tr } M(k^2))\sigma + \det M(k^2) = 0,$$

where

$$M(k^2) = \begin{pmatrix} f_u - D_u k^2 & f_v \\ g_u & g_v - D_v k^2 \end{pmatrix},$$

hence

$$\sigma = \frac{1}{2}\text{tr } M(k^2) \pm \sqrt{\left(\frac{1}{2}\text{tr } M(k^2)\right)^2 - \det M(k^2)}. \qquad (151)$$

Assume that

$$\det M(0) > \left(\frac{1}{2}\text{tr } M(0)\right)^2.$$

In this case, the imaginary part of the growth rate is nonzero for sufficiently long waves. When

$$\text{tr} M(0) = f_u + g_v$$

grows and crosses 0, a *long-wave oscillatory instability* is developed in the system.

1D complex Ginzburg-Landau equation

Near the threshold, one can derive an equation for a slowly changing in time and space amplitude function $A(t_2, x_1)$ [62], which is similar to the NWS equation discussed in Sec. 3. In the case of a supercritical Hopf bifurcation, the generic amplitude equation, after rescaling, reads (cf.(73)):

$$A_t = A + (1 + i\alpha)A_{xx} - (1 + i\beta)|A|^2 A. \qquad (152)$$

The latter equation is called the *complex Ginzburg-Landau equation*. Using the notation $A = R\exp(i\theta)$, one can rewrite (152) as a system of two real equations

$$R_t = R + \left(R_{xx} - R\theta_x^2\right) - \alpha\left(2R_x\theta_x + R\theta_{xx}\right) - R^3, \qquad (153)$$

$$R\theta_t = (2R_x\theta_x + R\theta_{xx}) + \alpha\left(R_{xx} - R\theta_x^2\right) - \beta R^3. \qquad (154)$$

A vast literature is devoted to the investigation of properties of the complex Ginzburg-Landau equation (see the review paper [65]). Here we discuss only a few of the most basic topics.

Periodic waves.   The system of equation (153), (154) has a one-parameter family of solutions

$$R = R_0(K), \ \theta = Kx - \Omega(K)t, \ |K| < 1,$$

where

$$R_0(K) = \sqrt{1 - K^2}, \; \Omega(K) = \beta + (\alpha - \beta)K^2,$$

which correspond to spatially periodic solutions $A(x,t)$ of equation (152). Similar to the stationary roll solutions studied in Sec. 3, these solutions may be stable or unstable, depending on the parameters K, α and β.

Stability of uniform oscillations. For the sake of simplicity, we will consider here only the stability of uniform oscillations, $K = 0$, $R_0(0) = 1$, $\Omega(0) = \beta$. For normal disturbances

$$(\tilde{R}, \tilde{\theta}) \sim e^{\sigma t + i\tilde{K}x},$$

one obtains the following linearized problem:

$$\sigma \tilde{R} = \tilde{R}(1 - \tilde{K}^2) + \alpha \tilde{K}^2 \tilde{\theta} - 3\tilde{R},$$

$$-\beta \tilde{R} + \sigma \tilde{\theta} = -\tilde{K}^2 \tilde{\theta} - 3\beta \tilde{R} - \alpha \tilde{K}^2 \tilde{R},$$

hence

$$\begin{vmatrix} -2 - \tilde{K}^2 - \sigma & \alpha \tilde{K}^2 \\ -2\beta - \alpha \tilde{K}^2 & -\tilde{K}^2 - \sigma \end{vmatrix} = 0,$$

or

$$\sigma^2 + 2\left(1 + \tilde{K}^2\right)\sigma + 2\tilde{K}^2(1 + \alpha\beta) + (1 + \alpha^2)\tilde{K}^4. \tag{155}$$

If $1 + \alpha\beta < 0$, the uniform oscillations are unstable with respect to modulations with the wavenumbers in the interval

$$0 < \tilde{K}^2 < K_m^2 = -\frac{1 + \alpha\beta}{1 + \alpha^2}.$$

This instability is known as the *Benjamin-Feir instability*. Regions of the Benjamin-Feir instability in (α, β) parameter plane are shown in Fig.19.

Nonlinear phase equation. If K_m^2 is small:

$$1 + \alpha\beta = -\epsilon^2, \; \beta = -\frac{1}{\alpha} - \frac{\epsilon^2}{\alpha}, \; \epsilon \ll 1,$$

the nonlinear evolution of the Benjamin-Feir instability can be studied by means of a nonlinear phase equation (cf. Sec. 4.3.1) [63].
The linear theory predicts $\tilde{K} \sim \epsilon$, $\sigma \sim \epsilon^4$. Therefore, we assume

$$R = R(X, T), \; \theta = -\beta t + \epsilon^2 \vartheta(X, T); \; X = \epsilon x; \; T = \epsilon^4 t$$

and obtain:

$$\epsilon^4 R_T = R + \epsilon^2 \left[\left(R_{XX} - \epsilon^4 R \vartheta_X^2\right) - \alpha\epsilon^2 \left(2R_X \vartheta_X + R \vartheta_{XX}\right) - R^3 \right], \tag{156}$$

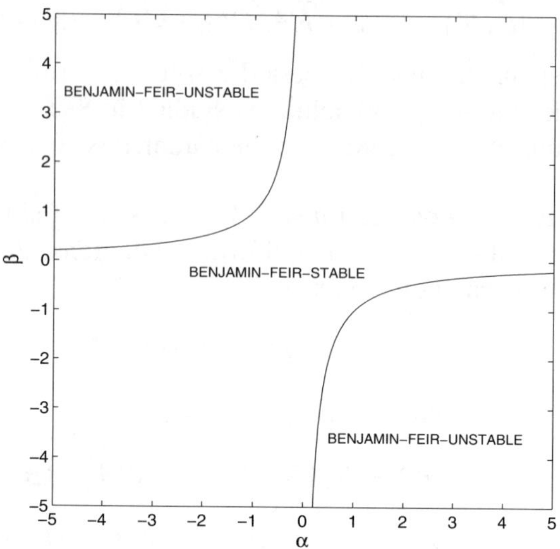

Figure 19.   Regions of the Benjamin-Feir instability.

$$\left(\frac{1}{\alpha} + \frac{\epsilon^2}{\alpha}\right) R + \epsilon^6 R\vartheta_T = \epsilon^2 \left[\left(2R_X\vartheta_X + R\vartheta_{XX}\right)\epsilon^2\right.$$

$$\left. + \alpha\left(R_{XX} - \epsilon^4 R\vartheta_X^2\right)\right] + R^3\left(\frac{1}{\alpha} + \frac{\epsilon^2}{\alpha}\right). \tag{157}$$

Let us construct the solution in the form of an asymptotic series in powers of ϵ^2. We find that the amplitude R is slaved to the phase ϑ_0:

$$r = 1 - \frac{1}{2}\epsilon^4\alpha\vartheta_{XX}^{(0)} + \frac{1}{2}\epsilon^6\left(-\frac{1}{2}\vartheta_{XXXX}^{(0)} - (\vartheta_X^{(0)})^2 - \alpha\vartheta_{XX}^{(2)}\right) + \ldots.$$

The leading order equation governing the evolution of the phase is the *Kuramoto-Sivashinsky equation*:

$$\vartheta_T^{(0)} = -\vartheta_{XX}^{(0)} - \frac{1}{2}\left(1 + \alpha^2\right)\vartheta_{XXXX}^{(0)} - \left(\alpha + \alpha^{-1}\right)\left(\vartheta_X^{(0)}\right)^2. \tag{158}$$

Equation (158) is a paradigmatic model for studying spatio-temporal chaos [64]. It exhibits solutions in the form of spatially-irregular "cells" splitting and merging in a chaotic manner in time. An example of spatio-temporal behavior of a chaotic solution of the Kuramoto-Sivashinsky equation (158) as well as a snapshot of this solution at a particular moment of time are shown in Fig.20.

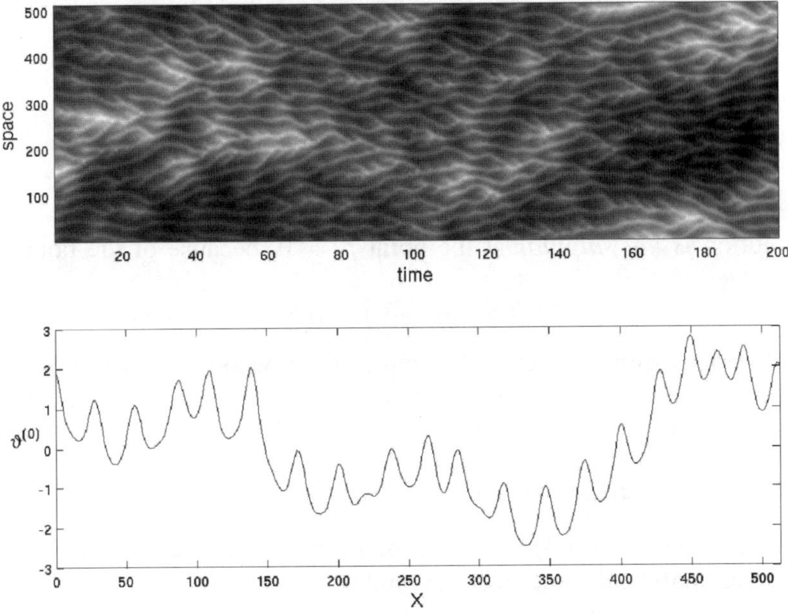

Figure 20. Numerical solutions of eq.(158): spatio-temporal diagram showing spatio-temporal chaos (upper figure) and a snapshot of the solution at a particular moment of time.

2D complex Ginzburg-Landau equation: spiral wave

In the 2D case, the complex Ginzburg-Landau equation reads:

$$A_t = A + (1 + i\alpha)\nabla^2 A - (1 + i\beta)|A|^2 A, \tag{159}$$

or ($A = R\exp(i\theta)$)

$$R_t = R + \nabla^2 R - R(\nabla\theta)^2 - \alpha(2\nabla R \cdot \nabla\theta + R\nabla^2\theta) - R^3, \tag{160}$$

$$R\theta_t = 2\nabla R \cdot \nabla\theta + R\nabla^2\theta + \alpha(\nabla^2 R - R(\nabla\theta)^2) - \beta R^3. \tag{161}$$

The 2D complex Ginzburg-Landau equation is characterized by extremely diverse behavior (see [65]). Here we will discuss the most remarkable objects typical for this equation, spiral waves.

A spiral wave is a solution of the type

$$R = R(\rho), \ \theta(\rho, \varphi, t) = m\varphi + \psi(\rho) + \Omega t, \tag{162}$$

where (ρ, φ) are polar coordinates in the plane (x, y). For the sake of simplicity, we shall take $\alpha = 0$, and consider only waves with $m = 1$. The problem is governed by the following system of equations:

$$R'' + \frac{1}{\rho}R' + R\left[1 - R^2 - (\psi')^2 - \frac{1}{\rho^2}\right] = 0, \tag{163}$$

$$q' + \frac{1}{\rho}q + \frac{2R'}{R}q - \beta R^2 = \Omega. \tag{164}$$

Here $'$ denotes differentiation with respect to ρ; $q = \psi'$.

The solution *is not singular* at the point $\rho = 0$, because of the boundary conditions

$$R(0) = 0, \; q(0) = 0. \tag{165}$$

At large distances from the center the spiral wave becomes indistinguishable from a plane wave with a certain wavenumber k_∞, which is the eigenvalue of the nonlinear problem (163), (164):

$$R(\infty) = \sqrt{1 - k_\infty^2}, \; q(\infty) = k_\infty. \tag{166}$$

The general problem (163)-(166) can be solved numerically. Here we will present a semi-analytical solution in the limit of small β: $\beta = -\epsilon, |\epsilon| \ll 1$ [66], [67]. In this limit, one can distinguish between the core region ($\rho = O(1)$) and far field region ($\rho = O(1/\epsilon)$), where the asymptotic expansions are different.

Inner expansion.   In the region $\rho = O(1)$, we seek the solution to the system (163)-(166) in the form

$$R(\rho) = R_0(\rho) + \epsilon R_1(\rho) + \ldots, \; q = \epsilon q_1 + \ldots.$$

Also, we assume that $|k_\infty| \ll 1$.

At leading order, we obtain the following nonlinear problem for $R_0(\rho)$:

$$R_0'' + \frac{1}{\rho}R_0' + \left(1 - \frac{1}{\rho^2} - R_0^2\right)R_0 = 0; \; R_0(0) = 0; \; |R_0(\infty)| < \infty. \tag{167}$$

This problem was studied earlier in the context of the vortex core for the Gross-Pitaevskii equation which describes a superfluid flow. It is known that the solution of this problem exists and is unique. The solution can be found only numerically. For large ρ, the asymptotics of the solution is $R_0^2 \sim 1 - 1/\rho^2 + \ldots$.

For the local wavenumber q_1, we obtain the linear problem:

$$q_1' + \left(\frac{1}{\rho} + \frac{2R_0'}{R_0}\right)q_1 = 1 - R_0^2, \; q(0) = 0. \tag{168}$$

Its solution can be written explicitly as:

$$q_1(\rho) = \frac{1}{\rho R_0^2} \int_0^\rho d\rho' \rho' R_0^2(\rho')[1 - R_0^2(\rho')]. \tag{169}$$

For large ρ, the solution (169) behaves as

$$q_1(\rho) \sim \frac{1}{\rho}(\ln \rho + C + \ldots), \tag{170}$$

hence

$$\psi'(\rho) \sim \frac{\epsilon}{\rho}(\ln \rho + C + \ldots). \tag{171}$$

A numerical evaluation of the constant C gives $C \approx -0.098$.

Outer expansion. For the construction of the outer expansion, it is better to return to the original system of equations (160)-(161) and take into account that for $\rho \gg 1$ the spatial derivatives of the fields are small. We find that the amplitude field R is slaved to the phase field θ:

$$R^2 \sim 1 - (\nabla\theta)^2.$$

Taking into account that $\Omega = \epsilon(1 - k_\infty^2)$, we obtain the following nonlinear phase equation:

$$\nabla^2\theta = \epsilon\left[k_\infty^2 - (\nabla\theta)^2\right] = 0. \tag{172}$$

This is the *Burgers equation* which can be linearized by means of the *Hopf-Cole transformation*:

$$\theta = -\frac{1}{\epsilon}\ln F, \tag{173}$$

$$\nabla^2 F - \epsilon^2 k_\infty^2 F = 0. \tag{174}$$

Introduce the scaled variable $s \equiv \epsilon k_\infty \rho$. The spiral-wave solution $\theta = \varphi + \psi(s)$ is transformed to

$$F = e^{-\epsilon\varphi}H(s),$$

where $H(s)$ satisfies the equation

$$\frac{d^2 H}{ds^2} + \frac{1}{s}\frac{dH}{ds} - \left(1 - \frac{\epsilon^2}{s^2}\right)H = 0. \tag{175}$$

The appropriate solution is a Bessel function with imaginary index: $H = \text{const } K_{i\epsilon}(s)$, hence

$$F = \text{const } e^{-\epsilon\varphi}K_{i\epsilon}(s), \quad \theta = \varphi - \frac{1}{\epsilon}\ln K_{i\epsilon}(\epsilon k_\infty \rho);$$

$$\psi(\rho) = -\frac{1}{\epsilon}\ln K_{i\epsilon}(\epsilon k_\infty \rho).$$

The derivative $\psi'(\rho)$ should be matched to the expression (171) obtained from the inner solution.

Matching. Using the asymptotics of the Bessel function for small $\epsilon k_\infty \rho$,

$$K_{i\epsilon}(\epsilon k_\infty \rho) \sim \sin\left(\epsilon \ln \frac{\epsilon k_\infty \rho}{2}\right) + \epsilon\gamma \cos\left(\epsilon \ln \frac{\epsilon k_\infty \rho}{2}\right),$$

where $\gamma \approx 0.577$ is the Euler constant, we find:

$$-\frac{1}{\rho}\frac{\cos\left(\epsilon \ln \frac{\epsilon k_\infty \rho}{2}\right) - \epsilon\gamma \sin\left(\epsilon \ln \frac{\epsilon k_\infty \rho}{2}\right)}{\sin\left(\epsilon \ln \frac{\epsilon k_\infty \rho}{2}\right) + \epsilon\gamma \cos\left(\epsilon \ln \frac{\epsilon k_\infty \rho}{2}\right)} \sim \frac{\epsilon}{\rho}(\ln \rho + C).$$

The matching can be performed if

$$\epsilon \ln \frac{\epsilon k_\infty \rho}{2} = -\frac{\pi}{2} + \delta, |\delta| \ll 1,$$

so that

$$\cos\left(\epsilon \ln \frac{\epsilon k_\infty \rho}{2}\right) \sim \delta, \ \sin\left(\epsilon \ln \frac{\epsilon k_\infty \rho}{2}\right) \sim -1,$$

and the asymptotics of the outer solution is given by the expression

$$\psi'(\rho) \sim \frac{1}{\rho}(\delta + \epsilon\gamma) = \frac{1}{\rho}\left(\frac{\pi}{2} + \epsilon \ln \frac{\epsilon k_\infty \rho}{2} + \epsilon\gamma\right). \tag{176}$$

Comparing (176) and (171), we find the matching condition:

$$\frac{\pi}{2} + \epsilon \ln \frac{\epsilon k_\infty}{2} + \epsilon\gamma = \epsilon C.$$

Thus, the selected wavenumber k_∞ is determined by the formula

$$k_\infty = \frac{2}{\epsilon} \exp\left(-\frac{\pi}{2\epsilon} - \gamma + C\right). \tag{177}$$

Fig.21 shows an example of a spiral-wave solution of eq.(159) obtained numerically. The left figure is a snap-shot corresponding to a particular moment of time. The spiral is rotating counterclockwise with constant frequency. One can see that far from the spiral core the wavenumber is constant and so is the amplitude $|A|$ that is related to the wavenumber by eq.(166). In the core center $A = 0$.

7.   Conclusions

We have discussed universal features of pattern formation in several systems that are often encountered at nanometer scales. These systems can be divided in two large classes: variational (potential) systems and non-variational (non-potential) systems. Variational systems are characterized by a free energy

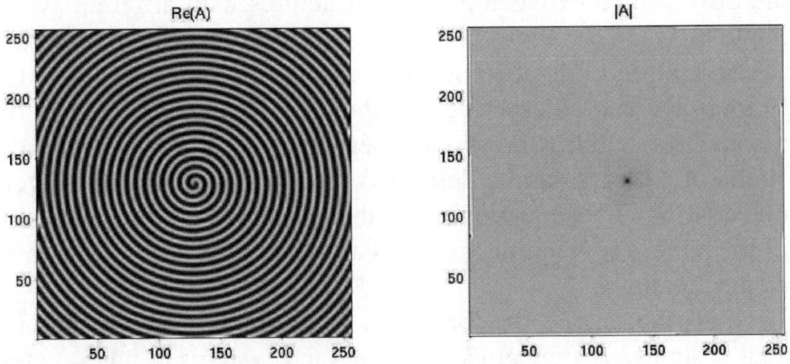

Figure 21.   Numerical solution of the complex Ginzburg-Landau equation (159) in the form of a spiral wave.

functional and the system evolution leads to the steady state with the minimum free energy. Non-variational systems can exhibit more complex, nonstationary behavior.

One of the main physical reasons for the formation of patterns in variational systems is phase separation. We have shown that, depending on the type of interactions in a particular system phase separation can result from a long-wave instability and lead to the formation of irregular domains of different phases that coarsen in time (as in binary alloys), or can result from a short-wave instability and lead to self-assembly of stable, spatially periodic structures (as in diblock-copolymer systems). In the latter case the pattern formation phenomenon is similar to that in macroscopic systems, like Rayleigh-Benard convection, and can be described by some generic equations, particularly by the Swift-Hohenberg equation. We have discussed the selection of patterns with the two most common symmetries: roll patterns and hexagonal patterns. We have shown that their dynamics and the selection process is described by a system of Landau equations and the competition between the patterns is determined by the Landau nonlinear interaction coefficients. Near the instability threshold, patterns in large aspect-ratio systems can undergo spatio-temporal modulations. We have shown that these modulations are described by the Newel-Whithead-Segel (NWS) equation that can describe two basic modulational instabilities of spatially periodic patterns: the Eckhaus instability and the zig-zag instability. These two instabilities are associated with the mechanism of the wavenumber selection. They are also associated with the formation of the two most common types of defects in spatially periodic patterns: dislocations and domain walls between the domains of patterns with different wavenumbers. The structure of these defects is described by the nonlinear phase diffusion equations that follow from the NWS equation. Moving fronts are also formed between periodic structures with different symmetries, for example between hexagons and rolls, and serve as a vehicle for pattern selection. The motion of such fronts is similar to the motion of a particle in a potential well and can be described by the corresponding Lagrange function. Farther from the instability threshold, where the NWS equation is not valid, the modulations of patterns can be described by a nonlinear Cross-Newell phase diffusion equation. This equation is capable of describing large but slow variations of the pattern wavenumber as well as another type of defects in periodic patterns: disclinations.

We have also discussed the formation of spatio-temporal patterns in non-variational systems. A typical example of such systems at nano-meter scales is reaction-diffusion systems that are ubiquitous in biology, chemical catalysis, electrochemistry, etc. These systems are characterized by the energy supply from the outside and can exhibit complex nonlinear behavior like oscillations and waves. A macroscopic example of such a system is Rayleigh-Benard convection accompanied by mean flow that leads to strong distortion of periodic patterns and the formation of labyrinth patterns and spiral waves. Similar nano-meter scale patterns are observed during phase separation of di-block copolymer films in the presence of hydrodynamic effects. The pattern's nonlinear dynamics in both macro- and nano-systems can be described by a Swift-Hohenberg equation coupled to the non-local mean-flow equation.

A typical feature of a non-potential systems is the non-stationary oscillatory behavior that usually manifests itself in the propagation of waves. We have shown that the nonlinear evolution of waves near the instability threshold is described by the complex Ginzburg-Landau (CGL) equation. This equation is capable of describing various kinds of instabilities of wave patterns, like the Benjamin-Feir instability. In two dimensions, the CGL equation describes the formation of spiral waves that are observed in many biological and chemical systems characterized by the interplay of diffusion and chemical reactions at nano-scales.

The material of this chapter can be useful for understanding other chapters of the present book. For a subsequent reading, the review papers [1], [65] and the book [2] are also recommended.

Acknowledgement

This work was supported by the NATO ASI grant 980684, and by the US Department of Energy grant #DE-FG02-03ER46069.

References

[1] M.C. Cross and P.C. Hohenberg, Rev. Mod. Phys. **65** (1993) 851.

[2] D. Walgraef, *Spatio-Temporal Pattern Formation with Examples from Physics, Chemistry and Materials Science*, Springer-Verlag, Berlin, 1997.

[3] C. Bowman and A.C. Newell, Rev. Mod. Phys. **70** (1998) 289.

[4] H. Benard, Ann. Chim. Phys. **7**, Ser. 23 (1900) 62.

[5] Lord Rayleigh, Proc. Roy. Soc. London, Ser. A **93** (1916) 148.

[6] M. Assenheimer and V. Steinberg, Nature (London) **367** (1994) 345.

[7] E. Bodenschatz, W. Pesch, and G. Ahlers, Annu. Rev. Fluid Mech. **32** (2000) 709.

[8] E.L. Koschmieder, Adv. Chem. Phys. **26** (1974) 177.

[9] J.D. Murray, *Mathematical Biology*, Springer-Verlag, Berlin, 1989.

[10] T. Ackemann and T. Lange, Appl. Phys. B: Lasers Opt. **72** (2001) 21.

[11] Q. Ouyang and H.L. Swinney, Nature (London) **352** (1991) 61.

[12] A.A. Abrikosov, Sov. Phys. JETP **5** (1957) 1174.

[13] M. Seul and R. Wolfe, Phys. Rev. Lett. **68** (1992) 2460.

[14] F.S. Bates, M.F. Schultz, A.K. Khandpur, S. Foster, J.H. Rosedale, K. Almdahl, and K. Mortensen, Faraday Discuss. **98** (1994) 7.

[15] G. Springholtz, V. Holy, M. Pinczolits and G. Bauer, Science **282** (1998) 734.

[16] H. Masuda and K. Fukuda, Science **268** (1995) 1466.

[17] A.Huczko, Appl. Phys. A **70** (2000) 365.

[18] Y. Chen and A. Pepin, Electrophoresis **22** (2001) 187.

[19] A. Joets and R. Ribotta, J. Phys. (Paris) **47** (1986) 595.

[20] W.S. Edwards and S. Fauve, J. Fluid Mech. **278** (1994) 123.

[21] P. Kolodner, C.M. Surko, and H. Williams, Physica D **37** (1989) 319.

[22] C.H. Blohm and H.C. Kuhlmann, J. Fluid Mech. **450** (2002) 67.

[23] P. Le Gal, A. Pocheau, and V. Croquette, Phys. Rev. Lett. **54** (1985) 2501.

[24] A.M. Rucklidge, Proc. Roy. Soc. Lond. A **453** (1997) 107.

[25] J.W. Cahn and J.E. Hilliard, J. Chem. Phys. **28** (1958) 258.

[26] J.S. Langer, Ann. Phys. (New York) **65** (1971) 53.

[27] K. Kawasaki and T. Ohta, Physica A **116** (1982) 573.

[28] H. Calisto, M. Clerc, R. Rojas, and E. Tirapegui, Phys. Rev. Lett. **85** (2000) 3805.

[29] I.M. Lifshitz and V.V. Slyozov, J. Phys. Chem. Solids **19** (1961) 35.

[30] A. Chakrabarti, R. Toral, and J.D. Gunton, Phys. Rev. E **47** (1993) 3025.

[31] L. Ratke and P.W. Voorhees, *Growth and Coarsening: Ostwald Ripening in Material Processing (Engineering Materials)*, Springer, 2002.

[32] A.J. Bray, Adv. Phys. **51** (2002) 481.

[33] L. Leibler, Macromolecules **13** (1980) 1602.

[34] T. Ohta and K. Kawasaki, Macromolecules **19** (1986) 2621.

[35] Y. Oono and Y. Shiwa, Mod. Phys. Lett. B **1** (1987) 49.

[36] J.J. Christensen and A.J. Bray, Phys. Rev. E **58** (1998) 5364.

[37] J. W. Swift and P.C. Hohenberg, Phys. Rev. A **15** (1977) 319.

[38] V.E. Zakharov, V.S. L'vov, and G. Falkovich, *Kolmogorov Spectra of Turbulence*, Springer-Verlag, Berlin, 1992.

[39] A.C. Newell and Y. Pomeau, J. Phys. A **26** (1993) L429.

[40] A. Schluter, D. Lortz, and F. Busse, J. Fluid Mech. **23** (1965) 129.

[41] V.L. Gertsberg and G.I. Sivashinsky, Prog. Theor. Phys. **66** (1981) 1219.

[42] G.Z. Gershuni and E.M. Zhukhovitsky, *Convective Stability of Incompressible Fluid*, Keter, Jerusalem, 1976.

[43] F.H. Busse, J. Fluid Mech. **30** (1967) 625.

[44] L.D. Landau and E.M. Lifshitz, *Statistical Physics*, Pergamon Press, Oxford, 1980.

[45] J. Kevorkian and J.D. Cole, *Multiple Scale and Singular Perturbation Methods*, Springer, New York, 1996.

[46] A.C. Newell and J.A. Whitehead, J. Fluid Mech. **38** (1969) 279.

[47] L.A. Segel, J. Fluid Mech. **38** (1969) 203.

[48] W. Eckhaus, *Studies in Non-linear Stability Theory*, Springer-Verlag, Berlin, 1965.

[49] E.D. Siggia and A. Zippelius, Phys. Rev. A **24** (1981) 1036.

[50] A.A. Nepomnyashchy and L.M. Pismen, Phys. Lett. A **153** (1991) 427.

[51] B.A. Malomed, A.A. Nepomnyashchy and M.I. Tribelsky, Phys. Rev. A **42** (1990) 7244.

[52] Y. Pomeau, Physica D **23** (1986) 3.

[53] M.C. Cross and A.C. Newell, Physica D **10** (1984) 299.

[54] T. Passot and A.C. Newell, Physica D **74** (1994) 301.

[55] H.S. Greenside and M.C. Cross, Phys. Rev. A **31** (1985) 2492.

[56] S.W. Morris, E. Bodenschatz, D. Cannel, and G. Ahlers, Phys. Rev. Lett. **71** (1993) 2026.

[57] L.M. Pismen, Phys. Lett. A **116** (1986) 241.

[58] H.-W. Xi, J.D. Gunton, and J. Vinals, Phys. Rev. Lett. **71** (1993) 2030.

[59] M.C. Cross and Y. Tu, Phys. Rev. Lett. **75** (1995) 834.

[60] R. Schmitz, W. Pesch, and W. Zimmermann, Phys. Rev. E **65** (2002) 037302.

[61] Y. Shiwa, Phys. Rev. E **61** (2000) 2924.

[62] Y. Kuramoto and T. Tsuzuki, Progr. Theor. Phys. **55** (1976) 356.

[63] T. Yamada and Y. Kuramoto, Progr. Theor. Phys. **56** (1976) 681.

[64] T. Bohr, M.H. Jensen, G. Paladin, and A. Vulpiani, *Dynamical System Approach to Turbulence*, University Press, Cambridge, 1998.

[65] I.S. Aranson and L. Kramer, Rev. Mod. Phys. **74** (2002) 99.

[66] P.S. Hagan, SIAM J. Appl. Math. **42** (1982) 762.

[67] L.M. Pismen and A.A. Nepomnyashchy, Physica D **54** (1992) 183.

CONVECTIVE PATTERNS IN LIQUID CRYSTALS DRIVEN BY ELECTRIC FIELD

An overview of the onset behavior

Agnes Buka
Research Institute for Solid State Physics and Optics of the Hungarian Academy of Sciences
H-1525 Budapest, POB. 49, Hungary
ab@szfki.hu

Nándor Éber
Research Institute for Solid State Physics and Optics of the Hungarian Academy of Sciences
H-1525 Budapest, POB. 49, Hungary
eber@szfki.hu

Werner Pesch
Institute of Physics, University of Bayreuth
D-95440 Bayreuth, Germany
Werner.Pesch@uni-bayreuth.de

Lorenz Kramer
Institute of Physics, University of Bayreuth
D-95440 Bayreuth, Germany

Abstract   A systematic overview of various electric-field induced pattern forming instabilities in nematic liquid crystals is given. Particular emphasis is laid on the characterization of the threshold voltage and the critical wavenumber of the resulting patterns. The standard hydrodynamic description of nematics predicts the occurrence of striped patterns (rolls) in five different wavenumber ranges, which depend on the anisotropies of the dielectric permittivity and of the electrical conductivity as well as on the initial director orientation (planar or homeotropic). Experiments have revealed two additional pattern types which are not captured by the standard model of electroconvection and which still need a theoretical explanation.

Keywords: Pattern formation, instabilities, liquid crystals, electroconvection

55

A. A. Golovin and A. A. Nepomnyashchy (eds.),
Self-Assembly, Pattern Formation and Growth Phenomena in Nano-Systems, 55–82.
© 2006 *Springer. Printed in the Netherlands.*

Introduction

Patterns formed in non-equilibrium systems are fascinating objects, which arise in many physical, chemical and biological systems [1]. *Liquid crystals*, the other key word in the title, are substances with captivating properties [2–4] and their study has made amazing progress – both in basic research and applications - in the past couple of years. How do these two subjects join and *why are liquid crystals especially attractive for studying pattern formation?* An attempt will be made in this tutorial review to answer these questions to some extent.

Pattern forming phenomena in liquid crystals can be divided into two groups. *In the first group*, liquid crystals replace isotropic fluids in the study of well known classical phenomena like Rayleigh-Benard convection, Taylor vortex flow, viscous fingering, free solidification from melt, directional solidification etc. [5]. This gives a possibility to extend the investigations from simple isotropic systems to more complex, partially ordered media. As liquid crystals are intrinsically anisotropic substances, unusual nonlinear couplings (e.g. electro-mechanical, thermo-mechanical, rotation-flow) of the hydrodynamic variables become possible. They induce various focussing effects (heat, light or charge) which in many cases give rise to considerably lower values of the external control parameters at the onset of the instability.

In the second group we find pattern forming phenomena based on new instability mechanisms arising from the specific features of liquid crystals, which have no counterpart in isotropic fluids or at least are difficult to assess. Some examples are shear (linear, elliptic, oscillatory, etc.) induced instabilities, transient patterns in electrically or magnetically driven Freedericksz transitions, structures formed in inhomogeneous and/or rotating electric or magnetic fields, electroconvection (EC), etc. [5–7].

Liquid crystals have become an important paradigm to study generic aspects of pattern forming mechanisms. Besides their stability the large aspect ratios of the typical convection cells allows the observation of extended regions with regular roll patterns. The patterns are easy to visualize by exploiting the bire-fringence of liquid crystals. It is convenient, that the number of accessible control parameters is larger than in standard isotropic systems. For instance, one can easily tune magnetic fields or the amplitude and the frequency of an applied voltage. It is also not difficult to change the symmetry of a convection layer via the boundary conditions or the amount of anisotropy by changing the temperature, in order to observe the effect of a transition from an isotropic to an anisotropic pattern forming system. In general one might state that liquid crystals have just the right amount and right kind of complexity and non-linearity, to make them so attractive. The understanding of patterns in liquid crystals has immensely benefitted from a close collaboration between experimentalists and theoreticians. Though the hydrodynamic description of liquid crystals may look prohibitively complex in fact a quantitative description of experiments has been achieved in many cases. In addition universal aspects of pattern formation can be addressed more easily in some cases (e.g. in terms of amplitude equations) than in isotropic systems [8].

The overview is organized as follows. In section **1**, we outline briefly the relevant properties of liquid crystals and sketch the theoretical description. In section **2**, we discuss electroconvection for different material parameter sets and geometries focusing mainly on the onset of convection. A summary concludes the paper.

1. Physical properties of nematics

The term *liquid crystals* denotes a family of mesophases, which consist of elongated (or sometimes oblate) molecules. They are characterized by a long range *orientational order* of the molecular axes. The resulting preferred direction in the system is described by the director field $\mathbf{n}$ (with $\mathbf{n} \cdot \mathbf{n} = 1$). The various mesophases differ in the positional order of the constituent molecules. Quite often one finds with decreasing temperature a multi-step transition from the fluid-like, random positional ordering of nematic liquid crystals (nematics) through several, layered structures to smectic liquid crystals (smectics) possessing short range crystalline order. In the following, we will constrain ourselves to the highest-symmetry liquid crystalline phase – the nematic – which is the simplest representative of anisotropic uniaxial liquids.

The thermodynamical equilibrium of nematics would correspond to a spatially uniform (constant $\mathbf{n}(\mathbf{r})$) director orientation. External influences, like boundaries or external fields, often lead to spatial distortions of the director field. This results in an *elastic* increment, f_d, of the volume *free energy* density which is quadratic in the director gradients [2, 3]:

$$f_d = \frac{1}{2}K_1(\nabla \cdot \mathbf{n})^2 + \frac{1}{2}K_2(\mathbf{n} \cdot (\nabla \times \mathbf{n}))^2 + \frac{1}{2}K_3(\mathbf{n} \times (\nabla \times \mathbf{n}))^2. \quad (1)$$

Here K_1, K_2 and K_3 are elastic moduli associated with the three elementary types of deformations; splay, twist and bend, respectively. Though the three elastic moduli are of the same order of magnitude; the ordering $K_2 < K_1 < K_3$ holds for most nematics. As a consequence of the orientational elasticity a local restoring torque (later referred to as elastic torque) acts on the distorted director field which tends to reduce the spatial variations.

In most experiments (and applications) a nematic layer is sandwiched between two solid (glass) surfaces supporting transparent electrodes. Special *surface* coatings and/or treatments allow control of the director *alignment* at the bounding plates. There are two basic geometries: the *planar* one where $\mathbf{n}$ is parallel to the surfaces (usually along $\hat{\mathbf{x}}$) and the *homeotropic* one where $\mathbf{n}$

is normal to the surfaces (along $\hat{\mathbf{z}}$). In most cases, the interaction between the liquid crystal and the surface is strong enough to inhibit a change of the direction of $\mathbf{n}$ at the boundaries (strong anchoring) despite director gradients in the bulk. The surface treatments combined with the elastic torques originating from Eq. (1) ensure the initial homogeneous (i.e. no spatial variations in the plane of the layer) director alignment of liquid crystal cells.

Due to their orientational order nematics are *anisotropic* substances. Therefore, in contrast to isotropic fluids, many physical quantities of nematics must be described using tensors [2, 3]. As nematics are non-chiral they exhibit an inversion symmetry as well as a cylindrical symmetry around the director. In addition, they are characterized by a non-polar molecular packing; thus the nematic phase is invariant under the transformation $\mathbf{n} \rightarrow -\mathbf{n}$. These symmetries imply that the dielectric susceptibility ϵ, the electrical conductivity σ, and magnetic susceptibility χ tensors each have only two different components in their principal-axis system: $\epsilon_\parallel$, $\sigma_\parallel$, $\chi_\parallel$ and $\epsilon_\perp$, $\sigma_\perp$, $\chi_\perp$, respectively. For instance, the dielectric displacement $\mathbf{D}$ induced by an electric field $\mathbf{E}$ is given as $\mathbf{D} = \epsilon_\perp \mathbf{E} + \epsilon_\mathbf{a}\mathbf{n}(\mathbf{n} \cdot \mathbf{E})$. Analogous relations connect the electric current $\mathbf{j}$ to $\mathbf{E}$, and the magnetization with a magnetic field, respectively. The difference $\epsilon_a = \epsilon_\parallel - \epsilon_\perp$ defines the anisotropy of the dielectric susceptibility. Substances both with $\epsilon_a > 0$ and with $\epsilon_a < 0$ can be found among nematics, moreover, the sign may change with the frequency and/or the temperature in some compounds (at optical frequencies always $\epsilon_a > 0$). Though liquid crystals are intrinsically insulators, they usually contain (or are intentionally doped with) some ionic impurities which lead to a finite electric conductivity. In most cases the anisotropy of the electrical conductivity $\sigma_a = \sigma_\parallel - \sigma_\perp$ is positive; in other words charges are more easily transported parallel to the mean orientation (director $\mathbf{n}$) of the elongated nematic molecules than perpendicular. In the layered smectic phases, on the contrary, typically $\sigma_a < 0$. In some liquid crystals with a nematic-to-smectic phase transition, when decreasing the temperature, pre-transitional fluctuations induce a sign change of σ_a already in the nematic temperature range. The anisotropy of the magnetic susceptibility $\chi_a = \chi_\parallel - \chi_\perp$ is positive for the majority of nematics due to saturated aromatic rings as main building blocks of the constituent molecules. The few exceptions with $\chi_a < 0$ are composed of exclusively non-aromatic (e.g. cyclohexane) rings.

The sign of the anisotropies ϵ_a and χ_a governs the behavior of the liquid crystal in an electric ($\mathbf{E}$) or a magnetic field ($\mathbf{H}$) field via an *electromagnetic* contribution, f_{em}, to the *orientational free energy* density:

$$f_{em} = -\frac{1}{2}\epsilon_o\epsilon_a(\mathbf{n} \cdot \mathbf{E})^2 - \frac{1}{2}\mu_o\chi_a(\mathbf{n} \cdot \mathbf{H})^2. \tag{2}$$

As a result, for $\epsilon_a > 0$ or $\chi_a > 0$ the electromagnetic torque tends to align the director along the fields, while in the case of $\epsilon_a < 0$ or $\chi_a < 0$ an orientation perpendicular to the field directions is preferred [2, 3, 9]. This behavior establishes the basic working principles of most liquid crystalline electro-optic devices (displays). Though Eq. (2) indicates a similarity between the behavior in electric and magnetic fields, one crucial difference should be emphasized. Since χ_a is of the order of 10^{-6} in SI units, distortions of the constant applied magnetic field when the director varies in space, can safely be neglected. In the electric case, however, ϵ_a is usually of the order of unity, and then the electric field distortions have to be taken into account.

Though nematics are non-polar substances, a polarization may emerge in the presence of director gradients, even in the absence of an electric field. This *flexoelectric polarization* [2, 3]

$$\mathbf{P}_{fl} = e_1 \mathbf{n}(\nabla \cdot \mathbf{n}) - e_3 \mathbf{n} \times (\nabla \times \mathbf{n}) \tag{3}$$

originates in the shape anisotropy of the molecules. Since the flexoelectric coefficients e_1 and e_3 of rod-like nematic molecules are usually quite small, their contribution

$$f_{fl} = -\mathbf{P}_{fl} \cdot \mathbf{E} \tag{4}$$

to the free energy density is negligible in the majority of cases or is only detectable under special conditions.

Though the free energy considerations introduced above are sufficient to describe static orientational deformations in nematics, they cannot provide information about the dynamical properties of the system (e.g. the rate of reorientation upon a change of an external field). Usually dynamics involves *material flow* which couples to the director field. In the standard nemato-electrohydrodynamic theory the flow field $\mathbf{v}$ is described by a Navier-Stokes equation, which, besides the elastic and viscous stresses, also includes the Coulomb force of an electric field on charges present. The nematic anisotropy is manifested in a complex form of the viscous stress tensor, such that the effective viscosity depends on the director orientation and the gradients of the velocity field components, which appear for instance in the strain tensor $\nabla \otimes \mathbf{v}$. The *dynamics of the director* $\mathbf{n}$ in liquid crystals is governed by a balance-of-torques principle, which involves the elastic and electromagnetic torques, as well as additional viscous torques in the presence of shear flow. It follows directly from standard symmetry arguments [2, 3, 10] that the complicated viscous behavior of nematics (reorientation of the director induces flow and vice versa, flow aligns the director) can be described by eight phenomenological transport coefficients - the Ericksen-Leslie *viscosity coefficients* $\alpha_1, ..., \alpha_6$, and the rotational viscosities γ_1 and γ_2 - though in fact only 5 of them are independent (for their definitions and the relations among them refer to e.g. [2, 3, 10]). In the nemato-electrohydrodynamic model nematics are treated as ohmic conductors. To describe the *dynamics of the charge density* $\rho = \nabla \cdot \mathbf{D}$, which is coupled via $\mathbf{D} = \epsilon \cdot \mathbf{E}$ to the electric field $\mathbf{E}$, the quasi-static approximation of the Maxwell equations is sufficient in our case, which is equivalent to the charge conservation:

$$\frac{d\rho}{dt} + \nabla \cdot \mathbf{j} = 0 \qquad (5)$$

with the electric current $\mathbf{j} = \sigma \cdot \mathbf{E}$.

The *optical properties* of nematics correspond to those of uniaxial crystals [11]. The director defines the local optical axis. The most obvious indication of the anisotropy of nematics is their birefringence. Since nematics are composed of elongated molecules their extraordinary refractive index n_e is always larger than the ordinary one n_o, i.e. nematics have a positive optical anisotropy $n_a = n_e - n_o$. When light is passing through a nematic layer, an optical path difference

$$\Delta s = \int \left[n_{eff}(z) - n_o \right] dz \qquad (6)$$

between the ordinary and extraordinary rays builds up, which depends on the local director orientation. The effective refractive index $n_{eff}(z)$ for the light of extraordinary polarization decreases with increasing angle between the director and the light polarization ($n_o \leq n_{eff}(z) \leq n_e$). If placed between crossed polarizers, variation of Δs results in changes of color and/or the intensity of the transmitted light. This feature makes the polarizing microscope a standard tool for studying the textures of nematic liquid crystals.

There are, however, conditions where modulation of the optical properties can be detected with a single or without any polarizer. This occurs if the director has a spatial tilt modulation periodic in a direction perpendicular to the light path (e.g. in electroconvection). The extraordinarily polarized light then senses the ensuing periodic modulation of the refractive index n_{eff}, so the sample acts like an array of lenses. The illuminating light is focused and defocussed, resulting in a sequence of alternating dark and bright stripes in a properly adjusted microscope. This technique, which is used in the majority of experiments, is known as the *shadowgraphy* [12, 13]. In this setup of course no intensity modulation exists when illuminated with light of ordinary polarization (n_o is constant per definition). Under unpolarized light the intensity modulations remain visible; however, the contrast is reduced since only part of the light with the appropriate polarization will contribute.

Besides microscopy, *diffraction* is another possibility for analyzing periodic patterns. The modulation of the refractive index is equivalent to an optical grating, thus illuminating with a laser beam the fringe pattern of the diffraction can be detected at a distant screen. This allows determination of the pattern wavelength as well as monitoring the pattern amplitude via the fringe intensities (see [14] and references therein).

All physical parameters mentioned above are material specific and temperature dependent (for a detailed discussion of the material properties of nematics, see for instance [4]). Nevertheless, some general trends are characteristic for most nematics. With the increase of temperature the absolute values of the anisotropies usually decrease, until they drop to zero at the nematic-isotropic phase transition. The viscosity coefficients decrease with increasing temperature as well, while the electrical conductivities increase. If the substance has a smectic phase at lower temperatures, some pre-transitional effects may be expected already in the nematic phase. One example has already been mentioned when discussing the sign of σ_a. Another example is the divergence of the elastic modulus K_2 close to the nematic-smecticA transition since the incipient smectic structure with an orientation of the layers perpendicular to **n** impedes twist deformations.

2. Electroconvection

Convection instabilities driven by temperature gradients are common in nature. They are, for instance, crucial ingredients for the dynamics of our atmosphere and drive the earth dynamo. They present an intensively studied paradigm for the dynamics of extended nonlinear systems with many degrees of freedom and show clearly the typical bifurcation sequences: spontaneous pattern formation by destabilization of the *homogeneous basic state* → *complex patterns (secondary bifurcations)* → *chaos/turbulence*. The understanding of such systems has been promoted, in particular, by laboratory experiments in the buoyancy-driven, classical Rayleigh-Benard convection in a layer of a simple fluid heated from below [15]. Already this experiment allows a wide range of possible modifications, such as rotating or inclining the experimental setup or the use of more complex working fluids such as binary fluids or electrically conducting liquid metals. The wealth of phenomena are still far from being exhausted, either from the experimental or from the theoretical point of view.

Electrically driven convection in nematic liquid crystals [6, 7, 16] represents an alternative system with particular features listed in the Introduction. At onset, EC represents typically a regular array of convection rolls associated with a spatially periodic modulation of the director and the space charge distribution. Depending on the experimental conditions, the nature of the roll patterns changes, which is particularly reflected in the wide range of possible wavelengths λ found. In many cases λ scales with the thickness d of the nematic layer, and therefore, it is convenient to introduce a dimensionless wavenumber as $q = \frac{2\pi}{\lambda}\frac{d}{\pi}$ that will be used throughout the paper. Most of the patterns can be understood in terms of the Carr-Helfrich (CH) mechanism [17, 18] to be discussed below, from which the standard model (SM) has been derived

[19]. Some scenarios fall outside that frame and need obviously new, other mechanism(s) for description.

Experiments and theoretical considerations have shown that the key parameters are the symmetry of the system (planar or homeotropic boundary conditions), the dielectric and the conductivity anisotropies. It is therefore convenient to categorize the various combinations of parameters as listed in Table 1. In the last column the structures predicted and/or observed are summarized and will be discussed below systematically.

Table 1. Eight different combinations (labelled **A** to **H**) of initial director alignments and the sign of anisotropies ϵ_a, σ_a. The EC pattern species are characterized in the last column: CH stands for patterns, which are compatible with the Carr-Helfrich mechanism, in contrast to the remaining, nonstandard ones (ns-EC).

Case	Alignment	ϵ_a	σ_a	Type of transition
A	planar	< 0	> 0	direct CH, ns-EC (prewavy)
B	homeotropic	> 0	< 0	direct CH
C	homeotropic	< 0	> 0	secondary CH, ns-EC (prewavy)
D	planar	> 0	< 0	secondary CH
E	planar	> 0	> 0	direct CH, Freedericksz
F	homeotropic	> 0	> 0	direct CH (α-induced)
G	planar	< 0	< 0	direct CH (α-induced), ns-EC (longitudinal)
H	homeotropic	< 0	< 0	direct CH, Freedericksz, ns-EC (longitudinal)

First we discuss configurations which can be described by the standard model where patterns appear either directly or as a secondary instability (section **2.1**). Then we discuss briefly EC phenomena not covered by the standard model (section **2.2.**).

2.1 Standard EC based on the Carr-Helfrich mechanism

EC is typically driven by an ac voltage. Its amplitude is used as the main control parameter, while the driving frequency provides a convenient secondary control parameter. Two types of modes, the *conductive* and the *dielectric*, are allowed by symmetry (see the generic stability diagram sketched in Fig. 1). In the low-frequency, conductive regime the director and the flow field are practically time independent while the electric field follows the external driving in time; in the high-frequency dielectric regime (for frequencies above the so-called cut-off frequency f_c, which increases with decreasing charge relaxation time $\tau_q = \frac{\epsilon_o \epsilon_\perp}{\sigma_\perp}$) the situation is reversed. In the following we will mostly focus on the conductive regime.

EC occurs in a layer (parallel to the $x - y$ plane) of homogeneously aligned nematics in the presence of an electric field across the layer (along the $\hat{z}$ axis).

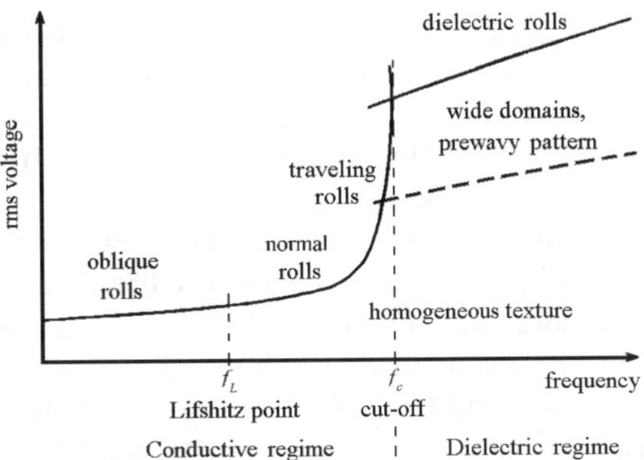

Figure 1.   Schematic morphological phase diagram in the $U - f$ plane. Solid lines correspond to the threshold voltage of standard electroconvection, the dashed line denotes the threshold of the prewavy patterns or wide domains (see later). For details see [16].

To describe the convecting state, one needs the velocity field $\mathbf{v}(\mathbf{r}, t)$, the director field $\mathbf{n}(\mathbf{r}, t)$ and the charge distribution $\rho(\mathbf{r}, t)$ (or the electric potential $\phi(\mathbf{r}, t)$ inside the layer), which are available from the nemato-hydrodynamic equations described in section **1**. These coupled partial differential equations cannot be solved analytically with realistic rigid boundary conditions (vanishing $\mathbf{v}$, fixed $\mathbf{n}$ and ϕ at the confining plates). Typically investigations of EC start with a linear stability analysis of the basic homogeneous state, which yields the threshold voltage $U_c(f)$ and the critical wavenumber $q_c(f)$ as function of frequency. Much insight into the mechanisms of EC has been obtained by deriving approximate, analytical expressions for the critical quantities U_c and q_c. This development started with the seminal work of Carr [17] and Helfrich [18] in the planar geometry. They extracted the basic positive feedback mechanism responsible for EC, which is now called the 'Carr-Helfrich (**CH**) mechanism' in the literature: any director fluctuation leads to charge separation, flow is excited due to the Coulomb force in the Navier-Stokes equation. The flow exerts a viscous torque on the director reinforcing its initial fluctuation and thus the charge density. The mechanism is opposed by viscous damping of the flow and the elastic torques, such that EC appears only above a certain threshold voltage.

The original, so called 1-d formula of Carr and Helfrich, was later refined and generalized into a 3-d theory capable of calculating the wavevector and describing real, three dimensional patterns (like normal or oblique rolls), other geometries and the dielectric regime [16].

Approximate analytical threshold formulas (examples are shown in Eqs. (7) and (8)) have been very useful not only to interpret specific experiments but also to get insight into general trends. For instance, it can be shown that the CH mechanism remains unaltered if all three key parameters – the initial director alignment, ϵ_a and σ_a – are "reversed" simultaneously, i.e. the change planar $\rightarrow$ homeotropic is combined with the sign reversal of ϵ_a and σ_a. In fact, pairs of systems connected by this reversal transformation (cases **A**$\leftrightarrow$**B**, **C**$\leftrightarrow$**D**, **E**$\leftrightarrow$**H** and **F**$\leftrightarrow$**G** in Table 1) show close similarities.

In the following, we first discuss the situations where EC occurs as a primary forward bifurcation and where the standard model is directly applicable (cases **A** and **B**). Then we discuss configurations where EC sets in as a secondary instability upon an already distorted Freedericksz ground state and compare it with experiments (cases **C** and **D**). Note that in this case the linear analysis based on the standard model already becomes numerically demanding. Finally, we address those combinations of parameters where a direct transition to EC is not very robust, since it is confined to a narrow ϵ_a range around zero. For cases **E** and **H** this range may be accessible experimentally while for cases **F** and **G** it is rather a theoretical curiosity only.

Case A: planar alignment, $\epsilon_a < 0$ & $\sigma_a > 0$.   This is the most studied, classical case, since the conductivity anisotropy of usual nematics (substances without a smectic phase) is typically positive. As for ϵ_a, there is a wide range of materials with negative dielectric anisotropy.

The starting point is the analytical expression for the *neutral curve* in planar alignment with realistic rigid boundary conditions in the conductive regime [20]:

$$U^2(q, f) = \frac{\pi^2 K^{eff}}{\epsilon_o \epsilon_a^{eff} + I_h(\alpha_3/q^2 - \alpha_2)\tau_q \sigma_a^{eff}/\eta^{eff}}, \qquad (7)$$

with the overlap integral $I_h = 0.97267$, and the charge relaxation time, $\tau_q = \epsilon_o \epsilon_\perp / \sigma_\perp$, that decreases with increasing $\sigma_\perp$. The effective material parameters $K^{eff} > 0$, $\epsilon_a^{eff} < 0$, $\sigma_a^{eff} > 0$, $\eta^{eff} > 0$ are proportional to the corresponding physical quantities (elastic moduli, dielectric and conductivity anisotropies and viscous damping coefficients, respectively). The effective values depend on the frequency (f) and the wavenumber (q) (for the complete expressions see [20]). Eq. (7) has been derived using a truncated Galerkin expansion: each variable is represented by *one test function* with respect to z which satisfies the boundary conditions and has the appropriate symmetry. The minimum of $U(q, f)$ with respect to q yields the critical wavenumber $q_c(f)$ that determines the threshold voltage $U_{th}(f) = U(q_c, f)$.

One can easily identify the effect of various terms in Eq. (7): with the increase of the orientational elasticity (K^{eff}), the dielectric torques (ϵ_a^{eff}) (which both tend to turn back the director to the initial homogeneous alignment) or the viscous damping of the flow (η^{eff}), the threshold increases. The only destabilizing term, which is responsible for the onset of the patterning ($q \neq 0$) instability, is the **CH** term, $(\alpha_3/q^2 - \alpha_2)\tau_q\sigma_a^{eff}$ (note $\alpha_2 < 0$ and $|\alpha_3| \ll |\alpha_2|$).

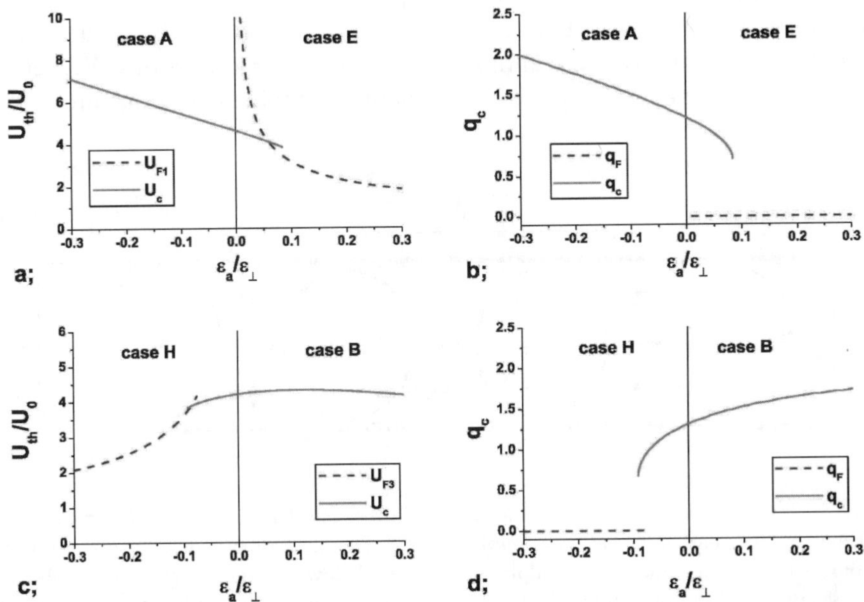

Figure 2. Threshold voltage U_{th}/U_0 and the critical wavenumber q_c versus the dimensionless dielectric anisotropy $\varepsilon_a/\varepsilon_\perp$ calculated from Eqs. (7) and (8). a; b; Planar alignment with $\sigma_a > 0$, c; d; homeotropic alignment with $\sigma_a < 0$. Dashed lines correspond to the Freedericksz transition, solid lines to the direct EC transition.

The important role played by the electrical conductivity anisotropy σ_a is evident: if σ_a decreases, $U^2(q, f)$ increases and diverges at a small positive value of σ_a when the two terms in the denominator of Eq.(7) cancel each other. If σ_a approaches zero or becomes negative (case **F**), the **CH** term vanishes or acts as a stabilizing term, respectively; thus convection is not expected. The role of ϵ_a is somewhat different: an EC transition exists for vanishing and even for positive ϵ_a (case **E**).

When analyzing the frequency dependences $U_{th}(f)$ and $q_c(f)$, it is obvious from Eq. (7) and in agreement with general symmetry arguments that both quantities start with zero slope at $f = 0$. They increase monotonically with f and diverge at the cutoff frequency f_c where the dielectric regime takes over. For simplicity, we restrict the detailed presentation of the threshold behavior to the limit $f \to 0$ and to MBBA material parameters [2, 4, 19] except that we allow for variations of ϵ_a (while keeping $\epsilon_\perp$ constant) and for reverse of the sign of σ_a for the cases **B**, **D**, **G** and **H**.

In Fig. 2, the results for the critical voltage U_{th} (left panels, in units of $U_0 = \sqrt{\pi^2 K_1/(\epsilon_o \epsilon_\perp)}$, $U_0 = 1.19$V for MBBA parameters) and the corresponding critical wavenumber q_c (right panels) are summarized as functions of $\epsilon_a/\epsilon_\perp$. The data are barely distinguishable from the results of a rigorous linear stability analysis based on the full standard model [21].

In case **A** (left panels of Fig. 2a-b), $U(q)$ has only one minimum at a finite q_c; both $U_c(\epsilon_a)$ and $q_c(\epsilon_a)$ are almost linearly decreasing functions. EC sets in in the whole range $\epsilon_a < 0$, at a few Volts applied to the convection cell.

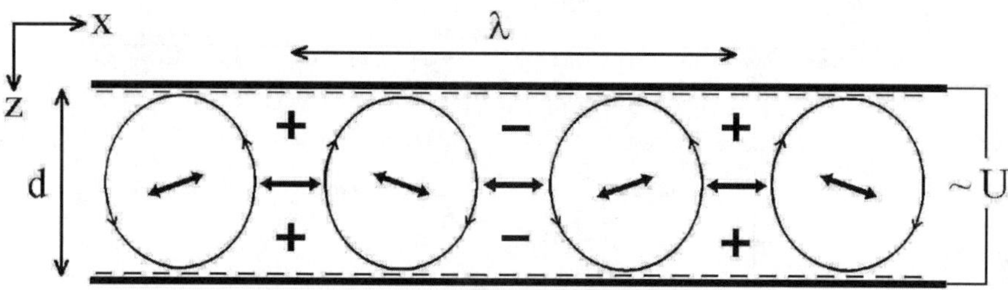

Figure 3. Cross section of a roll pattern at the direct onset to EC in the planar geometry (indicated by the small dashes at the confining plates). Double arrows denote the director modulations, which are maximal at the midplane. The lines follow the stream lines. The symbols + and − denote the sign of the induced charges, shown at a phase of the applied ac voltage where the electric field points downward.

Figure 3 exhibits the director and charge distribution as well as the velocity field in the $x - z$ plane at onset of electroconvection, where the $\hat{x}$ direction is parallel to the initial (planar) director alignment and λ is the pattern wavelength.

Experiments carried out on MBBA, I52 and Merck Phase 4 and 5 [22–25] match very well the quantitative calculations of the stability diagram. Often oblique rolls (see Fig. 4a), where the wavevector of the striped patterns makes a nonzero angle with the basic director alignment, appear in the conductive regime below a Lifshitz point ($f < f_L$), and normal rolls (Fig. 4b) appear above it. Their wavelengths λ are of the order of the cell thickness d. The dielectric rolls (Fig. 4c) appearing at frequencies above the cut-off ($f > f_c$) can be normal or oblique. However, their wavelength is independent of d and is usually $3 - 4\mu$m. The patterns are regular (the snapshot in Fig. 4c was taken at a higher voltage above threshold in order to have a higher contrast to the expense of producing defects due to secondary bifurcations) and have a large aspect ratio.

For completeness, we note that in many cases in experiments traveling rolls (Hopf bifurcation) have been observed at onset, usually at the high frequency

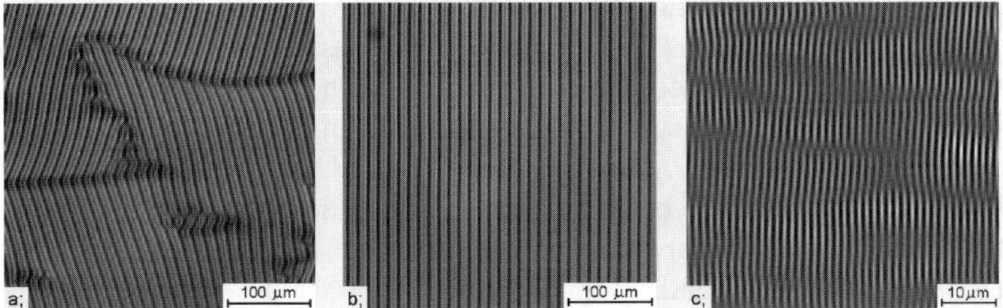

Figure 4. Snapshots of EC patterns slightly above onset for case **A** taken in a polarizing microscope with a single polarizer (shadowgraph images, Phase 5, $d = 9\mu$m). a; Oblique rolls, b; normal rolls, c; dielectric rolls (Note the difference in magnification.). The initial director orientation is horizontal. The contrast was enhanced by digital processing.

end of the conductive range (see Fig. 1). The phenomenon lacked theoretical understanding for a long time, until the standard model was generalized in a way that the simple ohmic conductivity was replaced by ionic mobilities. The resulting weak electrolyte model (WEM) [26, 27] provided the explanation and a good quantitative agreement with experiments in MBBA [28], I52 [24] and Phase 5 [29].

The dashed line in Fig. 1 is the experimental threshold curve for prewavy patterns or wide domains ($\lambda \approx 4 - 10d$) that also represent electroconvecting structures though not captured by the standard model (see later in Section **2.2.**).

Case B: homeotropic alignment, $\epsilon_a > 0$ & $\sigma_a < 0$. Here the initial director orientation and the sign of both anisotropies are reversed as compared to case **A**.

Recent work showed [30] that in this case EC sets in with a continuous transition from the homogeneous state directly, similarly to the classical configuration of case **A**. It has been shown that the pattern forming mechanism, including the role of the elastic, dielectric, viscous and **CH** terms, is analogous to that of case **A**, and the standard model is applicable. The analytical one-mode neutral-curve expression for homeotropic initial alignment [31] is as follows:

$$U^2(q, f) = \frac{\pi^2 K^{eff}}{-\epsilon_o \epsilon_a^{eff} - I_h(\alpha_3 - \alpha_2/q^2)\tau_q \sigma_a^{eff}/\eta^{eff}} . \qquad (8)$$

Eq. (8) is similar to Eq. (7); however, the effective quantities and τ_q in Eq. (8) are defined differently. In fact, they can be transformed into each other by interchanging the subscripts $\| \leftrightarrow \perp$ and the material parameters $K_1 \leftrightarrow K_3$, $\eta_1 \leftrightarrow \eta_2$ and $\alpha_2 \leftrightarrow -\alpha_3$. These transformations are natural consequences of switching the boundary conditions between planar and homeotropic cases.

The right hand side of Eq. (8) is positive for $\epsilon_a > 0$ and $\sigma_a < 0$ (case **B**). One expects for $U_c(\epsilon_a)$ and $q_c(\epsilon_a)$ similar functions (mirror images) to those of case **A**. However, the corresponding plots shown in Fig. 2 look somewhat different. This is due to the asymmetry of the scaling factors (in all cases, we plot ϵ_a in units of $\epsilon_\perp$ and U_{th} in units of $U_0 = \sqrt{\pi^2 K_1/\epsilon_o\epsilon_\perp}$).

The essential difference between cases **A** and **B** lies in the symmetry of the system. In case **A** the planar geometry is anisotropic, and the wavevector direction is selected by the boundary conditions. In contrast, the homeotropic alignment in case **B** provides isotropic conditions in the plane of the patterns, and the direction of the wavevector of the striped patterns is chosen accidentally at threshold, which corresponds to a spontaneous breaking of the rotational symmetry.

The director field and charge distribution, as well as the velocity fields are sketched in Fig. 5 for the homeotropic case. The director tilt angle is determined by the applied voltage, but, as already stressed before, the azimuthal angle is not selected in this isotropic configuration. As will be demonstrated below, this freedom leads easily to disordered patterns with slow variations of the azimuthal director component in the weakly nonlinear regime.

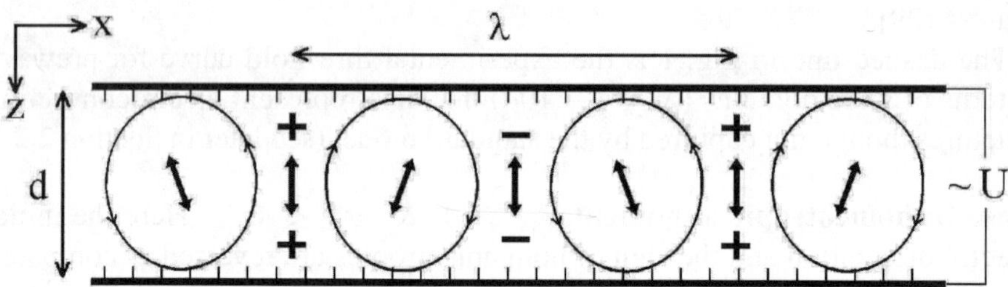

Figure 5. Cross section of a roll pattern at the direct onset of EC in the homeotropic geometry (indicated by the small dashes at the confining plates). Double arrows denote the director modulations, which are maximal at the midplane. Thin lines are for the flow field. The symbols $+$ and $-$ denote the sign of the induced charges.

Experiments have been carried out on p-(nitrobenzyloxy)-biphenyl [30] and typical patterns in the conductive range at onset are shown in Fig. 6. At low frequencies disordered rolls without point defects have been observed with a strong zig-zag (ZZ) modulation (see Fig. 6a) which can be interpreted as the isotropic version of oblique rolls. Above a critical frequency, a square pattern is observed which retains the ZZ character because the lines making up the squares are undulated. At onset the structure is disordered; however, after a transient period defects are pushed out and the structure relaxes into a nearly defect-free, long-wave modulated, quasi-periodic square pattern (see Fig. 6b).

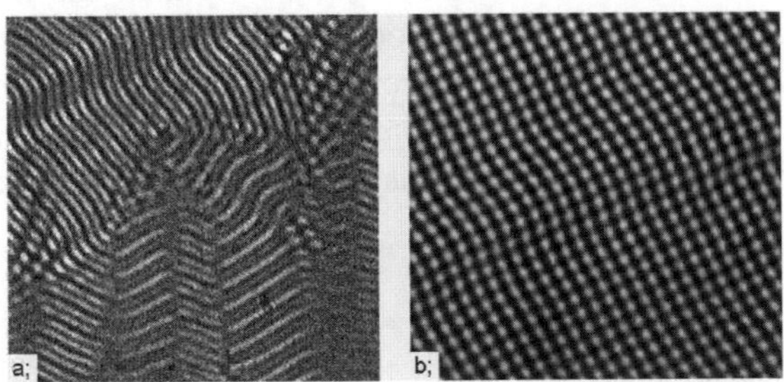

Figure 6.   Snapshots of EC patterns slightly above onset for case **B**. a; ZZ modulated disordered rolls, b; undulated (soft) squares.

The $U_c - f$ phase diagram is similar to Fig. 1 and can be reproduced quantitatively by the standard model. The dielectric regime has not been observed experimentally for, most likely, purely technical reasons (there is no reason to exclude the dielectric regime). As for the various patterns in the nonlinear regime, they have been well reproduced by a suitable weakly nonlinear analysis [32, 33]. In contrast to the generalized Ginzburg-Landau amplitude equations usually used in the anisotropic regimes, a generalized Swift-Hohenberg model had to be constructed here.

Case C: homeotropic alignment, $\epsilon_a < 0$ & $\sigma_a > 0$.   In this combination of the material parameters, the linear stability analysis of the basic state does not predict a direct transition to EC since the resulting expression for $U^2(q)$ in Eq. (8) is negative for all $q \neq 0$ (except for ϵ_a in the immediate vicinity of zero, see below). The reason is that the two terms in the denominator act differently compared to the case **B** ($\epsilon_a > 0$, $\sigma_a < 0$) described in the previous subsection. The Carr-Helfrich torque is now stabilizing while the dielectric torque ($\propto \epsilon_a^{eff}$) is destabilizing. At $q = q_F = 0$, this term dominates and describes, at the threshold U_{F3} (see Fig. 7a), the continuous bifurcation to the Freedericksz distorted state of homogeneous (along the $\hat{x}$ direction) bend (see Fig. 8a).

However, in the vicinity of $\epsilon_a \approx 0$, the destabilizing influence of the **CH** term is restored if it becomes comparable with the ϵ_a^{eff} term. The inspection of Eq. (8) shows that this may occur at a very large q: the $(\alpha_3 - \alpha_2/q^2)$ term, which is large and positive for usual q ($q \approx 1$), decreases with increasing q and becomes eventually negative for materials with $\alpha_3 < 0$ if $q^2 > \alpha_2/\alpha_3$ (≈ 100 for MBBA). This results in a patterning mode with $q = q_\alpha$ against which the basic homogeneous homeotropic configuration becomes unstable. This mode is only activated if its threshold U_α becomes lower than U_{F3}. U_α and U_{F3} intersect at a very small, negative $\epsilon_a/\epsilon_\perp \approx -5 \cdot 10^{-5}$ (the intersection point is not resolvable even in Fig. 7c). To the right of the intersection, q_α becomes in fact the fastest growing mode [34]. Note that the "α-induced" direct transition to EC is only an interesting theoretical possibility with no experimental relevance, since one would need a nematic with ϵ_a almost exactly zero, and one should be able to detect patterns with extremely large q at very high U_α.

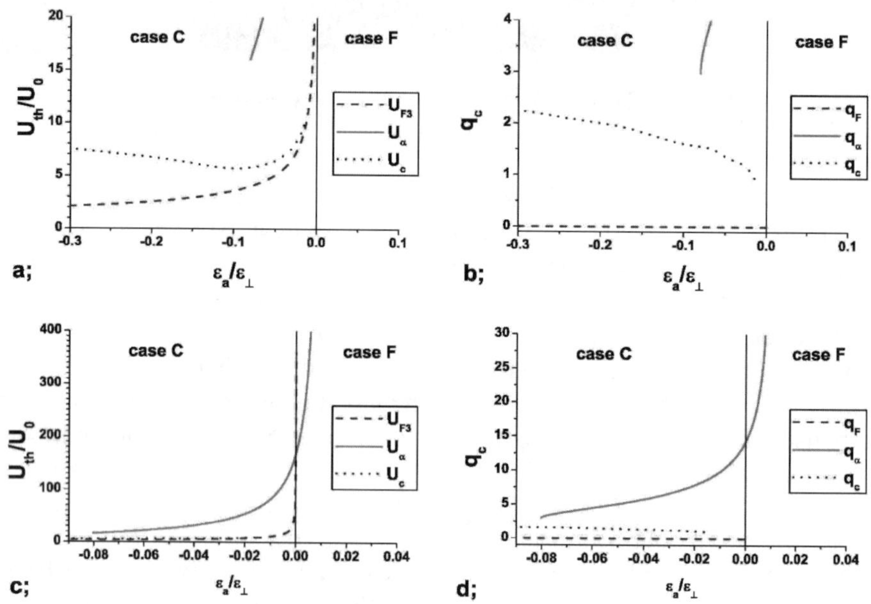

Figure 7.   Threshold voltages U_{th}/U_0 and the critical wavenumber q_c versus the relative dielectric anisotropy $\varepsilon_a/\varepsilon_\perp$ calculated from Eq. 8. Homeotropic alignment with $\sigma_a > 0$. The upper (a; b;) and lower (c; d;) plots differ only in the axis scales. Dashed lines correspond to the Freedericksz transition, solid lines correspond to the direct transition to an ("α-induced") EC patterned state, dotted lines represent a secondary transition to EC.

Above the Freedericksz threshold U_{F3}, the tilt angle with respect to $\hat{z}$ increases with increasing voltage such that eventually one arrives at a practically planarly aligned nematic layer at the midplane. Consequently, the planar CH mechanism is expected to be activated. Convection rolls are now to be superimposed on the elastically pre-distorted Freedericksz state (Fig. 8b), and there is no simple analytical threshold formula. Thus, one has to rely on a numerical linear stability analysis of the Freedericksz state in the framework of the standard model. Since boundary layers have to be resolved, one needs more $(6 - 8)$ Galerkin modes than typically required in the standard planar case. These numerical calculations have achieved the same good agreement with the experiments in MBBA or Phase 5 [35, 36], as previously in the case of the primary bifurcations (cases **A** and **B**). U_c in Fig. 7 (a,c) and q_c in Fig. 7 (b,d) (dotted lines) represent the results of such calculations that match the experimental data well [9]. Note that the U_c and q_c curves should continue to almost zero ϵ_a; however, we have been unable to access this regime numerically since the minimum of the neutral curve becomes too shallow. In the range $-0.2 < \epsilon_a/\epsilon_\perp < -0.03$, a bifurcation to oblique (instead of normal) rolls occurs at threshold. This manifests itself as little wiggles in the slope of the modulus of the critical wavevector plotted in Fig. 7b. Recent experiments [36] have shown, however, that at certain combinations of the material parameters (e.g. for Phase 5/5A) an unusual situation occurs, namely, oblique rolls become restricted to a finite frequency interval, $f_{L1} < f < f_{L2}$. Below f_{L1}, normal rolls appear, and reappear above f_{L2}; thus there are two Lifshitz points in this case.

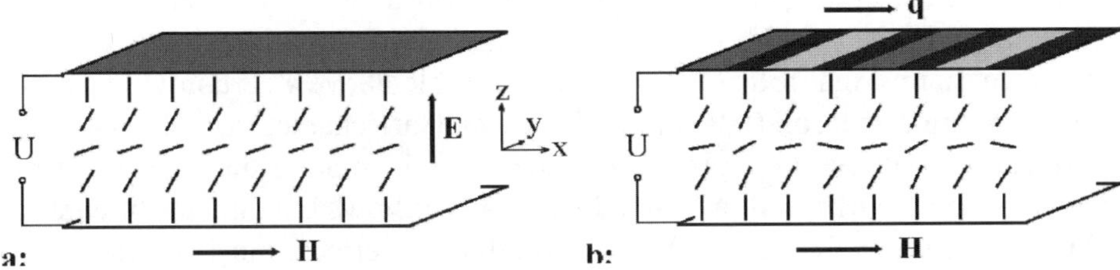

Figure 8. Schematic director profile in case **C**. a; Freedericksz distorted state, b; with superposed electroconvection pattern.

Though the initial homeotropic state is isotropic (as in case **B**), the isotropy in the plane is spontaneously broken due to the Freedericksz transition. Consequently, the EC pattern is formed on an anisotropic background with a preferred direction in the $x - y$ plane (as in case **A**). The local azimuthal angle of the Freedericksz tilt direction is singled out by coincidence; thus, it may vary in space as well as in time, representing a soft Goldstone mode coupled to the EC patterning mode. As a result, the patterns at onset – oblique rolls (Fig. 9a) or normal rolls (Fig. 9b) depending on the frequency – are disordered and correspond to a special manifestation of spatio-temporal chaos, the soft mode turbulence [37–39].

The chaotic behavior reflected in the disordered patterns can be suppressed if the initial isotropy of the homeotropic alignment is broken by applying a magnetic field **H** parallel to $\hat{\mathbf{x}}$ as shown in Fig. 8 [40]. The azimuthal angle of **n** is then singled out by the magnetic field, and the patterns become nicely ordered and exhibit similar morphologies as shown in Fig. 4 for the case **A** (e.g. the disordered pattern of Fig. 9a becomes similar to that in Fig. 4a).

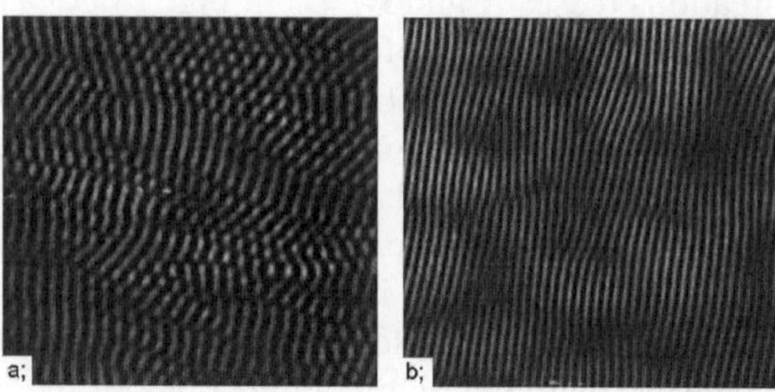

Figure 9. Snapshots of electroconvection patterns superposed on the Freedericksz state in case **C**. a; oblique rolls, b; normal rolls.

The homeotropic geometry offers some advantages in observing certain phenomena in the (weakly) nonlinear regime of EC (at voltages above U_c). It is known that normal rolls (NR) become unstable above a certain voltage with respect to abnormal roll (AR) modes [41] which are characterized by a rotation of the director in the $x - y$ plane while the roll axis remains unchanged. In planar cells, the polarization of light adiabatically follows the director orientation (Mauguin's principle) [2, 3]. Weak non-adiabatic effects must be resolved in this case to detect possible in-plane rotations of the director which are maximal near the mid-plane and vanish at the boundaries.

In homeotropic cells, however, in-plane rotations of the director are reflected in a net azimuthal rotation of the optical axis (and the light polarization) across the cell which has allowed a detailed exploration of the characteristics of the NR-AR transition. Experiments have shown an excellent agreement with the predictions of generalized Ginzburg-Landau models [36].

Another example is related to the motion of defects (dislocations in the roll pattern) which constitutes the basic mechanism of wavevector selection. In the normal-roll regime, the stationary structure is characterized by the condition $\mathbf{q} \parallel \mathbf{H}$. However, when changing the field direction one can easily induce a temporary wavevector mismatch $\Delta\mathbf{q} = \mathbf{q}_{new} - \mathbf{q}_{old}$ which relaxes via a glide ($\mathbf{v} \parallel \mathbf{q}$) motion of defects. Experiments have confirmed the validity of detailed theoretical predictions, both with respect to the direction ($\mathbf{v} \perp \Delta\mathbf{q}$) and the magnitude (consistent with logarithmic divergence at $|\Delta\mathbf{q}| \rightarrow 0$) of the defect velocity $\mathbf{v}$ [42].

The homeotropic geometry also allows for the appearance of structures with a secondary spatial periodicity – chevrons – in the conductive regime at voltages considerably larger than U_c [43]. Such type of chevrons, which are characterized by a periodic arrangement of defect chains, have been observed before exclusively in the dielectric regime.

Case D: planar alignment, $\epsilon_a > 0$ & $\sigma_a < 0$. This configuration is realized by "reversing" the sign of the anisotropies and the initial director orientation compared to case **C**. Thus, cases **D** and **C** belong to the same "family", analogous to cases **B** and **A** discussed before. The standard model applies and the theoretical analysis can be based on Eq. (7) for the neutral curve $U^2(q)$.

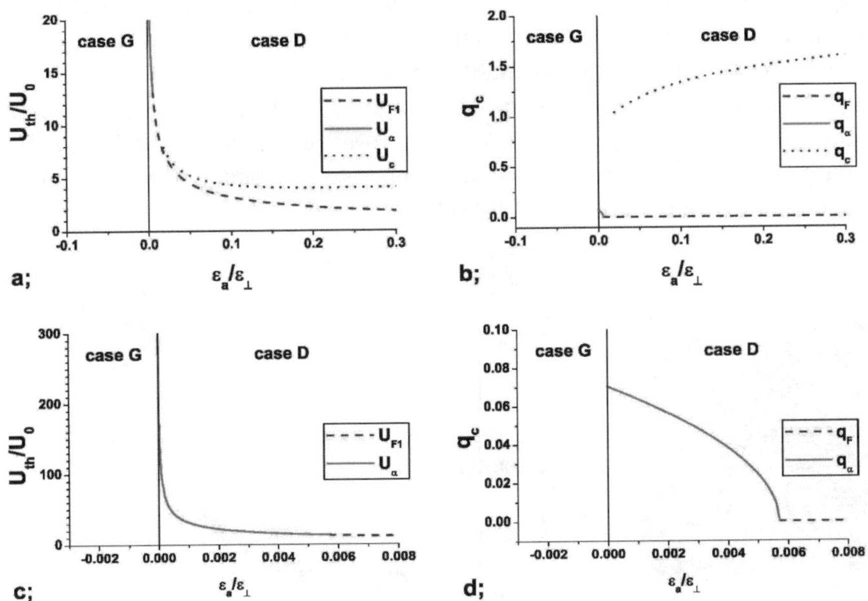

Figure 10. Threshold voltages U_{th}/U_0 and the critical wavenumber q_c versus the relative dielectric anisotropy $\varepsilon_a/\varepsilon_\perp$ calculated from Eq. (7). Planar alignment with $\sigma_a < 0$. The upper (a; b;) and lower (c; d;) plots differ only in the axis scales. Dashed lines belong to the Freedericksz transition, solid lines to the direct transition to an ("α-induced") EC patterned state. Dotted lines represent a secondary transition to EC.

Upon increasing the voltage the first transition is usually to the homogeneous ($q = q_F = 0$) splay Freedericksz state with a frequency-independent threshold voltage U_{F1} (Fig. 10a), i.e the absolute minimum of the neutral curve is at $q = 0$. However, very near to $\epsilon_a = 0$ the absolute minimum appears at a finite q. Thus we have the planar counterpart of the "α-induced" EC described in case **C**. The Freedericksz threshold smoothly transforms into an EC threshold U_α at $\epsilon_a/\epsilon_\perp = 0.0057$ (see Fig. 10c-d). Below this, a direct transition to EC is predicted with q_α growing continuously from zero and remaining extremely small. This transition seems to be a better candidate for experimental observation than its homeotropic counterpart, because both U_α and q_α are substantially lower than in case **C**.

For $\epsilon_a/\epsilon_\perp > 0.0057$ EC occurs superimposed onto the Freedericksz state (secondary instability) at a higher voltage $U_c > U_{F1}$. The standard model can also be applied here by carrying out numerical linear stability analysis of the Freedericksz distorted state, and one is faced with similar modifications and difficulties as mentioned in case **C** before. The ϵ_a-dependent U_c and q_c, presented by the dotted lines in Fig. 10(a-b), have been calculated numerically. It should be noted that the convection rolls are now oriented parallel to the initial director alignment, contrary to the normal rolls in case **A** or **C**.

Measurements in the only available substance have revealed well-aligned rolls in the whole conductive frequency range [30] similarly to case **A** (compare Fig. 11a with Fig. 4b), indeed with $\mathbf{q} \perp \mathbf{n}$ as predicted by the theory. The wavenumber scales as d^{-1}. The calculations above provided a good quantitative agreement with experiments for both $U_c(f)$ and $q_c(f)$.

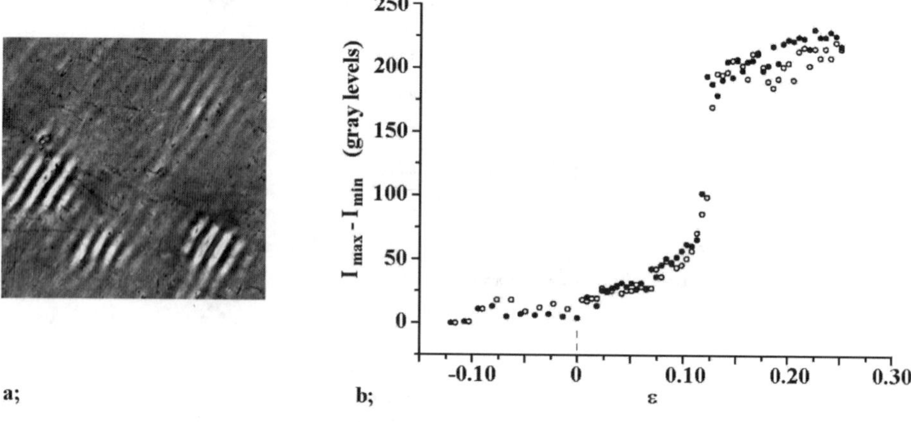

a;   b;

Figure 11.   a; Snapshot of EC pattern in the Freedericksz distorted planar geometry of case **D**. b; Voltage dependence of the contrast (the difference of the maximum I_{max} and minimum I_{min} intensities) of the EC pattern in case **D**. $\varepsilon = (U^2 - U_c^2)/U_c^2$ is a dimensionless control parameter.

The transition was found to be mediated by nucleation and traveling of sharp fronts (Fig. 11a) that indicates a backward bifurcation, although hysteresis has not been identified directly. Rather, a sharp jump in the contrast (pattern amplitude) with increasing voltage has been detected, with some indications that a low contrast pattern already arises at voltages before the jump occurs in Fig. 11b. A preliminary, weakly non-linear analysis has exhibited a bifurcation, which is in fact weakly supercritical at low frequencies. If small changes of the parameters and/or additional effects are included (e.g. flexoelectricity and weak-electrolyte effects) the bifurcation could become a more expressed subcritical one [32, 33].

Case E: planar alignment, $\epsilon_a > 0$ & $\sigma_a > 0$. Here the magnitude of ϵ_a plays a role. Starting from negative values (case **A**) a direct EC threshold persists for zero and positive ϵ_a (see Eq. (7) and Fig. 2a-b). The upper limit in ϵ_a is set by the onset of the homogeneous ($q = q_F = 0$) splay Freedericksz transition at a threshold U_{F1}. The intersection of U_c and U_{F1} occurs about $\epsilon_a/\epsilon_\perp = 0.06$, where the Freedericksz transition starts preceding the EC bifurcation. The local minimum on the neutral curve still exists at a finite q_c up to $\epsilon_a/\epsilon_\perp = 0.09$ where q_c discontinuously drops to zero. In the parameter range where $U_c \approx U_{F1}$, interesting scenarios are expected as a result of the competition between the homogeneous and the convective mode [44, 45]. As in cases **C** and **D**, we have performed a linear stability analysis of the Freedericksz distorted state but did not find any secondary EC threshold. To understand this behavior we note that the dielectric destabilizing torque of the basic state preferring $q = 0$ prevails over the CH torque resulting in the Freedericksz transition at U_{F1}. With the increase of the voltage U above U_{F1}, the tilt-angle of the director increases and reduces further the destabilizing effect of the CH term so that an even higher voltage would be required for EC onset. At large enough director tilt, the CH term becomes even stabilizing and the EC threshold "runs away". In this case EC does not superimpose onto the Freedericksz state contrary to cases **C** and **D** discussed above.

Case F: homeotropic alignment, $\epsilon_a > 0$ & $\sigma_a > 0$. This is one of the rather uninteresting situations. The only transition predicted is the "α-induced" high q instability discussed in case **C** which is the primary transition in a very narrow $\epsilon_a > 0$ interval (see Fig. 7a-d). No transition has been observed experimentally for any wavenumber at any voltage.

Case G: planar alignment, $\epsilon_a < 0$ & $\sigma_a < 0$. This is the counterpart of case **F**. Thus, theoretically, the "α-induced", low-q instability discussed in case **D** persists for $\epsilon_a < 0$ but the existence range is much smaller than in case **F**. The U_α curve diverges here at $\epsilon_a/\epsilon_\perp \approx -10^{-5}$ that cannot be resolved in Fig. 10c and has not been observed experimentally.

Case H: homeotropic alignment, $\epsilon_a < 0$ & $\sigma_a < 0$. Here one expects a qualitative behavior similar to case **E**. A direct EC transition is predicted for negative ϵ_a down to $\epsilon_a/\epsilon_\perp \approx -0.1$ (see Fig. 2c-d), though there is no experimental evidence yet. For more negative ϵ_a, the Freedericksz transition takes over, above which no secondary EC threshold is expected. The discussion and arguments given for case **E** apply here as well.

2.2 Non-standard EC excluded by the CH mechanism

The standard model is very powerful. As demonstrated before, it describes quantitatively the EC structures at onset and also quite deeply in the nonlinear regimes. Nevertheless, there are some situations (one of them is the prewavy structure mentioned under case **A**) where the standard model does not predict a bifurcation to EC, although experimental observations have clearly identified pattern formation accompanied by convection (ns-EC) in the presence of electric fields.

Cases A and C: $\epsilon_a < 0$ **&** $\sigma_a > 0$. As discussed above, compounds with $\epsilon_a < 0$ and $\sigma_a > 0$ are the most common examples of substances to exhibit the CH electroconvection. In substances with higher electrical conductivity, however, another periodic stripe pattern - the *prewavy pattern* - arises occasionally at voltages below the EC threshold U_c. This pattern has been reported for homeotropic [46] (case **C**) as well as for planar cells (case **A**) and has been called wide domains [47, 48, 9]). It is characterized by a wavelength λ much larger than the sample thickness ($\lambda \approx 4 - 10d$) (Fig. 12) as well as by much longer relaxation times than those of standard EC patterns. The stripes run perpendicular to the director, i.e. in the same direction as the conductive normal rolls. The prewavy pattern does not produce (at least near threshold) a shadowgraph image. It can be made visible only using crossed polarizers. This implies that the pattern is due to an azimuthal modulation of the director which is associated with flow vortices parallel to the surfaces [49] (i.e. in the $x - y$ plane, in contrast to the normal rolls in CH mechanism when both the director modulation and the flow occur in the $x - z$ plane). While the azimuthal rotation of the director is easily detectable in homeotropic samples for any d, its visualization in planar samples requires the detection of non-adiabatic corrections to the light propagation, which restricts the sample thickness.

Measurements have shown that the prewavy pattern appears in a forward bifurcation [50]. Its threshold voltage U_{pw} has a weak, nearly linear frequency dependence. It usually occurs at higher frequencies (see Fig. 1). Conductive normal rolls, dielectric rolls and the prewavy pattern may follow each other with increasing f (dielectric rolls may be skipped in compounds with higher conductivity). Near the crossover frequency f_c, the conductive (or dielectric) rolls can coexist with the prewavy pattern resulting in the defect-free chevron structure [51].

The characteristics of the prewavy pattern which clearly differ from those of the classical EC patterns cannot be explained using the standard model. The underlying mechanism is not yet known. One proposed interpretation – the inertial mode of EC [9] – fails to predict the correct direction of the stripes. The observation that U_{pw} seems to remain continuous and that the flow survives

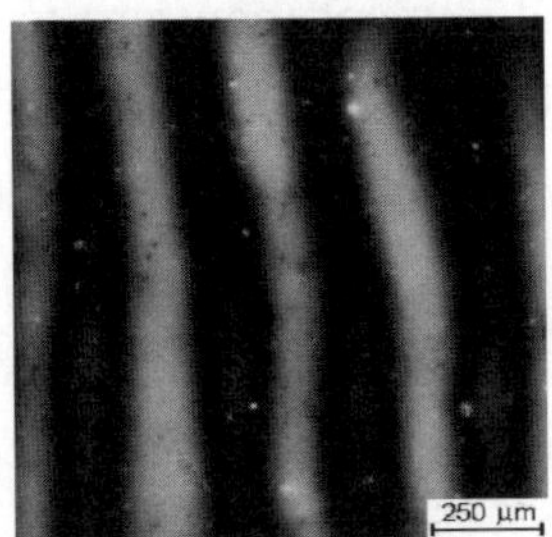

Figure 12. The prewavy pattern in homeotropic MBBA. $d = 50\mu$m.

when passing the nematic-to-isotropic transition with increasing temperature may suggest that the prewavy pattern could correspond to the chevron structure of a not yet detected primary pattern created by an isotropic mechanism already at voltages below U_{pw}. However, there are no direct experimental or theoretical proofs for this idea.

Case G: planar alignment, $\epsilon_a < 0$ & $\sigma_a < 0$. Standard EC (based on the **CH** mechanism) cannot occur for the material parameter combination $\epsilon_a < 0, \sigma_a < 0$ [2] except the "α induced" pattern type. Nevertheless, convection associated with roll formation has been observed in ac electric field in the homologous series of N-(p-n-alkoxybenzylidene)-n-alkylanilines, di-n-4-4'-alkyloxyazoxybenzenes and 4-n-alkyloxy-phenyl-4-n'alkyloxy-benzoates [52–54]. The characteristics of the patterns: the orientation of the rolls, contrast, frequency dependence of the wavevector and the threshold, director variation in space and time etc. – are substantially different from those observed in the standard EC. Since this roll formation process falls outside of the frame of the standard model, it has been called *nonstandard electroconvection* (ns-EC).

The main characteristics of these ns-EC patterns that differ from those of standard EC are:

- The overall contrast of the pattern is low compared to the standard EC structures. Near onset, the ns-EC pattern is not visible with the conventional shadowgraph method; crossed polarizers are needed to detect the pattern. Thus, the director field has no z component, i.e. the director is only modulated in the $x - y$ plane. This feature also explains the low contrast.

- The threshold voltage scales with the cell thickness; thus, the onset is characterized by a threshold field (not a voltage).

- The critical wavevector is perpendicular to, or subtends a large angle with the initial director alignment (contrary to normal rolls); thus, the rolls are parallel (longitudinal) or strongly oblique (see Fig. 13).

- The critical wavelength is comparable or larger than the cell thickness.

- The director field oscillates with the driving frequency, similar to the dielectric regime of standard EC.

- The threshold is a linear function of the driving frequency.

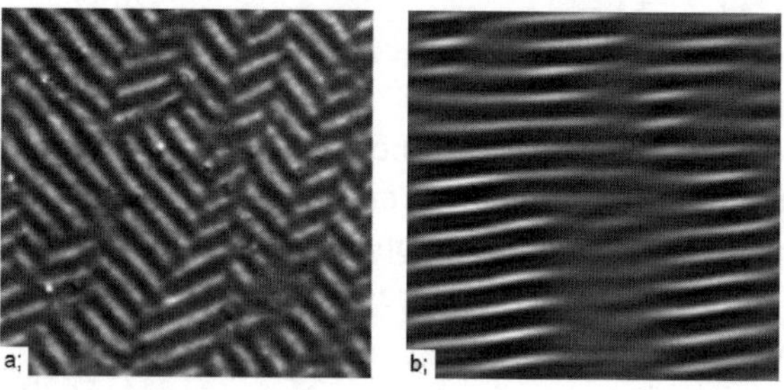

Figure 13.   Snapshots of nonstandard electroconvection pattern in case **G** taken with crossed polarizers. a; Oblique rolls, b; parallel rolls. Contrast was enhanced by digital processing. The initial director orientation is horizontal. The depicted image is $0.225 \times 0.225\text{mm}^2$, $d = 11\mu\text{m}$.

As possible explanations, several ideas have been proposed: a hand-waving argument based on "destabilization of twist fluctuations" [52], a possibility of an isotropic mechanism based on the non-uniform space charge distribution along the field [53] and the flexoelectric effect [55–57].

Case H: homeotropic alignment, $\epsilon_a < 0$ & $\sigma_a < 0$.   Above the Freedericksz transition where no standard EC is predicted, convection (**ns-EC**) builds up with properties similar to those listed for case **G**. The patterns are disordered (see Fig. 14) as expected for an initial homeotropic alignment.

Summary

In this paper we have reviewed the structures appearing at onset of electro-convection in nematic liquid crystals. The influence of the relevant material parameters (ϵ_a and σ_a) and the role of the initial director alignment were explored. Our calculations using a linear stability analysis of the standard model of electroconvection (performed for zero frequency) revealed that four different scenarios characterized by different ranges of the wavenumber q can be identified: (**1**) the $q_F = 0$ mode (a homogeneous deformation known as the *Freedericksz transition*) predicted and observed in cases **C**, **D**, **E** and **H**, which is

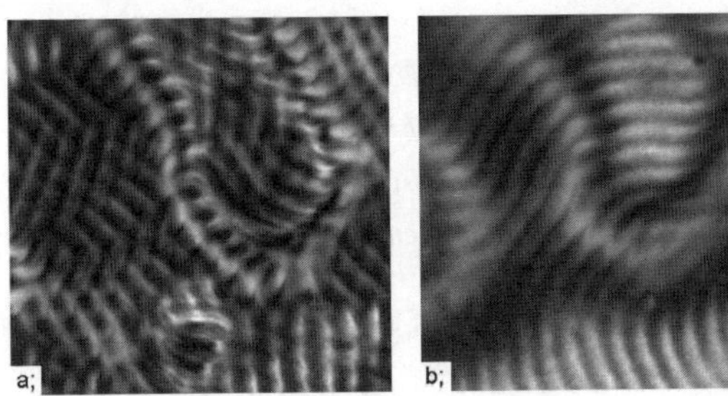

Figure 14. Snapshots of nonstandard electroconvection pattern in case **H** taken with crossed polarizers. a; Oblique rolls, b; parallel rolls. The initial director orientation is horizontal. The depicted image is $0.45 \times 0.45 \text{mm}^2$, $d = 20\mu\text{m}$.

actually a convection-free state; (**2**) the $q_c \approx 1$ mode (wavelength in the range of d) which is the *classical electroconvection*, appearing either as a primary or a secondary bifurcation; both have been detected and explained quantitatively in cases **A**, **B** and **C**, **D**; (**3**) a short-wavelength structure with $q_\alpha > 15$ in case **F**; and (**4**) a long-wavelength structure with $q_\alpha < 0.07$ in case **D**. The scenarios (**3**) and (**4**) are only theoretically predicted (*"α-induced" EC*) for substances with $\alpha_3 < 0$. With extending the consideration from zero to low frequencies, no qualitative change of the mode classification is expected, although some morphological changes of patterns can occur (oblique versus normal rolls).

At high frequencies – in the dielectric regime – Eqs. (7) and (8) do not apply. For completeness, we mention that independent calculations for this range would invoke a different q-mode (**5**), the *dielectric rolls*, with the wavelength that does not scale with d.

Experiments, however, have proved that other stripe patterns occur that do not yet have an unambiguous, widely accepted interpretation. These patterns can be classified in two additional q-modes: (**6**) the *prewavy* pattern (or wide domains) with $0.2 < q < 0.5$ observed in cases **A** and **C** and replacing modes (**2**) and/or (**5**) or coexisting with them; and (**7**) the parallel (*longitudinal*) rolls that are observed in cases **G** and **H**, with $q_c \approx 1$ as for mode (**2**), but otherwise having different characteristics.

Acknowledgment

The authors wish to thank E. Kochowska and A. Cauquil-Vergnes for the snapshots of ns-EC patterns. This work was supported by the EU grant EU-HPRN-CT-2002-00312, the NATO ASI 980684, CRG.LG 973103 and the Hungarian Research Fund OTKA T-037336.

References

[1] M.C. Cross and P.C. Hohenberg, *Rev. Mod. Phys.* **65**, 851 (1993).

[2] P.G de Gennes and J. Prost, *The Physics of Liquid Crystals*, Clarendon Press, Oxford, 1993.

[3] S. Chandrasekhar, *Liquid Crystals*, University Press, Cambridge, 1992.

[4] *Physical Properties of Liquid Crystals: Nematics*, editors: D.A. Dunmur, A. Fukuda and G.R. Luckhurst, Inspec, London, 2001

[5] *Pattern Formation in Liquid Crystals*, editors: A. Buka and L. Kramer, Springer, New York, 1996.

[6] W. Pesch and U. Behn, Electrohydrodynamic Convection in Nematics. In *Evolution of Spontaneous Structures in Dissipative Continuous Systems*, editors: F.H. Busse and S.C. Müller, pages 335–383, Springer, New York, 1998.

[7] L. Kramer and W. Pesch, Electrohydrodynamics in Nematics. In *Physical Properties of Liquid Crystals: Nematics*, editors: D.A. Dunmur, A. Fukuda and G.R. Luckhurst, pages 441–454, Inspec, London, 2001

[8] W. Pesch and L. Kramer, General Mathematical Description of Pattern-Forming Instabilities. In *Pattern Formation in Liquid Crystals*, editors: A. Buka and L. Kramer, pages 69–90, Springer, New York, 1996.

[9] L.M. Blinov and V.G. Chigrinov, *Electrooptic Effects in Liquid Crystal Materials*, Springer, New York, 1994.

[10] H. Pleiner and H. Brandt, Hydrodynamics and Electrohydrodynamics of Liquid Crystals. In *Pattern Formation in Liquid Crystals*, editors: A. Buka and L. Kramer, pages 15–68, Springer, New York, 1996.

[11] M. Born and E. Wolf, *Principles of Optics*, Pergamon, Oxford, 1996.

[12] S. Rasenat, G. Hartung, B.L. Winkler and I. Rehberg, *Exp. Fluids* **7**, 412 (1989).

[13] S. P. Trainoff and D. Canell, *Physics of Fluids* **14**, 1340 (2002).

[14] N. Éber, S.A. Rozanski, Sz. Németh, Á. Buka, W. Pesch and L. Kramer, *Phys. Rev. E* **70**, 61706 (2004).

[15] E. Bodenschatz, W. Pesch and G. Ahlers, *Annu. Rev. Fluid Mech.* **32** , 709-778 (2000).

[16] L. Kramer and W. Pesch, Electrohydrodynamic Instabilities in Nematic Liquid Crystals. In *Pattern Formation in Liquid Crystals*, editors: A. Buka and L. Kramer, pages 221–255, Springer, 1996.

[17] E.F. Carr, *Mol. Cryst. Liq. Cryst.* **7**, 253 (1969).

[18] W. Helfrich, *J. Chem. Phys.* **51**, 4092 (1969).

[19] E. Bodenschatz, W. Zimmermann and L. Kramer, *J. Phys. (Paris)* **49**, 1875 (1988).

[20] See equation (6.5) on p. 221 in [16]. Note that there $I_h = 0.97267$ has been replaced by one and that $\alpha_3/q^2 - \alpha_2$ has been approximated by $|\alpha_2|$.

[21] Numerical codes for the linear analysis of the full standard model can be obtained upon request from the authors.

[22] A. Joets and R. Ribotta, *J. Phys. (Paris)* **47**, 595 (1986).

[23] S. Kai, N. Chizumi and M. Kokuo, *Phys. Rev. A* **40**, 6554 (1989).

[24] M. Dennin, M. Treiber, L. Kramer, G. Ahlers and D. Cannell, *Phys. Rev. Lett.* **76**, 319 (1995).

[25] S. Rasenat, V. Steinberg and I. Rehberg, *Phys. Rev. A* **42**, 5998 (1990).

[26] M. Treiber and L. Kramer, *Phys. Rev. E* **58**, 1973 (1989).

[27] M. Treiber and L. Kramer, *Mol. Cryst. Liq. Cryst.* **261**, 951 (1995).

[28] I. Rehberg, S. Rasenat and V. Steinberg, *Phys. Rev. Lett.* **62**, 756 (1989).

[29] M. Treiber, N. Éber, Á. Buka and L. Kramer, *J. Phys. II (Paris)*, **7**, 649 (1997).

[30] Á. Buka, B. Dressel, W. Otowski, K. Camara, T. Tóth-Katona, L. Kramer, J. Lindau, G. Pelzl and W. Pesch, *Phys. Rev. E* **66**, 051713/1-8 (2002).

[31] See equation (6.29) on p.244 in [16], which has been rewritten. The definition of $\bar\sigma_a$ in equation (6.31) contains a misprint. The correct equation reads:
$$\bar\sigma_a(eff) = \frac{\sigma_a(\epsilon_\perp/\epsilon_\parallel - \epsilon_a\sigma_\perp/(\epsilon_\parallel\sigma_a))}{(1+\omega'^2)\bar S}.$$

[32] Á. Buka, B. Dressel, L. Kramer and W. Pesch, *Phys. Rev. Lett.* **93(4)**, 044502/1-4 (2004).

[33] Á. Buka, B. Dressel, L. Kramer and W. Pesch, *Chaos* **14**, 793-802 (2004).

[34] L. Kramer, A. Hertrich and W. Pesch, Electrohydrodynamic Convection in Nematics: the Homeotropic Case. In *Pattern Formation in Complex, Dissipative Systems*, editor: S. Kai, pages 238–246, World Scientific, Singapore (1992).

[35] A. Hertrich, W. Decker, W. Pesch and L. Kramer, *Phys. Rev. E* **58**, 7355 (1998).

[36] A.G. Rossberg, N. Éber, Á. Buka and L. Kramer, *Phys. Rev. E* **61**, R25 (2000).

[37] S. Kai, K. Hayashi and Y. Hidaka, *J. Phys. Chem.* **100**, 19007 (1996).

[38] Y. Hidaka, J.-H. Huh, K. Hayashi, M. Tribelsky and S. Kai, *J. Phys. Soc. Jpn.* **66**, 3329 (1997).

[39] P. Tóth, Á. Buka, J. Peinke and L. Kramer, *Phys. Rev. E* **58**, 1983 (1998).

[40] H. Richter, N. Kloepper, A. Hertrich and Á. Buka, *Europhys. Lett.* **30**, 37 (1995).

[41] H. Richter, Á. Buka and I. Rehberg, *Phys. Rev. E* **51**, 5886 (1995).

[42] P. Tóth, N. Éber, T.M. Bock, Á. Buka and L. Kramer, *Europhys. Lett.* **57**, 824 (2002).

[43] Á. Buka, P. Tóth, N. Éber and L. Kramer, *Physics Reports* **337**, 157 (2000).

[44] B. Dressel and W. Pesch, *Phys. Rev. E.* **67**, 031707 (2003)

[45] B. Dressel, L. Pastur, W. Pesch, E. Plaut and R. Ribotta, *Phys. Rev. Lett.* **88** , 024503 (2002)

[46] S. Kai and K. Hirakawa, *Solid State Commun.* **18** 1573 (1976).

[47] P. Petrescu and M. Giurgea, *Phys. Lett.* **59A**, 41 (1976).

[48] A.N. Trufanov, M.I. Barnik and L.M. Blinov, A Novel Type of the Electrohydrodynamic Instability in Nematic Liquid Crystals. in *Advances in Liquid Crystal Research and Application*, editor: L. Bata, pages 549–560, Akadémiai Kiadó - Pergamon Press (1980).

[49] J.-H. Huh, Y. Yusuf, Y. Hidaka, and S. Kai, *Mol. Cryst. Liq. Cryst.* **410**, 39 (2004).

[50] J.-H. Huh, Y. Hidaka, Y. Yusuf, N. Éber, T. Tóth-Katona, Á. Buka and S. Kai, *Mol. Cryst. Liq. Cryst.* **364**, 111 (2001).

[51] J.-H. Huh, Y. Hidaka, A.G. Rossberg and S. Kai, *Phys. Rev. E* **61**, 2769 (2000).

[52] M. Goscianski and L. Léger, *J. Phys. (Paris)* **36**, N.3, C1-231 (1975).

[53] L.M. Blinov, M.I. Barnik, V.T. Lazareva and A.N. Trufanov, *J. Phys. (Paris)* **40**, N.4, C3-263 (1979).

[54] E. Kochowska, S. Németh, G. Pelzl and Á. Buka, *Phys. Rev. E* **70**, 011711 (2004).

[55] N.V. Madhusudana and V.A. Raghunathan, *Mol. Cryst. Liq. Cryst. Lett.* **5**, 201 (1988).

[56] N.V. Madhusudana and V.A. Raghunathan, *Liquid Crystals*, **5**, 1789 (1989).

[57] M.I. Barnik, L.M. Blinov, A.N. Trufanov and B.A. Umanski, *J. Phys. (Paris)* **39**, 417-422 (1978).

DYNAMICAL PHENOMENA IN NEMATIC LIQUID CRYSTALS INDUCED BY LIGHT

Dmitry O. Krimer,[1] Gabor Demeter,[2] and Lorenz Kramer[1]

[1]*Physikalisches Institut der Universitaet Bayreuth, D-95440 Bayreuth, Germany*

[2]*Research Institute for Particle and Nuclear Physics of the Hungarian Academy of Sciences, Konkoly-Thege Miklos ut 29-33, H-1121 Budapest, Hungary*

Abstract   A wide range of interesting dynamical phenomena have been observed in nematic liquid crystals, that are induced by strong laser radiation. We review the latest theoretical advances in describing and understanding these complex phenomena.

Keywords: Nematic liquid crystals, Dynamical phenomena, Optically induced instabilities

Introduction

The optics of liquid crystals has been a widely investigated subject for decades, whose importance stems from the enormous range of technological applications where liquid crystals are utilized for their optical properties. The most important advantage of liquid crystals is that these optical properties can be changed and controlled quickly and easily. The basic physical origin of these phenomena is the subject of a number of review papers [1, 2] and monographs [3, 4].

A very interesting group of phenomena is that associated with the so-called light-induced director reorientation in nematic liquid crystals (or nematics for short). Liquid crystals consist of elongated molecules (rod shaped, or disc shaped) which have anisotropic polarizability. In a nematic liquid crystal phase, where the molecular orientation is ordered, the propagation of light waves is then governed by an anisotropic dielectric tensor. The optical axis of the nematic is aligned along the local direction of the molecular axis which is called the director. On the other hand, the anisotropic polarizability also means that an orienting torque is exerted on the molecules by any electrical field, including the electric field of the light. Thus an intense light, whose electric field is strong enough to reorient the molecules (i.e. turn the director), alters the optical properties of the medium it propagates through. This leads to a large

A. A. Golovin and A. A. Nepomnyashchy (eds.),
Self-Assembly, Pattern Formation and Growth Phenomena in Nano-Systems, 83–122.
© 2006 *Springer. Printed in the Netherlands.*

variety of nonlinear optical responses of the liquid crystal medium. Some of these phenomena are rather spectacular (such as the ring pattern due to light self-phase modulation) and are sometimes collectively referred to as the Giant Optical Nonlinearity of liquid crystals [1, 2].

Among these phenomena, there are a number of situations where a constant illumination of the liquid crystal leads to persistent oscillation of the molecules and sometimes even to chaotic behavior. There has been considerable effort recently to observe, describe and understand these phenomena. A lot of experiments have been performed, and theoretical models have been proposed. In some cases, there is sufficient agreement between theory and experiment, in other cases not. In all cases it is clear that the dynamical behavior excited by light in nematics is very rich - numerous bifurcations, transitions and regimes have been predicted and observed. In this paper, we review recent theoretical developments in modeling, simulating and understanding the physical origins of these phenomena. We put special emphasis on explorations in the "nonlinear domain", where considerable progress has been made during the past few years. Numerous instabilities and bifurcations were found, and they helped interpret experimental observations a great deal. By no means is the task accomplished however - there is a number of experimental situations which have been considered theoretically only very superficially, or not at all. Understanding these complex phenomena is, as yet, far from complete.

1.   Simple setups - complicated phenomena

The basic experimental setup that can be used to generate the complex nonlinear behavior in nematics is deceptively simple. A thin cell is made from two parallel glass plates enclosing the nematic layer, whose thickness is the order of $L = 10 - 100\mu m$. The glass plates are coated with some chemical surfactant to achieve a fixed orientation of the nematic director at the interface. The cell is irradiated with a continuous laser beam, and the reorientation of the molecules induced by the light is monitored by measuring the changes in intensity and polarization of the outcoming light. Sometimes a separate probe beam is used whose changing polarization and intensity supplies information on director reorientation (Fig. 1). The key properties of the setup are the thickness of the nematic layer, the orientation of the director at the boundaries, the polarization, angle of incidence, and intensity of the light.

Depending on the relative orientation of the vector of polarization of light and the initial director orientation, we can distinguish two different cases. In one case, the two directions enclose some angle $0 < \beta < \pi/2$, and so an orienting torque acts, that turns the director for arbitrarily small light intensity - there is no threshold intensity required (though considerable intensity may be needed for appreciable changes in the director orientation, as the elasticity

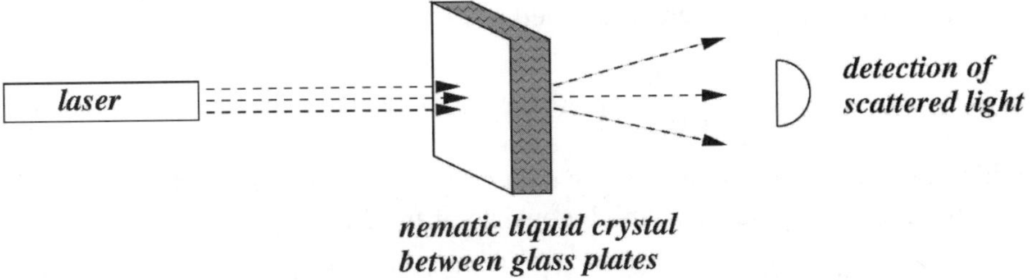

Figure 1.　　Typical experimental setup: a thin layer of nematic liquid crystal is sandwiched between two glass plates and irradiated with laser light. The scattered light (or the light of a probe beam) is monitored behind the cell.

of the nematic counteracts the reorientation). In the other case, the director orientation is parallel or perpendicular to the polarization vector, and so the orienting torque acting on the director is zero. However, even though the configuration where the director is perpendicular to the polarization is always an equilibrium, it will be unstable above a certain light intensity. Thus, above a certain threshold intensity, director reorientation will take place. This is the so called Light-Induced Freedericksz Transition or Optically Induced Freedericksz Transition (OFT) [1, 3], which is analogous to the classical Freedericksz transition induced by static (or low frequency) electric or magnetic fields. This transition can already lead to time-dependent behavior in certain geometries. In other geometries the simple static reoriented state undergoes further bifurcations as the light intensity increases and produces complex, time dependent behavior.

2.　　Theoretical description

Basic equations

The starting point for the description of these complex phenomena is the set of hydrodynamic equations for the liquid crystal and Maxwell's equation for the propagation of the light. The relevant physical variables that these equations contain are the director field $\mathbf{n}(\mathbf{r}, t)$, the flow of the liquid $\mathbf{v}(\mathbf{r}, t)$ and the electric field of the light $\mathbf{E}_{light}(\mathbf{r}, t)$. (We assume an incompressible fluid and neglect temperature differences within the medium.) The Navier-Stokes equation for the velocity $\mathbf{v}$ can be written as [5]

$$\rho_m \left(\partial_t + \mathbf{v} \cdot \nabla \right) v_i = -\nabla_j (p\, \delta_{ij} + \pi_{ij} + T_{ij}^{visc}),　\tag{1}$$

where ρ_m and p are the density and the pressure of the nematic, respectively. π_{ij} is the Ericksen stress tensor defined as

$$\pi_{ij} = \frac{\partial F}{\partial(\partial_j n_k)} \cdot \partial_i n_k, \quad i = x, y, z, \tag{2}$$

(where the summation over doubly occurring indices is assumed.) In Eq. (2) F is the free energy density which consists of the elastic part

$$F^{(elast)} = \frac{K_1}{2} \left(\nabla \cdot \mathbf{n}\right)^2 + \frac{K_2}{2} \left(\mathbf{n} \cdot \nabla \times \mathbf{n}\right)^2 + \frac{K_3}{2} \left(\mathbf{n} \times \nabla \times \mathbf{n}\right)^2, \tag{3}$$

and the external part which, in our case, is

$$F^{(ext)} = -\frac{\varepsilon_a}{16\pi} \mid \mathbf{n} \cdot \mathbf{E}_{light} \mid^2 . \tag{4}$$

Here, K_1, K_2, K_3 are, respectively, the splay, twist and bend elastic constants [5], and $\mathbf{E}_{light}$ is the amplitude of the optical electric field. Any other external fields which act on the director (static or low frequency electric or magnetic fields for example) can be incorporated into $F^{(ext)}$ by adding similar terms. The viscous stress tensor T_{ij}^{visc} in Eq. (1) is written in terms of the six Leslie coefficients α_i [6],

$$-T_{ij}^{visc} = \alpha_1 n_i n_j n_k n_l A_{kl} + \alpha_2 n_j N_i + \alpha_3 n_i N_j + \tag{5}$$
$$\alpha_4 A_{ij} + \alpha_5 n_j n_k A_{ki} + \alpha_6 n_i n_k A_{kj} .$$

The symmetric strain-rate tensor A_{ij} and the vector $\mathbf{N}$, which gives the rate of change of the director relative to the fluid, are

$$A_{ij} = (\partial_i v_j + \partial_j v_i)/2 , \tag{6}$$
$$\mathbf{N} = (\partial_t + \mathbf{v} \cdot \nabla)\mathbf{n} - \boldsymbol{\omega} \times \mathbf{n} .$$

Here $\boldsymbol{\omega} = (\nabla \times \mathbf{v})/2$ is the local fluid rotation. The Leslie coefficients satisfy the Parodi relation $\alpha_2 + \alpha_3 = \alpha_6 - \alpha_5$ [7]. In addition, the assumption of incompressibility means that the density ρ_m is constant, and so $\nabla \cdot \mathbf{v} = 0$. The equation for the director $\mathbf{n}$ is

$$\gamma_1(\partial_t + \mathbf{v} \cdot \nabla - \boldsymbol{\omega} \times)\mathbf{n} = -\underline{\delta}^{\perp} \left(\gamma_2 \underline{\underline{A}} \mathbf{n} + \mathbf{h}\right), \tag{7}$$

where $\gamma_1 = \alpha_3 - \alpha_2$ is the rotational viscosity, and $\gamma_2 = \alpha_3 + \alpha_2$. $\mathbf{h}$ is the molecular field obtained from the variational derivatives of F:

$$h_i = \frac{\delta F}{\delta n_i} = \frac{\partial F}{\partial n_i} - \partial_j \left(\frac{\partial F}{\partial n_{i,j}}\right), \quad i = x, y, z . \tag{8}$$

and the projection operator $\delta_{ij}^{\perp} = \delta_{ij} - n_i n_j$ in Eq. (7) ensures conservation of the normalization $\mathbf{n}^2 = 1$.

In addition, the electric field must be obtained from Maxwell's equations for light propagation, which, for a nonmagnetic material in the absence of any currents and charges, can be written as:

$$\nabla \times \mathbf{H} = \frac{1}{c}\underline{\underline{\varepsilon}}\,\frac{\partial \mathbf{E}}{\partial t}, \qquad \nabla \cdot (\underline{\underline{\varepsilon}}\,\mathbf{E}) = 0,$$

$$\nabla \times \mathbf{E} = -\frac{1}{c}\frac{\partial \mathbf{H}}{\partial t}, \qquad \nabla \cdot \mathbf{H} = 0, \tag{9}$$

with the spatially dependent dielectric tensor

$$\varepsilon_{ij} = \left(\varepsilon_\perp + i\gamma_\perp\right)\delta_{ij} + \left(\varepsilon_a + i\gamma_a\right)n_i n_j, \tag{10}$$

where $\varepsilon_a = \varepsilon_\| - \varepsilon_\perp$ [$\gamma_a = \gamma_\| - \gamma_\perp$] is the real [imaginary] part of the dielectric anisotropy. The imaginary part which describes absorption is usually negligible in pure nematics but must be taken into account if the nematic has been doped by absorbing dyes.

The boundary conditions needed for an unambiguous solution of the PDEs are usually taken to be a fixed orientation of the director (strong anchoring) and vanishing velocity field (no-slip) at the nematic-glass interface.

The equations (1), (7) and (9) constitute the starting point for any theoretical description of dynamical phenomena induced by light in nematics. Clearly, light propagation is influenced by the spatial distribution of the director orientation through the dielectric tensor (10), and the electric field of the light influences the orientation of the director through the free energy (4) whose derivatives (8) enter the director equation (7). The fluid flow must also be added, as flow is coupled to the director, so any dynamical process that leads to director reorientation will also induce flow even in the absence of pressure gradients. This is the so-called backflow.

Some general remarks

The above equations contain three distinct timescales: the time it takes the light to traverse the cell $\tau_l = L/c$, the momentum diffusion time $\tau_{visc} = \varrho_m L^2/\gamma_1$ and the director relaxation time $\tau = \gamma_1 L^2/\pi^2 K_3$. These usually differ by many orders of magnitude, since typically $\tau_l \sim 10^{-13}s$, $\tau_{visc} \sim 10^{-6}\,s$ and $\tau \sim 1s$, so the slow variable of the system is clearly the evolution of the director which enslaves the other two modes. The electric field of the light can thus be expressed from Maxwell's equations as a function of the instantaneous value of $\mathbf{n}$ and can be considered as a self-consistency relation or a constraint. In a similar way, due to the vastly different magnitude of τ and τ_{visc}, inertial terms in the Navier-Stokes equation can be neglected, and the flow of the nematic can be expected to be determined entirely by the director components and their time derivatives.

These equations are a very complex set of nonlinear partial differential equations. The main difficulty lies in the fact that, even though $\mathbf{E}_{light}$ and $\mathbf{v}$ are theoretically defined by $\mathbf{n}$ at every instant t, in general it is very hard, if not impossible to express them by $\mathbf{n}$. Most often we can only say that "$\mathbf{E}_{light}$ and $\mathbf{v}$ are a solution of this and that", where "this and that" will be a partial differential equation. Even when $\mathbf{E}_{light}$ and $\mathbf{v}$ can be expressed by $\mathbf{n}$, the expressions will be complicated integral relationships. To gain any meaningful solutions (even using computers), one must resort to further assumptions and approximations. Analytic results are available only under the most restrictive approximations and simplest cases. Numerical solutions are obtainable under much less restrictive conditions, but yield proportionally less insight into the physical origin of the phenomena. A delicate balance is needed when applying restrictive assumptions to the solutions in order to obtain solvable equations, and at the same time to keep physical phenomena within grasp. Should the assumptions be too restrictive, we will readily obtain solutions that miss important aspects of the dynamics or have no connection with real physical processes at all. A constant comparison between theoretical results and experimental observations is needed to avoid pitfalls.

There are several major simplifications that can be used to tackle these equations. First of all, fluid flow is almost always neglected altogether. Then the Navier-Stokes equation is not needed at all, $\mathbf{v}$ is no longer a variable and we only need to solve the director equation (7) which will now be

$$\gamma_1 \partial_t \mathbf{n} = -\underline{\underline{\delta}}^{\perp} \mathbf{h}. \tag{11}$$

$\mathbf{h}$ will still contain the electric fields through (4) and (8), so (11) is still coupled to (9). This approximation is sometimes justified by arguments that flow plays only a passive role (backflow) and makes only a quantitative difference. Sometimes it is argued that when reorientation is small, the effect of flow can be included in a renormalized (reduced) rotational viscosity. This, however, turns out not always to be true. A renormalized viscosity is applicable strictly only in a linear approximation (even then not always), and backflow turns out to make a qualitative difference in some cases. Thus the real reason why flow is usually neglected is simply because it reduces the complexity of the equations a great deal. Explicit treatment of backflow has been attempted only in very few cases [8–10]. Even without flow, obtaining $\mathbf{E}$ from $\mathbf{n}$ remains a formidable task, and usually more approximations are needed.

Another frequently employed simplification is the 1D assumption according to which all variables depend only on one coordinate, namely, the one transversal to the plane of the nematic layer, say, the z coordinate. This means that the incident light should be an infinite plane wave (hence this approximation is often called the infinite plane wave approximation), and, by virtue of incompressibility and the boundary conditions, $\mathbf{v} = (v_x(z,t), v_y(z,t), 0)$ which grants an enormous simplification of the initial equations [9]. The application of this assumption is slightly controversial. Experimental setups rarely use laser beams whose width w_0 is so much larger than that of the cell L, that would justify it. Also, there is evidence that even if the laser beams were ideal plane waves, transverse degrees of freedom could not be neglected, for spontaneous pattern formation would occur [11, 12] (see the last section). However, current experimental evidence leads us to believe that the width of the laser beam w_0 plays a crucial role in the observed phenomena only if the cell width L is about two times larger ($w_0/L \approx 0.5$) [13]. Theoretical results derived using the 1D assumption compare remarkably well to experimental observations obtained using laser beams with $w_0/L \approx 1$. Due to this, the 1D assumption is relaxed very rarely.

In the 1D approximation, the director equations (7) reduce to [9]:

$$\gamma_1 \partial_t n_x + n_z \left[(\alpha_2 - \gamma_2 n_x^2) \partial_z v_x - \gamma_2 n_x n_y \partial_z v_y \right] = - \left[\underline{\underline{\delta}}^{\perp} \mathbf{h} \right]_x ,$$

$$\gamma_1 \partial_t n_y + n_z \left[(\alpha_2 - \gamma_2 n_y^2) \partial_z v_y - \gamma_2 n_x n_y \partial_z v_x \right] = - \left[\underline{\underline{\delta}}^{\perp} \mathbf{h} \right]_y . \quad (12)$$

Following the concept of adiabatic elimination, the velocity gradients $\partial_z v_x, \partial_z v_y$ can be expressed with the components of $\mathbf{n}$ from (1) and substituted into (12) [14, 9]. The procedure is straightforward, but expressions are complicated, so the equations obtained can be solved only numerically.

A further simplification comes from the typical boundary condition of strong anchoring that allows us to expand the z dependence of the director in terms of a full set of base functions, typically trigonometric functions. Since elasticity in nematics inhibits the growth of reorientation with a force proportional to k_i^2 (here k_i is the inverse wavelength of the i-th mode), high order modes will be strongly damped, and thus a truncation to a finite number of modes is possible. This way, spatial dependence with respect to z is described by a few mode functions and their amplitudes. If additionally the 1D assumption is also used (i.e. the physical quantities depend only on z), this procedure of expansion and projection of the equations onto the base functions can be used to get rid of spatial dependence altogether. The system reduces to a finite dimensional one, described by a set of complicated nonlinear ODEs for the mode amplitudes. This is favorable, as investigating the solutions of nonlinear ODEs is almost always much simpler, even if, in this case not necessarily faster. On the other hand, taking too few modes can easily result in a loss of physically important solutions. In any case, a correct choice of director description (Cartesian components, various angles, e.g. spherical) is essential. Choosing a representation that corresponds to the symmetries of the setup can simplify the equations drastically.

It is important to note that the inhibition of high order modes due to elasticity always holds – even when it is not easy or practical to utilize. Thus sometimes a numerical solution of the equations does not use such an expansion, but the results are projected onto a set of modes for analysis, as the time behavior of mode amplitudes is often more meaningful and telling than the time behavior of the director at any one point within the cell.

Equations for the light propagation

The most basic and unavoidable of problems when treating dynamical phenomena induced by light in nematics is the solution of Maxwell's equations for some distorted director configuration. Phenomena in the nonlinear domain are governed by an interplay between director reorientation – light torque change due to modification of light propagation, so the success or failure of a theory often depends on the suitably chosen representation and/or approximations used when dealing with this problem.

One simplification can almost always be applied: since director reorientation changes very slowly on the spatial scale of the wavelength of light, the electric field can be separated into fast phase exponentials and slow amplitudes (similarly to many problems of light propagation in anisotropic media). If additionally the 1D assumption can be applied, it is possible to write relatively simple coupled ODEs for the slow field amplitudes (the amplitudes will depend only on z - slow time dependence will result only through the time dependence of the director). One possible way to obtain equations for slow amplitudes in the 1D case is to use the so-called Berreman formalism [15]. The electric and magnetic fields of the light should be written in the form:

$$\mathbf{E}_{light}(\mathbf{r}, t) = \frac{1}{2}(\mathbf{E}(z, t)e^{i(k_x x + k_y y)}e^{-i\omega t} + c.c.), \tag{13}$$

with the possible $x - y$ dependence of the fields due to oblique incidence entirely incorporated into the fast exponentials. From (9) or the wave equation that can be obtained from it, it is straightforward to derive an equation for the amplitudes $\mathbf{E}(z, t), \mathbf{H}(z, t)$. A vector of four independent amplitudes describes the light field, and a set of linear, first order, ordinary differential equations governs its evolution:

$$\frac{d\bar{\Psi}}{dz} = ik_0 \mathrm{D}\bar{\Psi}, \tag{14}$$

where

$$\bar{\Psi} = \begin{pmatrix} E_x \\ H_y \\ E_y \\ -H_x \end{pmatrix}. \tag{15}$$

The matrix D depends on the director components [3], and the other two field components are defined by the above four unambiguously. By calculating the eigenvalues of D, one can separate the fast oscillations in z from the slow amplitudes. Another frequently used formalism is the separation of the electric field into an ordinary and an extraordinary wave component [3]. This separation depends on space and time through the director components, but the resulting equations are very useful (especially for the case of perpendicular incidence), as the slow amplitudes change only due to twist distortions of the nematic which is often small enough to allow a perturbative solution of the equations.

If the equations of motion (1,7 or 11) are to be integrated by computer, the equations for the slow field amplitudes can be solved numerically relatively easily at each step of the integration. Sometimes it is also possible to obtain an approximate expression for the fields as a function of the director components. This is not very easy, however, and great care must be taken. While the amplitudes of the electric field may change relatively slowly as the light traverses the cell, phase differences between various components acquire importance much faster. This means that integrals of the director components will appear in the expressions, and perturbation theory has a very limited validity.

3.   Obliquely incident, linearly polarized light

One of the most interesting and investigated geometries is when a linearly polarized light wave is incident at a slightly oblique angle on a cell of homeotropically aligned nematic. The direction of polarization is perpendicular to the plane of incidence in this setup, so the system is symmetric with respect to inversion over this plane [the $x - z$ plane, see Fig. 2 (a)]. Very interesting dynamical phenomena were observed in numerous experiments in this geometry. First it was noted that persistent oscillations are possible above a certain threshold intensity [16, 17]. Then various "competing" oscillatory states and

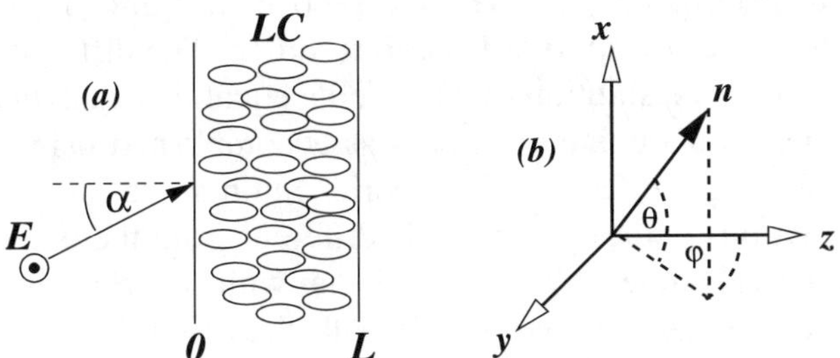

Figure 2.   Basic geometry of the setup. A linearly polarized light wave is incident upon a cell of homeotropically aligned nematic at a slightly oblique angle α. The direction of polarization is perpendicular to the plane of incidence (ordinary wave). The setup is symmetric with respect to the inversion $S : y \to -y$.

stochastic oscillations were found [18], and it was noted that the system eventually exhibits chaotic behavior [19]. Considerable effort went into exploring the various regimes of regular and stochastic oscillations and trying to identify the transitions between them [20, 21]. Without a detailed theory, however, interpretation of experimental observations remained inadequate. Initial theoretical investigations performed the linear stability analysis of the homeotropic state [16, 22]. These treatments used the 1D approximation, neglected flow, and, employing a mode expansion, managed to derive a solvable set of linear equations for small distortions around the homeotropic state. They showed that the first instability of the homeotropic state is a pitchfork bifurcation for small angles (the usual optical Freedericksz transition), but, above a certain angle of incidence α_{TB}, the primary instability becomes a Hopf bifurcation. However, since the angles at which regular oscillations and stochastic behavior were observed were smaller than this critical angle, it was obvious that one must go beyond the linear stability analysis and look for further bifurcations in the nonlinear regime.

The next step was the derivation of a "simple" model to try to describe these complex phenomena [23]. The starting point was again the same as that used previously for the linear stability analysis: a plane wave approximation for the light, no flow included and the assumption that reorientation is small around the homeotropic state, i.e. we are in the weakly nonlinear regime. The director was described in terms of two angles, as $\mathbf{n} = (\sin\theta, \cos\theta\sin\varphi, \cos\theta\cos\varphi)$ [see Fig. 2 b)], and, using the strong anchoring at the boundary, these were expanded as $\varphi(z,t) = \sum_n A_n(t)\sin(n\pi z/L)$, $\theta(z,t) = \sum_n B_n(t)\sin(n\pi z/L)$. The set of mode amplitudes $(A_1, .., B_1, ...)$ were truncated, and the resulting expression for the director was substituted into the equations of motion and projected onto these modes. The aim of this analysis was to derive a set of explicit first order ODEs for the time evolution of the mode amplitudes. Clearly, for a "minimal model" expected to be able to describe further bifurcations and possibly chaotic oscillations, one needs to keep at least three mode amplitudes and keep nonlinear terms up to at least third order. The difficult point of this analysis was finding the solution of Maxwell's equations analytically with the mode amplitudes as parameters which was accomplished using perturbation theory and exploiting the fact that the reorientation angles are expected to be small. (Note that this choice of angles is different from the one usually used, namely the spherical angles. The reason is that with spherical angles, only the polar coordinate can be assumed to be small, the azimuthal one not – thus a power expansion is not possible.) The general form of the equations obtained is:

$$\tau \dot{A}_i = \sum_j L^A_{ij} A_j + \sum_{j,k} P^A_{ijk} A_j B_k$$
$$+ \sum_{\substack{j,k,l \\ k \leq l}} Q^A_{ijkl} A_j B_k B_l + \sum_{j \leq k \leq l} R^A_{ijkl} A_j A_k A_l,$$

$$\tau \dot{B}_i = \sum_j L^B_{ij} B_j + \sum_{j \leq k} P^B_{ijk} A_j A_k + \sum_{\substack{j,k,l \\ k \leq l}} Q^B_{ijkl} B_j A_k A_l$$

$$+ \sum_{j \leq k \leq l} R^B_{ijkl} B_j B_k B_l . \tag{16}$$

The inversion symmetry with respect to the $x - z$ plane implies that the equations must be invariant under the transformation $S : \{A_i, B_i\} \rightarrow \{-A_i, B_i\}$, so that only odd powers of the A_i-s can appear in the first set of equations and only even powers in the second set. In a linear approximation, only the A-s have to be taken into account, as they are the ones driven by the light directly. Thus for a minimal model the three modes: A_1, A_2, B_1 have been kept. The resulting set of ODEs has been solved numerically, and the nature of the solutions has been analyzed as a function of the two control parameters of the problem: the angle of incidence α and the intensity of the light normalized by the OFT threshold ρ.

The numerical solution of the equations gave exciting results. In the region where the primary instability of the homeotropic state is a stationary instability, two new stationary states are born, which are mutual images under the symmetry transformation S. These then lose stability at some critical intensity in a Hopf bifurcation, where two limit cycles are born (again, mutual images under S). They are depicted in Fig. 3 a) where they are plotted in the phase space spanned by the three amplitudes $\{A_1, A_2, B_1\}$. This is very reassuring, since it accounts for the regular oscillating regime observed in the experiments for angles smaller than α_{TB}. As the intensity is increased, the symmetric limit cycles pass closer and closer to the origin (which, above the OFT threshold is a saddle), and at a certain intensity ρ_1, they become homoclinic trajectories to the origin (i.e. the homogeneous homeotropic state) [Fig. 3 b)]. Above ρ_1, the two limit cycles merge into one double-length limit cycle that is symmetric with respect to S, [Fig. 3 c)]. This bifurcation is called a homoclinic gluing, or a gluing bifurcation. A further increase in the light intensity then brings about another symmetry-breaking bifurcation, where the symmetric limit cycle gives way to two new asymmetric limit cycles, which are mutual images under S [Fig. 3 d)]. At a still higher intensity, these too become homoclinic to the origin [Fig. 3 e)] and merge into a quadruple-length symmetric limit cycle [Fig. 3 f)]. This sequence of splitting and re-merging of limit cycles continues

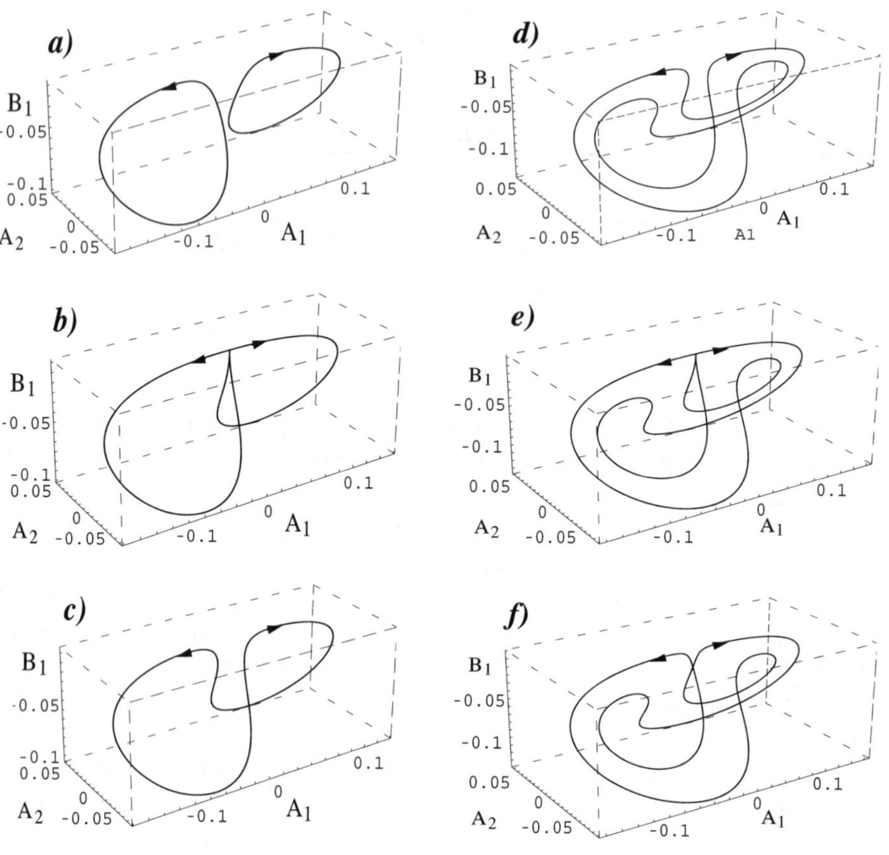

Figure 3. Limit cycles in phase space at various intensities as obtained from the simple model and $\alpha = 7°$. a) $\rho = 1.78$ b) $\rho = 1.80875$, c) $\rho = 1.85$, d) $\rho = 1.94$, e) $\rho = 1.9474$ f) $\rho = 1.96$.

ad infinitum, and the bifurcation thresholds ρ_i converge to certain value ρ_∞. Beyond this point the motion is chaotic, and the system moves along a strange attractor in phase space [Fig. 4]. The system exhibits typical signatures of low-dimensional deterministic chaos, such as great sensitivity to initial conditions and a positive Lyapunov exponent. The frequency spectrum of the mode amplitudes also shows this transition to chaos, by changing from a line spectrum (where all lines are integer multiples of the same fundamental frequency) to a continuous spectrum. We would like to emphasize: while this route to chaos involves the birth of double-length limit cycles at a sequence of points, it is very different from the usual period doubling scheme, as the stable homoclinic limit cycle at the bifurcation has an infinite period. This quite distinct route

to chaos was analyzed in a series of papers [24, 25], but to our knowledge has
never been observed in an experiment before.

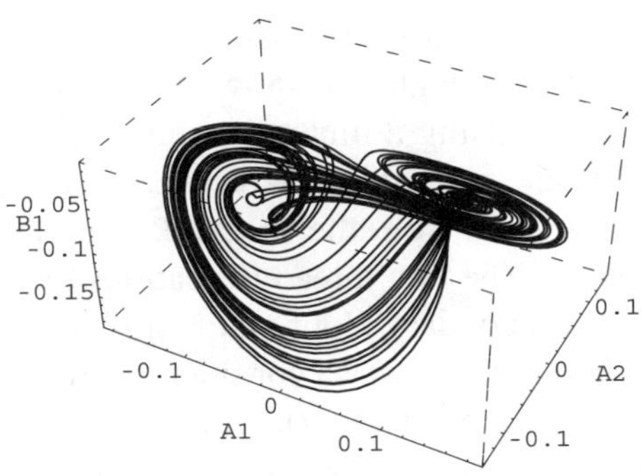

Figure 4.   Strange attractor in phase space at $\alpha = 7°$ and $\rho = 2.18$.

Apart from this peculiar route to chaos, there was a great variety of interesting nonlinear behavior found in this simple model in various domains of
the parameter plane spanned by the angle of incidence α and the intensity of
light ρ [23]. The most important result of the model is, however, that the first
region of regular oscillation - stochastic oscillation - regular oscillation which
was found in various experiments [19, 21] could be interpreted in terms of the
system passing through the first gluing bifurcation. As the system is in the
immediate vicinity of this bifurcation, and orbits are close to being homoclinic
trajectories to the origin, there are random transitions between orbits (or different parts of the same orbit), due to noise in the experiment. Thus two "competing" modes of oscillations exist with random transitions between them. Some
distance below and above the bifurcation, the limit cycles do not approach the
vicinity of the origin, and there are no random jumps. Therefore, the evolution
of the system seems regular above and below the bifurcation but is found to
be stochastic in the immediate vicinity. This interpretation is made even more
convincing by a reconstruction of the limit cycles from experimental data [26],
which shows clearly that the symmetry properties of the trajectories above and
below the bifurcation change precisely as expected. However, for higher intensities, the agreement between the predictions of the model and the observations
was not good. Experiments revealed what looked like another gluing bifurcation (whose nature was different from the one expected for the second gluing
in the model), another periodic regime, and then an abrupt transition to chaotic
behavior. This was found at intensities much higher than the intensities where
chaos exists in the model.

Thus, while the simple model was successful in identifying the first two bifurcations above the primary one, experimental results did not confirm the existence of the full cascade of gluing bifurcations leading to chaos. To gain further insight into dynamical phenomena, a numerical study of the equations was performed [27]. This study retained the assumption that light can be described by a plane wave, and it also neglected flow. The assumption that reorientation is small was relaxed. Using a finite differences algorithm, the equations were first solved, and mode amplitudes characterizing the motion of the system in phase space were extracted from the solution. The result of this treatment was somewhat surprising. The first three bifurcations the system goes through (Freedericksz transition, secondary Hopf and first gluing) were the same as in the simple model (although there was a considerable difference in the bifurcation threshold for the gluing bifurcation). For higher intensities however, the cascade of gluing bifurcations was not found. Furthermore, the observed dynamical scenarios were different from that observed experimentally, and chaos was not found in the simulations for the parameters with which the experiments were performed. This made it clear that, although higher order nonlinearities that were neglected in the simple model do play a major role in the dynamics of the system, further assumptions must be discarded for a correct description of the phenomena.

Clearly, the next step in refining theoretical description of these phenomena had to be the relaxation of one of the major assumptions and either include flow in the equations or discard the plane wave approximation and consider narrow beams. Since other experimental works that investigated dynamical phenomena induced by narrow laser beams indicated that the beam width becomes an important parameter only when it is about two times smaller than the cell width [13], the choice was to include flow and treat light as a plane wave. Thus another numerical study of the system was performed, along similar lines as the previous one. This time, the full nematodynamical equations including flow were solved [10]. The results of this calculation were much more satisfying. The first three bifurcations (primary, secondary Hopf and first gluing) occurred in the same way as in the simple model and the first simulation. There were only quantitative differences of the bifurcation thresholds (Fig. 5 shows the lines of the most important bifurcations on the $\rho - \alpha$ plane. The lines of the primary Hopf bifurcation and the first gluing bifurcation are shown as calculated from both simulations, so that the difference between the bifurcation thresholds caused by the inclusion of flow can be seen.) As the intensity is increased, the bifurcation scenario is qualitatively different for the calculation with flow. There is a second gluing bifurcation as observed in the experiments, but its nature is different from that suggested by the model. It is actually the inverse of the first one - the symmetric limit cycle breaks up into two small asymmetric limit cycles, like those depicted in Fig. 3 a). After

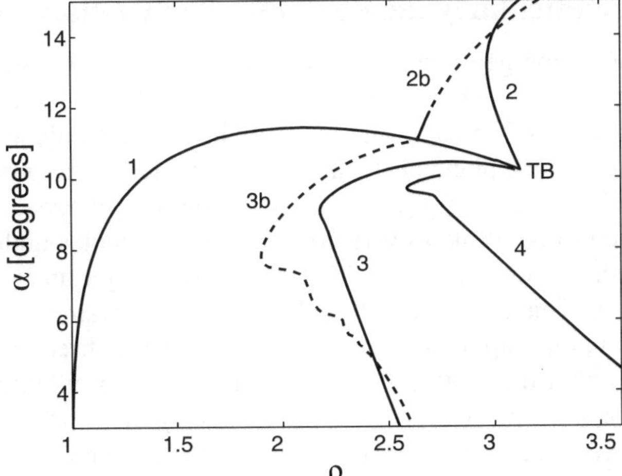

Figure 5.   Bifurcation diagram on the plane of the two control parameters ρ and α. The solid lines 1 and 2 mark the primary instability, where the homogeneous homeotropic orientation becomes unstable. At 1, the bifurcation is a stationary (pitchfork) bifurcation, and a Hopf one at 2. The two lines connect in the Takens-Bogdanov (TB) point. The solid lines 3 and 4 mark the first gluing bifurcation and the second gluing bifurcation respectively. The dashed lines 2b and 3b mark the lines of the primary Hopf bifurcation and the first gluing bifurcation when calculated without the inclusion of flow in the equations.

this second gluing, a strange attractor appears abruptly as the intensity is increased. This sequence now agrees with the dynamical scenarios observed in the experiments, so it appears that the calculation that includes flow is capable of interpreting experimental observations at all intensities. It also proves that flow plays a major part in the dynamics when the light intensity is high enough, so it is imperative to take flow into account if observations are to be interpreted correctly. This supports the assumption that finite beam size effects can be neglected, even though these experiments too were performed with $w_0/L \approx 1$.

Another exciting result of the calculation with flow was, that the route to chaos via a cascade of gluing bifurcations was actually located in the parameter plane, close to the Takens-Bogdanov point. Since bifurcation lines are nearly parallel to the ρ axis in this region, they can be traversed by keeping the intensity fixed and decreasing the angle of incidence. The scenario can be found in a region which was not explored by experiments, so the existence of this very peculiar route remains to be confirmed by a further experiment.

4.   Perpendicularly incident, circularly polarized light

Another intriguing geometry to which much attention was attracted during the last two decades is when a circularly polarized beam is incident perpendicularly on a layer of nematic that has initially homeotropic alignment. The light is polarized in the plane of the layer (the x-y plane) and propagates along the positive z axis (see Fig. 6). In this case, the optical Freedericksz transition is observed to be weakly hysteretic, and the molecules undergo a collective rotation above threshold [28] (that corresponds to a uniform precession of the director). This effect is well understood in the frame of a purely classical (hydrodynamic) approach [28]. The fact that the director rotates above the transition rather than settling to some stationary or oscillating state can be explained by symmetry. Contrary to the geometry discussed in the previous section, the present one possesses isotropic symmetry in the plane of the layer. Thus the only states allowed are rotating ones (the peculiar case of stationary distortion can be regarded as a rotation whose frequency becomes zero). Another explanation in terms of ordinary (o) and extraordinary (e) waves is also possible. When the director has homeotropic alignment, the phase speeds of e and o waves are the same, so the phase difference $\alpha(z)$ between e and o waves remains $\pi/2$. Thus light polarization is unchanged and remains circular when it propagates through the nematic layer. When the director reorients, $\alpha(z)$ changes because the phase speed of the e-wave traveling across the layer depends on z. Thus the polarization becomes elliptic inside the layer. The light-torque acting on the director tries to turn it towards the major axis of polarization, leading to precession.

The subcritical nature of the Freedericksz transition can be explained as follows. When the director settles to the precession state, light becomes elliptically polarized inside the nematic. On the other hand, it is known that the Freedericksz transition for elliptically polarized light depends on ellipticity and

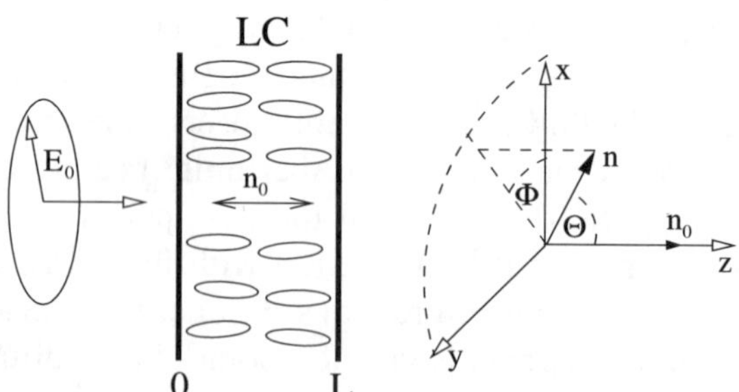

Figure 6.   Geometry of the setup: circularly polarized light incident perpendicularly on a nematic layer with the director $n_0 \parallel z$ (homeotropic state). The components of the director n are described in terms of the angles Θ, Φ ($\Theta = 0$ in the homeotropic state).

occurs at some intensity between the Freedericksz threshold for linearly polarized light and the one for circularly polarized light (see the next section). (The threshold for OFT for circularly polarized light is two times higher than that for a linearly polarized one.) Thus, once reorientation takes place, the elliptically polarized light that develops inside the nematic can sustain the distortion even if the intensity is decreased slightly below the initial threshold.

The precession of the molecules can also be interpreted in a quantum picture as spin angular momentum transfer from the light to the medium and is called self-induced stimulated light scattering [29]. Since collective molecular rotation dissipates energy, the light beam has to transmit part of its energy to the medium. As the pure nematic LC is a transparent medium (no absorption), this energy loss leads to a red shift of part of the light beam [29]. The mechanism can be described as follows: each scattered photon has its helicity reversed and thus transfers an angular momentum of $2\hbar$ (that is perpendicular to the layer) to the medium. Moreover, its energy is lowered by an amount $\hbar\Delta\omega$. Thus, p photons produce a constant torque $\tau_z = 2\hbar p$ per unit time, that acts on the medium and induces a collective molecular precession. This torque is balanced by the viscous torque. The angular velocity of the uniform precession Ω is related to the red shift $\Delta\omega$ by the simple formula $\Delta\omega = 2\Omega$. This formula can be derived from energy conservation using that i) p photons lose the amount of energy $p\hbar\Delta\omega$ per unit time; ii) the work made by the torque τ_z on the director is $\tau_z\Omega = 2p\hbar\Omega = p\hbar\Delta\omega$. The fact that $\hbar$ disappears in the final relation shows that one can obtain this formula through a classical approach [30].

In general, the angular momentum of the light beam consists of two parts: a spin part associated with polarization [31] and an orbital part associated with spatial distribution [32]. However, if the spatial distribution in the plane of the layer is supposed to be homogeneous (i.e. when one deals with a plane wave approximation), then the orbital part is zero. In this context, it may be interesting to note that laser light with a Laguerre-Gaussian amplitude distribution can be shown to have a well-defined orbital angular momentum [33].

The theoretical description of the OFT in this geometry was reported in [34] where the importance of twist deformations of the director was pointed out, and the hysteretic nature of the OFT was explained. In [30] a theoretical and experimental investigation of the dynamical behavior of the system for the region of higher intensities was reported. The authors of [30] observed a further discontinuous transition with large hysteresis from a precession regime with small reorientation amplitude occurring above the OFT to one with large reorientation. The frequency of the large amplitude precession was found to be much smaller than the one just above the OFT and to exhibit rapid variations with the incident intensity reaching zero at roughly periodic intervals. In this work, the authors presented an approximate model that can describe qualitatively both regimes of uniform director precession and also presented clear experimental evidence of the frequency reduction in the second regime.

Again, one of the simplifications used in this model was the infinite plane wave approximation. Under this assumption, all relevant functions depend solely on the spatial coordinate z and the time t. Obviously, the representation adapted to this geometry is the one given in usual spherical angles $\Theta(z,t)$ and $\Phi(z,t)$ such that $\mathbf{n} = (\sin\Theta\cos\Phi, \sin\Theta\sin\Phi, \cos\Theta)$ [see Fig. 6]. The twist angle is written then as $\Phi = \Phi_0(t) + \Phi_d(z,t)$, where $\Phi_0(t)$ does not depend on z and describes a rigid rotation of the director around the z axis (no distortion), while $\Phi_d(z,t)$ contains twist distortion. Such a decomposition is not unique in the sense that any constant can be added to Φ_0 and then subtracted from Φ_d. The key point, however, is that Φ_0 depends on time only and can be unbounded, while Φ_d is required to remain bounded. (Note that nonzero Φ_d means that the instantaneous director profile is out of plane.) To construct a simple model, some further simplifying hypotheses are needed [30], namely i) the backflow is neglected; ii) the splay-bend distortion is small, i.e., $\Theta^2(z,t) \ll 1$; iii) a sine trial function for $\Theta(z,t)$ is used; iv) the twist distortions are small, i.e., $|\partial_z\Phi_d| \ll 1/L$; v) the slow-envelope approximation for Maxwell's equations can be used. Retaining terms up to third order in Θ and keeping the lowest order terms in $\partial_z\Phi_d$, the following expressions for the frequency of the uniform director precession $2\pi f_0 = d\Phi_0(t)/dt$ and the twist gradient $\partial_z\Phi_d$ have been obtained:

$$f_0 = \frac{\rho\,(1-\cos\Delta)}{2\pi\Delta}\,,\quad \partial_z\Phi_d = \frac{\pi\rho}{2\Delta}\,\frac{(1-\cos\Delta)v(z)-1+\cos[\Delta v(z)]}{\sin^2 z}\,,\quad (17)$$

where time and length are normalized as $t \to t/\tau$ and $z \to \pi z/L$, respectively. Here ρ is the normalized intensity such that $\rho = 1$ corresponds to the threshold for OFT, and $v(z) = (z - \sin z \cos z)/\pi$. Δ is the phase delay induced by the whole layer and is a global measure of the amplitude of reorientation. It has a direct experimental interpretation, since the quantity $\Delta/2\pi$ represents roughly the number of self-diffraction rings in the far field [35] and, under the approximation used, is proportional (with a large prefactor) to the square of the amplitude of the polar angle $\Theta_1^2(t)$. Finally, an analytic solution for $\Delta(\rho)$ has been found [30] which is given by a rather cumbersome formula. As is seen from Eq. (17), f_0 indeed becomes smaller and exhibits rapid variation with increasing Δ (i.e. with increase of the reorientation).

Even though the approximate model gave a satisfactory description of the observed phenomena, the nature of the transition from one regime to the other was not understood in this framework. Some years later, a qualitative mechanism based on non-uniform spin angular momentum deposition from the light to the nematic was introduced to explain the origin of such a transition [36]. A particular interest to this problem arose again more recently, when an additional continuous secondary instability between the OFT and the abrupt transition to the largely reoriented state was observed [37]. Although a preliminary description of the bifurcation scenario was reported in [38], the global scenario was still obscure. It became clear that a numerical study was needed to gain better insight to the problem [39, 40]. The original problem consisting of two PDEs for Θ and Φ was simplified by means of expansions of Θ and Φ_d with respect to z in systems of orthogonal functions which satisfy the boundary conditions: $\Theta = \sum_{n=1}^{\infty} \Theta_n(t) \sin nz$, $\Phi_d = \sum_{n=1}^{\infty} \Phi_n(t) \sin(n+1)z / \sin z$. We substitute these expansions into the director equations and project the equation for Θ onto the modes Θ_n and the equation for Φ onto Φ_n (Galerkin method). This results in a set of coupled nonlinear ODEs for the modes Θ_n and Φ_n. In order to solve this set of ODEs, the two ODEs for the field amplitudes have to be integrated dynamically at each step of numerical integration for time t [40]. The infinite set of ODEs was reduced to a finite one by truncating the mode expansion for Θ and Φ_d. It is worth noting that for a state of uniform director precession (UP) (f_0 =const), the set of ODEs reduces to one of nonlinear algebraic equations for Φ_n and Θ_n which become constant in time. The results of this numerical study which explains the entire scenario is discussed in what follows.

In Fig. 7, the phase delay $\Delta/2\pi$ is plotted versus the normalized intensity ρ. The solid lines represent stable uniform precession (UP) states, while the dashed lines correspond to precession states that are unstable. The region in gray corresponds to a nonuniform precession (NUP) where nutation ($d\Delta/dt \neq 0$) is coupled to precession. In this regime, the lower and the upper lines that limit the region correspond to the minimum and maximum values taken by Δ during its oscillation.

The OFT occurs at $\rho = 1$ via a subcritical Hopf-type bifurcation where the system settles to a uniform precession state with a small reorientation amplitude ($\Delta \sim \pi$ so that $\Theta^2 \ll 1$) labeled UP1. Decreasing the intensity from the UP1 regime, the system switches back to the unperturbed state at $\rho = \rho_1^* \simeq 0.88$ where a saddle-node bifurcation occurs. The trajectory in the (n_x, n_y) plane is a circle whereas, in a coordinate system that rotates with frequency f_0 around the z axis, it is a fixed point. The time Fourier spectra of the director $\mathbf{n}$ have one fundamental frequency f_0, whereas Θ_n, Φ_n and Δ do not depend on time.

It is worth noting here that from the weakly nonlinear analysis, it follows [40] that the nature of the OFT is governed by the sign of the coefficient $C = K_1/K_3 - (9/4)\,(\varepsilon_a/\varepsilon_{||})$. $C < 0$ corresponds to a subcritical OFT, while $C > 0$ corresponds to a supercritical OFT. Incidentally, this criterion is identical to the one derived in the case of OFT under linearly polarized light [41]. In the present example, $C = 0.154$ and the OFT is actually supercritical. However, the solution branch turns over and becomes subcritical (and unstable) already

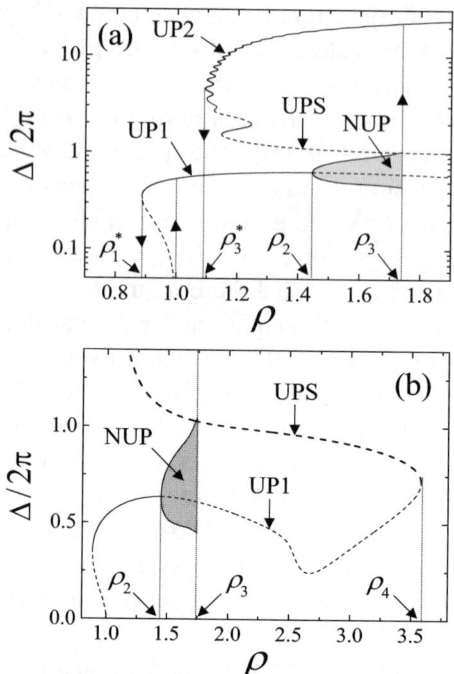

Figure 7. (a) $\Delta/2\pi$ on a log scale versus ρ for $\rho < 2$ and $\Delta < 50\pi$. (b) $\Delta/2\pi$ on a linear scale versus ρ for $\rho < 4$ and $\Delta < 3\pi$. Solid (dashed) curves correspond to stable (unstable) solutions.

at $\rho = 1 + \delta\rho$ where $\delta\rho \simeq 10^{-6}$. This explains why the OFT appears to be subcritical on the scale used in Fig. 7. In fact, although $\delta\rho$ increases when the cell thickness is decreased, even at $L = 10$ μm, the subcritical region is still too small to detect ($\delta\rho \simeq 10^{-4}$).

With further increase of the intensity, the UP1 state loses stability via a supercritical Hopf bifurcation at $\rho = \rho_2$ where the director starts to nutate (NUP regime). For the NUP state, all modes Θ_n and Φ_n with $n \geq 1$ are time dependent, and their Fourier spectrum contains frequencies mf_1, where m is an integer. The spectra of the phase delay Δ and the director $\mathbf{n}$ have contributions at frequencies given by the simple formulas: $\tilde{\Delta} = \{mf_1\}, \tilde{\mathbf{n}} = \{f_0, mf_1 \pm f_0\}$. In some narrow region around $\rho_3 \approx 1.75$, the period $T = 1/f_1$ of the NUP increases progressively with increasing light intensity and indeed appears to diverge logarithmically at ρ_3, as shown in Fig. 8. Thus as ρ approaches ρ_3, the NUP limit cycle collides with the unstable UPS branch which is a saddle. In fact, we deal here with a homoclinic bifurcation of the simplest type, where a limit cycle collides with a saddle point having only one

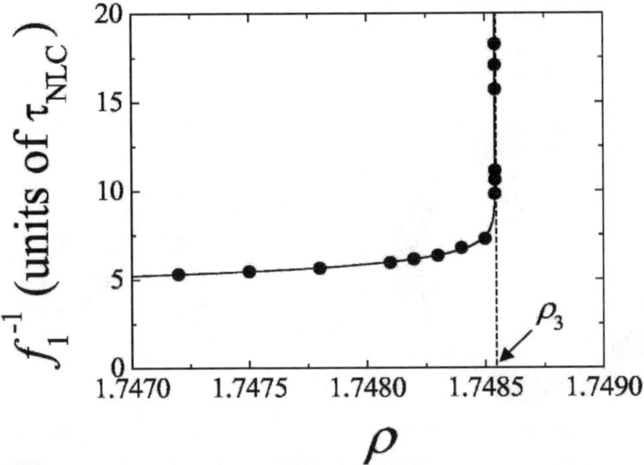

Figure 8. Characterization of the homoclinic bifurcation $f_1^{-1}(\rho) = \mathcal{O}[\ln(\rho_3 - \rho)]$ near ρ_3 ($\rho_3 = 1.748542389055$). The solid line is the best fit to the theoretically calculated values (•).

unstable direction [42] (all the eigenvalues have negative real parts except one, which is real and positive).

At $\rho = \rho_3$, the system jumps to a new state of uniform precession of the director (UP2) with large reorientation ($\Theta \simeq 74^o$) and slow precession rate. As displayed in Fig. 7, starting from the stable UP2 branch above ρ_3 and lowering the excitation intensity, one finds a large and rather complicated hysteretic cycle, which eventually flips back to the UP1 solution at $\rho_3^* = 1.09$. This part of the UP2 branch consists of alternatively stable and unstable regions exhibiting a series of saddle-node bifurcations. Eventually, this branch connects with the UPS one which makes a loop and connects with the UP1 branch.

As a result of the appearance of the new frequency f_1 at $\rho = \rho_2$, the director motion becomes quasi-periodic characterized by the two frequencies f_0 and f_1. This is illustrated in Fig. 9(a), where the trajectory of the director in the (n_x, n_y) plane is plotted for $\rho = 1.55$ at some z inside the layer. This trajectory is not closed in the laboratory frame indicating quasi-periodicity of the director. In fact, the two independent motions, namely the precession (f_0) and the nutation (f_1), can be isolated by transforming to a frame that rotates with frequency f_0. In the rotating frame, the director performs a simple periodic motion with frequency f_1 as is seen in Fig. 9(b) with the arrow indicating the sense of rotation for the case where the incident light is left circularly polarized. (The sense of rotation is always opposite to that of the underlying precession [43].) As is seen from Fig. 9(c,d) starting from initial conditions near the unstable UP1 solution or the UPS one, the director eventually settles on the NUP solution, which is represented by a simple limit cycle.

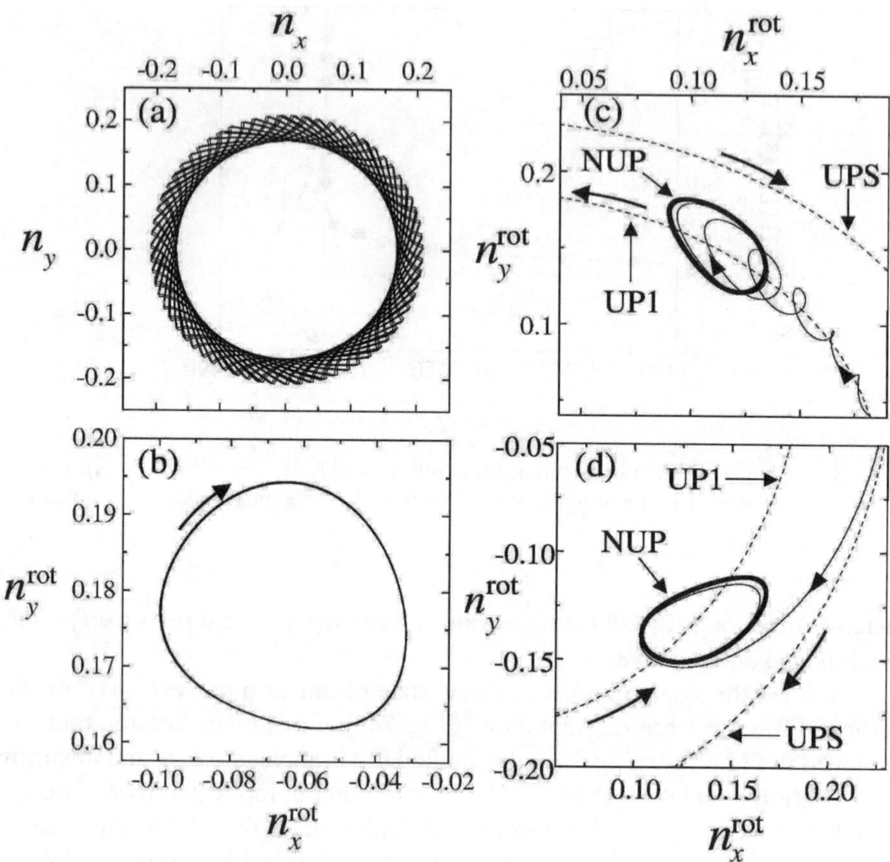

Figure 9. (a),(b): director trajectory at $\rho = 1.55$. (a) Quasiperiodic behavior in the laboratory frame (n_x, n_y). (b) Periodic limit cycle in the rotating frame $(n_x^{\mathrm{rot}}, n_y^{\mathrm{rot}})$. The arrow indicates the sense of rotation when the incident light is left circularly polarized.
(c),(d): director trajectory at $\rho = 1.55$ in the $f_0(\rho, \mathrm{NUP})$-rotating frame showing the instability of the UP1 and UPS solutions in the NUP regime. (c) Initial condition near the UP1 solution. (d) Initial condition near the UPS solution. The arrows indicate the sense of rotation of the corresponding trajectory when the incident light is left circularly polarized.

In fact, the unstable UPS branch represents the saddle point (or separatrix) that separates the regions of attraction of the NUP state (or, below ρ_2, the UP1 state) from that of the largely-reoriented UP2 state. At this point, it might also be interesting to note that the UP1 state represents a stable node at $\rho \sim \rho_1$ (the relevant stability exponents are real and negative). Then, between ρ_1 and ρ_2, it changes to a focus (the stability exponents become complex). At ρ_2, the real part of the complex pair of stability exponents passes through zero and then becomes positive.

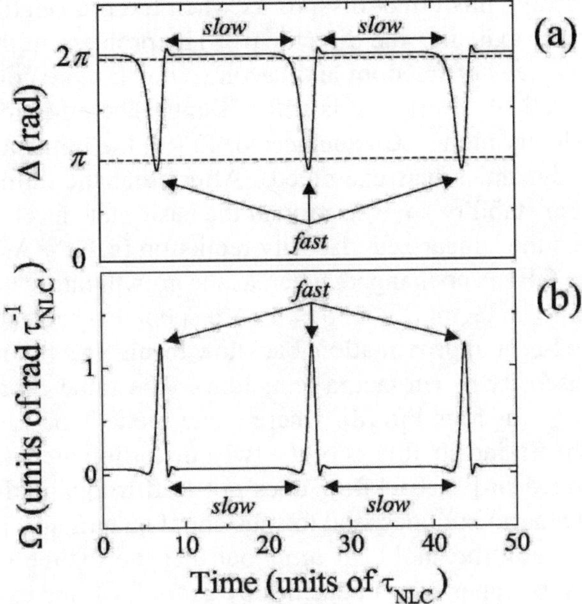

Figure 10. Calculated dynamics just below $\rho = \rho_3$. (a) Phase shift $\Delta(t)$. (b) Instantaneous angular velocity $\Omega(t) = d\Phi_0/dt$.

As ρ approaches the homoclinic bifurcation point, the trajectory of the director approaches the unstable UPS orbit for longer and longer intervals. The dynamics near ρ_3 possesses two time scales, a slow and a fast one, as expected from the homoclinic nature of the transition. Figure 10 emphasizes this point, where the phase shift $\Delta(t)$ and the instantaneous angular velocity $\Omega(t) = d\Phi_0/dt$ are plotted versus time.

The bifurcation scenario discussed above was actually observed in the experiment. Although a good qualitative agreement between theory and experiment was found [40], there are quantitative discrepancies. In the experiment, the measured onset of the nutation-precession motion turns out to be about 20% lower than predicted by theory. Moreover, the slope of the precession frequency versus intensity predicted by theory turned out to be different from that observed in the experiment. One of the two possible reasons could be the use of finite beam size in the experiment (that is typically of the order of the thickness of the layer), whereas in theory the plane wave approximation was assumed. Actually, the ratio δ between diameter of the beam and the width of the layer is another bifurcation parameter (in the plane wave approximation, $\delta \to \infty$) and was shown to play crucial role on the orientational dynamics [13]. There and in [44] the importance of the so called walk-off effect was pointed out which consists of spatial separation of Pointing vectors of the ordinary and extraordinary waves. (Ideally, this effect disappears when the propagation takes place along the principal axis, i.e. the director **n**.) The problem in this case has to include lateral degrees of freedom and becomes much more complicated. An appropriate theoretical description is still missing. The other reason could be that the fluid velocity in the LC is neglected. In [9], the influence of backflow on the director dynamics was examined. After adiabatic elimination of the flow field, a linear stability analysis around the basic state has been performed in order to assess the "linearized viscosity reduction factor". As expected, the threshold for the OFT is unchanged, whereas the growth rate $\sigma = (\rho-1)/(\tau\xi)$ acquires an additional factor $\xi < 1$ ($\xi = 1$ corresponds to neglect of backflow). Thus, within the linear approximation, backflow results in a renormalization of the rotational viscosity γ_1 (in fact, a reduction). The same expression for the reduction factor ξ was found in [8], where a one-mode approximation for the director components and smallness of the twist distortion were used.

As demonstrated in [9], backflow does not lead to qualitative changes in the dynamical scenario but does lead to substantial quantitative changes in the secondary bifurcation threshold. It turns out that the regime of nonuniform precession shifts to higher light intensities by about 20% and exists in a larger interval. In Fig. 11, $\Delta/2\pi$ is plotted versus the normalized intensity ρ. The phase delay Δ for the UP regimes is only slightly different from the case without backflow. However, the regime of nonuniform director precession shifts to higher intensities. As is seen from Fig. 11, the thresholds for the NUP and

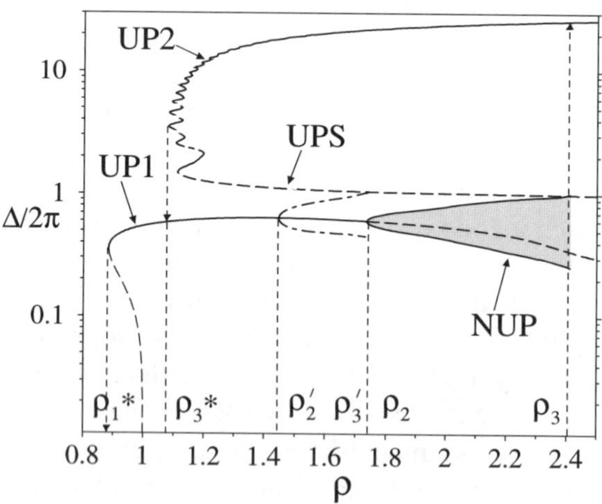

Figure 11. $\Delta/2\pi$ versus ρ in semi-logarithmic scale. Solid (dashed) curves correspond to stable (unstable) UP solutions. Gray region: nonuniform precession states of the director. Dash-dotted lines in the (ρ_2', ρ_3') interval: nonuniform precession states of the director when backflow is neglected. The fact that $\rho_3' \simeq \rho_2$ is accidental.

UP2 regimes turn out to be $\rho_2 = 1.75$ and $\rho_3 = 2.4$ instead of $\rho_2' = 1.45$ and $\rho_3' = 1.75$ when the backflow is neglected [39, 40]. In [9], it was also shown that the precession frequency f_0 for UP1 states actually increases when the backflow is included (as expected because γ_1 effectively decreases). An unanticipated spatial oscillations of the backflow in the UP2 regime were also found which results from spatial oscillations of the director twist $\partial_z \Phi$. They are a consequence of oscillations in the torque resulting from interference phenomena between ordinary and extraordinary light. The backflow behaves very differently for the three types of the director motion and thus can act as a sensitive diagnostic to distinguish them.

Thus, the inclusion of backflow makes the situation even worse, because the experimental values of the thresholds ρ_2 and ρ_3 are even smaller than that given by the theory without backflow [39, 40]. One is forced to conclude that the discrepancy between the theoretical predictions and the experiment is strongly affected by the fact that, in the experiments, the beam size is not large compared to the layer thickness. By using a large aspect-ratio geometry, one is now in a position to test the theoretical framework [9] quantitatively. This can be done by use of a dye-doped nematic, because the values of OFT in this case can be two orders of magnitude smaller than for a pure nematic (see [45, 46] and references therein). The low threshold intensity allows the spot size of the light to be much larger than the thickness of the layer, thus the plane wave approximation assumed in the theory might be better achieved in the experiment.

5.   Perpendicularly incident, elliptically polarized light

A natural generalization of the previous geometry is an elliptically polarized (EP) plane wave incident perpendicularly on a layer of nematic that initially has homeotropic alignment (see Fig. 6 in the previous chapter). The ellipticity $-\pi/4 \leq \chi \leq \pi/4$ is related to the ratio between the minor and the major axis of the polarization ellipse. The case $\chi = 0$ [$\chi = \pm\pi/4$] corresponds to a linearly [circularly] polarized light. The sign of χ determines the handedness of the polarization, thus it is sufficient to choose $\chi > 0$ only. The main difference with the CP light discussed in the previous section is the broken rotational invariance around the z-axis.

As shown in [47], the director is unperturbed (U) until the intensity reaches a critical value that depends on χ: $I_F^{EP} = I_F^{CP}/(1 + \cos 2\chi)$, where I_F^{CP} is the intensity for OFT of CP light. Thus we have two control parameters, the ellipticity χ and the incident intensity I. In what follows, the normalized intensity $\rho = I/I_F^{CP}$ is used.

For $\chi < \pi/4$ the OFT is a pitchfork bifurcation, and the reoriented state is a stationary distorted state (D). In [48], the oscillating states (O) were experimentally observed. The numerical analysis [48] of the basic equations indeed predicts the existence of such a state. It should be noted that reflection symmetry is spontaneously broken by the first bifurcation, so in the D and O states, one has two symmetry degenerate solutions related by $\{n_x \rightarrow -n_x, n_y \rightarrow -n_y\}$. In [49], a model was derived using the assumption of small director distortion, i.e. both the polar angle and the twist were assumed to be small ($\Theta^2 \ll 1$ and $\Phi_d \ll 1$). The director and the field equations were expanded with respect to these angles and only some significant nonlinear terms were kept. Then a mode expansion for Θ and Φ was used (the same as for CP light, see previous section), and only the first mode Θ_1 for the polar angle was retained. Within this approximation, the phase delay $\Delta \sim \Theta_1^2$. The equations for the Stokes vector (which determine the field amplitudes) were solved iteratively using Φ_d as a small parameter (actually the first iteration was taken). This allowed elimination of the field amplitudes from the director equations, which was finally reduced to a set of two ODEs for the phase delay Δ and the zeroth mode Φ_0 (which represents a rigid rotation). The twist modes $\Phi_{n \geq 1}$ were then assumed to adiabatically follow their steady state values and were shown to decrease rapidly with n, so only a few of them were kept. This relatively simple model was capable of predicting not only O states but also some other states occurring at higher intensities. It was demonstrated that with the increase of ρ, the transition from the D to O state takes place via Hopf bifurcation, while the transition from oscillation to rotation was shown to be the gluing of two symmetrical limit cycles (for a certain region of $\chi < \pi/4$). The hysteresis between rotations and oscillations at large ellipticity χ was also predicted. It should be noted that the experimental findings [49] are qualitatively reproduced by this model.

It was shown in the previous chapter that CP light induces quasiperiodic director rotation (QPR) if the incident intensity exceeds the one for OFT by about 40% (no backflow). This is already in a higher region of intensities than that considered in [49]. So the question was what happens if one mismatches slightly from the CP case at higher intensities? It became clear [50] that a full numerical analysis is needed to capture the QPR for the elliptic case, because i) the small distortion approximation failed to describe QPR for CP case; ii) higher order nonlinearities in twist terms [$\propto (\partial_z \Phi_d)^2$] are important. In [50], the QPR was found both theoretically and experimentally. Apart from this regime, other regimes of rotating, oscillating or stationary states with large director distortion and the transitions between them were predicted theoretically [50]. In what follows, we present a brief overview of the bifurcation scenario following [50], which is, in our opinion, complete for large and moderate values of ellipticity.

Figure 12 taken from [50] presents the different regimes that exist in the (χ, ρ) plane for $0.33 \leq \chi \leq \pi/4 \simeq 0.785$. Above the OFT threshold, several

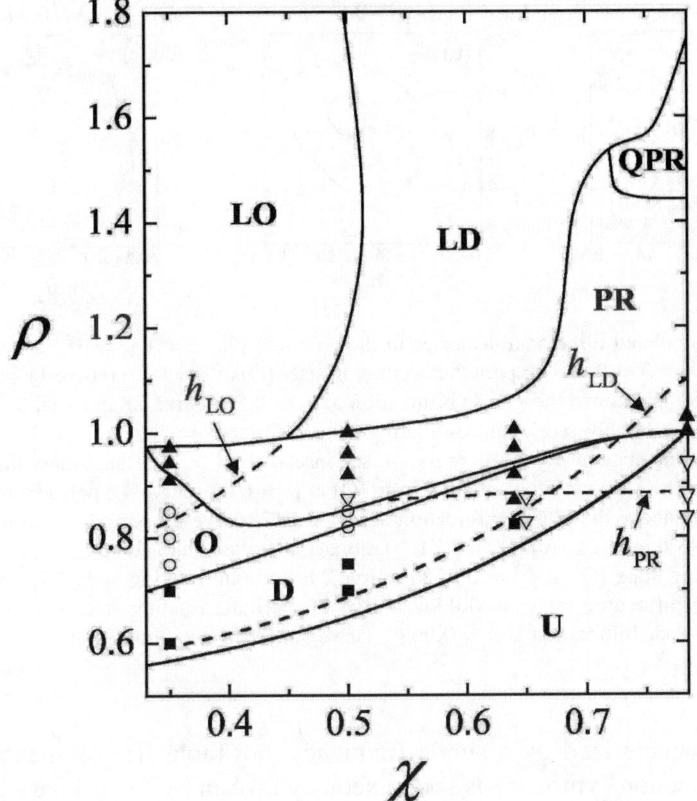

Figure 12. Phase diagram of the dynamical regimes in the parameter plane (χ, ρ). U: Undistorted state; D: stationary Distorted states; O: periodic Oscillating states; PR: Periodic Rotating states; QPR: Quasi-Periodic Rotating states; LD and LO: Large reorientation associated respectively with stationary Distorted and Oscillating states. The dashed lines h_{PR}, h_{LD} and h_{LO} correspond to the hysteretic region of the PR, LD and LO states, respectively. The points are experimental data extracted from [49] for D ($\blacksquare$), O ($\circ$), PR ($\blacktriangle$) and hysteretic PR ($\triangledown$).

regimes can exist depending on the values of χ and ρ: stationary distorted (D), oscillating (O), periodic rotating (PR), quasi-periodic rotating (QPR) and largely reoriented states ($\Theta \sim 1$), which may be stationary distorted (LD), oscillating (LO) or rotating (LR) states (LR states are not shown in Fig. 12, since they only arise in a narrow region $\Delta\chi \sim 10^{-2}$ near $\chi = \pi/4$). Keeping the ellipticity fixed and increasing the intensity, these regimes appear as a well-defined sequence of transitions (as summarized in Table I). The trajectories of the director in various regimes are shown in Fig. 13.

For $0.33 < \chi < 0.53$, the OFT is a pitchfork bifurcation, and the reoriented state is a D state [see the filled circles in Fig. 13(a)]. This state loses its stability through a supercritical Hopf bifurcation to an O state [curve 1 in

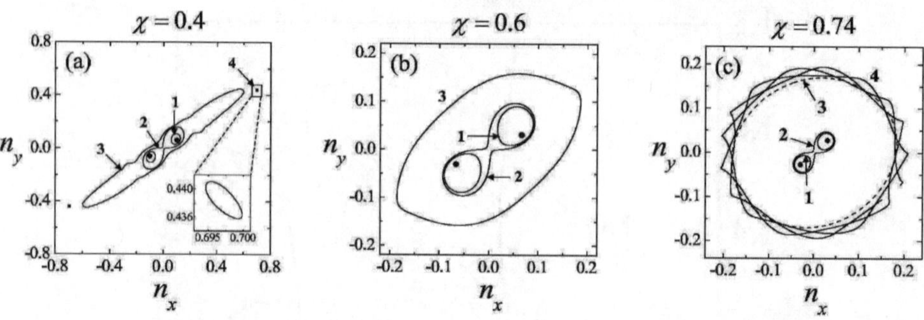

Figure 13.   Calculated director trajectories in the (n_x, n_y) plane. (a) $\chi = 0.4$: stationary distorted state (D) at $\rho = 0.72$ ($\bullet$); periodic oscillating state (O) at $\rho = 0.76$ (curve 1); periodic rotating state (PR) just above the gluing bifurcation at $\rho = 0.83$ (curve 2) and slightly below the transition to the largely reoriented oscillating state (LO) at $\rho = 0.97$ (curve 3); largely reoriented oscillating state at $\rho = 0.98$ (curve 4, see inset). (b) $\chi = 0.6$: stationary distorted state (D) at $\rho = 0.8$ ($\bullet$); periodic oscillating state (O) at $\rho = 0.91$ (curve 1); periodic rotating state PR_1 slightly above the gluing bifurcation at $\rho = 0.917$ (curve 2); periodic rotating state PR_2 at $\rho = 0.95$ (curve 3). (c) $\chi = 0.74$: stationary distorted state (D) at $\rho = 0.99$ ($\bullet$); periodic oscillating state (O) at $\rho = 0.9925$ (curve 1); periodic rotating state PR_1 slightly above the gluing bifurcation at $\rho = 0.9932$ (curve 2); periodic rotating state PR_2 slightly above the saddle-node bifurcation at $\rho = 0.9936$ (curve 3, dashed line); quasi-periodic rotating state at $\rho = 1.5$ (curve 4).

Fig. 13(a)] characterized by a single frequency f_0 (Table II). As mentioned above, the reflection symmetry is spontaneously broken by the primary bifurcation, so in D and O states one has two symmetry degenerate solutions. As ρ increases, these two limit cycles merge in a gluing bifurcation at the origin and restore the reflection symmetry. This leads to the appearance of a single double-length limit cycle that corresponds to the trajectory in the PR state [curve 2 in Fig. 13(a)]. A further increase of the intensity eventually leads to a discontinuous transition to a largely reoriented oscillating ($\chi < 0.45$) or stationary distorted ($\chi > 0.45$) state. In both cases, this transition is associated with a small relative jump of the director amplitude and corresponds to a homoclinic bifurcation. In fact, stable LO states exist until the intensity is decreased to a critical value represented by the hysteretic line h_{LO} in Fig. 12, below which the LO state becomes stationary distorted. This LD state finally vanishes when the intensity is decreased below the hysteretic line h_{LD}.

For $0.53 < \chi < 0.72$, one has the sequence U $\rightarrow$ D $\rightarrow$ O $\rightarrow$ PR as before [see Fig. 13(b)], however there is an additional bifurcation between PR states. In fact, the limit cycle amplitude of the PR regime, now labeled PR_1 [curve 2 in Fig. 13(b)], abruptly increases. This results in another periodic rotating regime labeled PR_2 with higher reorientation amplitude [curve 3 in Fig. 13(b)]. This is a hysteric transition connected to a double saddle-node structure with the (unstable) saddle separating the PR_1 and PR_2 branches as already found within the approximate model [49]. In that case, the system switches back to the O or D state at the line labeled h_{PR} in Fig. 12. In contrast, no hysteresis is observed when ρ is decreased starting from the PR_1 regime since the O $\rightarrow \mathrm{PR}_1$ transition is continuous. The $\mathrm{PR}_1 \rightarrow \mathrm{PR}_2$ transition is not shown in Fig. 12, because it is very near to the gluing bifurcation. At $\chi = 0.53$, the two saddle nodes coalesce. Finally, for high intensity the system switches abruptly from the PR_2 to the LD regime. The reorientation discontinuity associated with this transition is small for $\chi < 0.66$ and quite large for $\chi > 0.66$. This is due to the fact that, for $\chi < 0.66$, part of the limit cycle, associated with the PR state just below the transition, extends to large reorientations in the (n_x, n_y) plane. Consequently, it is already close to the largely reoriented states nearby [see e.g. curve 3 and 4 in Fig. 13(a)]. The transition to large reorientation is found to be a homoclinic bifurcation, and when the intensity is decreased, stable LD states exist until the line h_{LD} is reached.

For $0.72 < \chi < \pi/4$, the sequence U $\rightarrow$ D $\rightarrow$ O $\rightarrow \mathrm{PR}_1 \rightarrow \mathrm{PR}_2$ is observed as before [see Fig. 13(c)]. However, for higher values of ρ, a QPR regime is born through a secondary supercritical Hopf bifurcation, which introduces a new frequency f_1 into the system and transforms the dynamics into a quasi-periodic behavior [curve 4 in Fig. 13(c)]. As the intensity increases, the QPR state undergoes a homoclinic transition to a largely reoriented LD or LR regime, which are respectively represented by a stationary distorted or slowly rotating (close to $\chi = \pi/4$) state. This bifurcation is associated with a large discontinuity of the reorientation amplitude.

The signature of the anisotropy of incident light is visible in the director trajectories in the (n_x, n_y) plane. The PR trajectories are obviously non-circularly symmetric for $\chi = 0.4$ and $\chi = 0.6$ [see Figs. 13(a,b)], whereas PR_2 and QPR regimes are almost circularly symmetric when the polarization is almost CP [see Fig. 13(c)]. The spectral content of the variables $n_{x,y}$, Δ and $I_{x,y}$ for O, PR and QPR is listed in Table II.

Finally, one should mention the particular situation when the PR regime vanishes at $\chi = 0.33$ (see Fig. 12). There, a direct transition from the O to the LO regime occurs as the intensity is increased. The corresponding picture is this: as the intensity is increased, the limit cycle associated with the O regime collides with an unstable fixed point thus preempting the gluing at the origin. This dynamical sequence suppresses the appearance of the PR regime.

Starting from the PR or QPR regime and increasing the intensity, an instability eventually occurs at $\rho = \rho_L$ (which depends on χ, see Fig. 12), and the director settles to a largely reoriented oscillating (LO), stationary distorted (LD) or rotating (LR) state (the latter one exists in a narrow region $\Delta\chi \sim 10^{-2}$ around $\chi = \pi/4$). From Fig. 12, we see that the final state above ρ_L is a LO state if $\chi < 0.45$ and a LD state if $\chi > 0.45$. It was found that the transition from the PR or QPR regime to the largely reoriented states is related to

Table 1.   Calculated sequence of bifurcations as a function of the ellipticity χ of the incident light.

Ellipticity	Sequence of transitions	Bifurcation nature
$0.33 < \chi < 0.53$	Unperturbed $\rightarrow$ Distorted	Pitchfork
	Distorted $\rightarrow$ Periodic oscillation	Supercritical Hopf
	Periodic oscillation $\rightarrow$ Periodic rotation	Gluing
	Periodic rotation $\rightarrow$ Periodic oscillation or distorted	Homoclinic [a]
$0.53 < \chi < 0.72$	Unperturbed $\rightarrow$ Distorted	Pitchfork
	Distorted $\rightarrow$ Periodic oscillation	Supercritical Hopf
	Periodic oscillation $\rightarrow$ Periodic rotation-1	Gluing
	Periodic rotation-1 $\rightarrow$ Periodic rotation-2	Saddle-node
	Periodic rotation-2 $\rightarrow$ Distorted	Homoclinic [b]
$0.72 < \chi < \pi/4$	Unperturbed $\rightarrow$ Distorted	Pitchfork
	Distorted $\rightarrow$ Periodic oscillation	Supercritical Hopf
	Periodic oscillation $\rightarrow$ Periodic rotation-1	Gluing
	Periodic rotation-1 $\rightarrow$ Periodic rotation-2	Saddle-node
	Periodic rotation-2 $\rightarrow$ Quasi-periodic rotation	Supercritical Hopf
	Quasi-periodic rotation $\rightarrow$ Distorted or periodic rotation	Homoclinic [c]
$\chi = \pi/4$	Unperturbed $\rightarrow$ Periodic rotation	Subcritical Hopf
	Periodic rotation $\rightarrow$ Quasi-periodic rotation	Supercritical Hopf
	Quasi-periodic rotation $\rightarrow$ Periodic rotation	Homoclinic [c]

[a] small jump of the director amplitude
[b] small [large] jump of the director amplitude for $\chi < 0.66$ [$\chi > 0.66$]
[c] large jump of the director amplitude

Table 2.   Spectral content of the director components $n_{x,y}$, the output intensity components $I_{x,y}$ and the phase delay Δ for the different dynamical regimes for an elliptically polarized excitation.

Regime	$n_{x,y}$	$I_{x,y}$	Δ
Periodic oscillation (O)	nf_0	nf_0	nf_0
Periodic rotation (PR)	$(2n-1)f_0$	$2nf_0$	$2nf_0$
Quasi-periodic rotation (QPR)	$nf_1 \pm (2m+1)f_0$	$nf_1 \pm 2mf_0$	$nf_1 \pm 2mf_0$

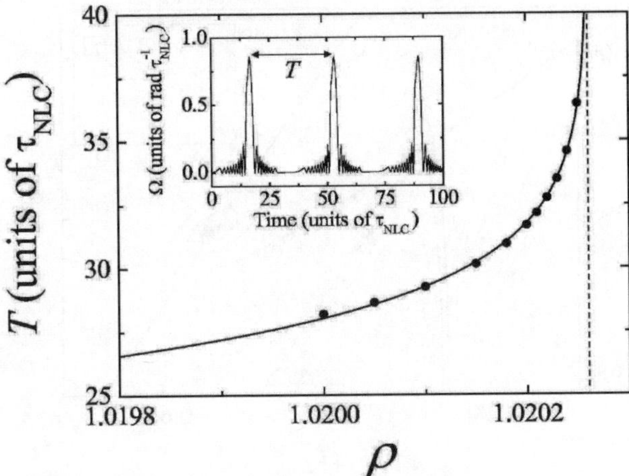

Figure 14. Characterization of the homoclinic bifurcation near ρ_L for $\chi = 0.57$, $T(\rho) = \mathcal{O}[\ln(\rho_L - \rho)]$, where T is the period of the instantaneous angular velocity $\Omega(\tau) = d\Phi_0/d\tau$. The solid line is the best fit to the theoretically calculated values ($\bullet$). Inset: time evolution of $\Omega(\tau)$ at $\rho = 1.02025$. A perfect agreement with the calculated values (filled circles) is obtained with the parameterizations $a + b\ln(\rho_L - \rho)$ for $a \simeq 8.232$ and $b \simeq -2.406$ (solid line).

an increase of the period of the corresponding limit cycle. More precisely, this period appears to diverge logarithmically at $\rho = \rho_L$. This behavior is illustrated in Fig. 14 for $\chi = 0.57$. In this figure, the period of the instantaneous angular velocity $\Omega(\tau) = d\Phi_0/d\tau$ is plotted as a function of ρ. The origin of this critical slowing down near the bifurcation point is illustrated in Fig. 15, where the director trajectory in the (n_x, n_y) plane is shown. Slightly below ρ_L ($\rho = 1.02025$, black solid line on the left), the trajectory approaches a saddle fixed point (open circle) during increasingly longer times; this corresponds to the "plateau" behavior when $\Omega \sim 0$ in the inset of Fig. 14. On the other hand, slightly above ρ_L ($\rho = 1.02026$, gray line in Fig. 15), the director eventually settles to a stable focus (filled circle) that corresponds to a LD state. (The stable LD state in the present example [just above ρ_L for $\chi = 0.57$] is represented by a stable focus, in that for lower values of χ the director settles to a LO state above ρ_L.) In fact, we deal at $\rho = \rho_L$ with a homoclinic bifurcation of the simplest type, where a limit cycle collides with a saddle point having only one unstable direction (all the eigenvalues have negative real parts except one, which is real and positive) [42].

As discussed in the previous section, a LR state appears at high intensities for the circular case ($\chi = \pi/4$). This slow dynamics is quite fragile and disappears for perturbations of the ellipticity as small as $\Delta\chi \sim 10^{-2}$, giving rise to a LD regime instead. The mechanism for the disappearance of the LR regime is the following. In the CP limit, the precession frequency associated with the

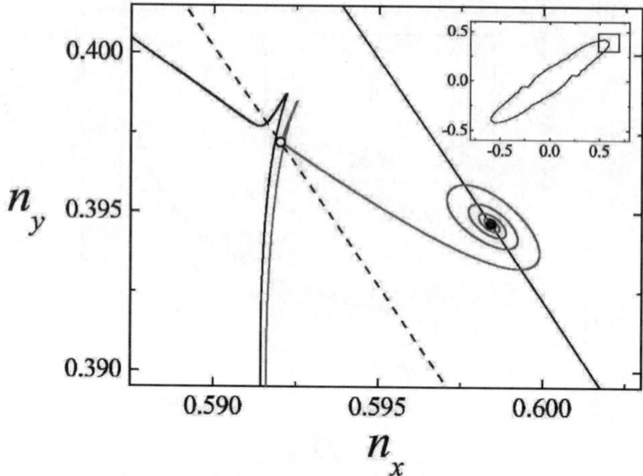

Figure 15.   Director trajectory in the (n_x, n_y) plane near the homoclinic bifurcation point ρ_L at $\chi = 0.57$. Inset: homoclinic PR trajectory slightly below ρ_L ($\rho = 1.02025$). Main graph: magnification of the region delimited by the box in the inset. The black solid line on the left is part of the PR trajectory at $\rho = 1.02025$, and the gray line is the transient trajectory converging to a stable fixed point just above ρ_L ($\rho = 1.02026$). The dashed line (solid line on the right) represents the location of the unstable (stable) fixed points in a small range of ρ centered around ρ_L. The open and filled circles represent the unstable and stable fixed points at $\rho = 1.02025$ and $\rho = 1.02026$, respectively.

LR state exhibits almost periodic modulation as a function of the intensity with zero minimum values. As soon as the ellipticity is reduced, the points of zero frequency are transformed into finite regions which continue to increase as χ is further decreased. Eventually, these regions join leading to the LR $\rightarrow$ LD transition.

In the region of largely reoriented states, the LO regime appears for $\chi < 0.51$ [see Fig. 12]. This state is characterized in the (n_x, n_y) plane by a limit cycle with a small radius [see curve 4 in Fig. 13(a)]. It was concluded [50] that the LD states [which were stable at $\rho = \rho_L$] lose their stability at some higher value of ρ in the range $0.45 < \chi < 0.51$, leading to a LO state. In fact, the transition LD $\rightarrow$ LO takes place via a Hopf bifurcation. Moreover, these unstable LD states eventually recover their stability at higher intensities, leading to the inverse transition LO $\rightarrow$ LD.

The observation of the QPR regime for ellipticities close to circular polarization gave results which agree qualitatively with the above theoretical findings [50]. To gain quantitative agreement, one again should refine the theory by i) inclusion of flow; ii) taking into account the effect of finite beam size. We think, however, that neither the former, nor the latter effect would lead to substantial differences (provided the beam is not several times narrower than the cell width).

6. Finite beam-size effects and transversal pattern formation

As mentioned in the introductory sections, using the 1D assumption to solve the relevant equations of motion is a simplification that is relaxed very rarely. Although a number of works dealt with the OFT in nematics using Gaussian beam profiles [51, 52], these works dealt exclusively with the primary instability and with the properties of the stationary reoriented state above the transition in various geometries. Dynamical phenomena were not considered in any theoretical works, even though there are experiments proving that, decreasing the width of the beam yields very interesting dynamical behavior and even chaos. One example is [13], where a circularly polarized light, incident perpendicularly on a cell of nematic was observed to induce various dynamical regimes. The ratio of the beam width to the cell width $\delta = w_0/L$ was treated as a control parameter alongside the intensity ρ, and the various bifurcation scenarios were compared as δ and ρ were changed. It was found that for $\delta \approx 0.3 - 0.4$, novel dynamical regimes and even chaotic oscillations could be observed. It was argued that a spatial mismatch between the ordinary and extraordinary waves that develops within the cell during propagation may have something to do with these phenomena. Another study [53] investigated the dynamics induced by a strongly astigmatic beam of circularly polarized light, again with normal incidence. The light of an astigmatic beam carries not just the usual spin angular momentum of circularly polarized photons, but also orbital angular momentum. It was shown in this study, that as the astigmatism of the beam is increased above a certain level (i.e. if the orbital component of the angular momentum reaches a certain ratio to the spin angular momentum component), chaotic rotation of the molecules can be observed. Both of these experiments emphasize that the laser beam shape can be an important control parameter, whose change gives rise to complex dynamics. A proper theory, however, that can account for the physical reasons or the nature of the transitions is missing altogether.

Another possibility to consider is, that even if the light is incident on the cell as a plane wave, pattern formation may occur in the plane of the layer spontaneously. In other words, the spatially homogeneous state may lose stability to a finite wavelength perturbation, and a dependence of the physical quantities on the transverse coordinates may develop. There are several arguments to suggest that such instabilities are to be expected. On one hand, it is known [54, 55] that periodic patterns can develop in the magnetic or electric field induced Freedericksz transition in nematics, if the anisotropy of the elastic constants reaches a certain value (i.e. if three elastic constants are sufficiently different). It is also known [56], that any nearly homoclinic limit cycle is generically unstable with respect to spatiotemporal perturbations. This instability is either a phase instability or a finite wavelength period doubling instability. This means that we can certainly expect very complicated behavior (probably spatio-temporal chaos) to develop in the vicinity of the homoclinic bifurcations that were found in the 1D calculations in various geometries. Solving the relevant PDEs in 2 or 3 dimensions to search for transversal pattern formation phenomena would be prohibitively difficult. However, investigating the stability of various spatially homogeneous states with respect to finite wavelength perturbations is much easier and has been performed in several cases.

In [11], the simplified models of director dynamics in two geometries were examined. One was the director dynamics induced by obliquely incident, linearly polarized light. The simple model of this geometry [23] (see section 3) was generalized to include a slow $x - y$ dependence of the amplitudes A_1, A_2, B_1. Performing a linear stability analysis of the basic state with this extended model, it was found that the undistorted homeotropic state always loses stability in a spatially homogeneous bifurcation, i.e. $\vec{k}_c = 0$. (In the course of any linear stability analysis, one considers spatially periodic perturbations, and the wave vector $\vec{k}$ of the mode that destabilizes the stationary state is called the critical wave vector $\vec{k}_c$ - if this is zero, the instability is said to be homogeneous.) This is true for both the stationary OFT (curve 1 on Figure 4) and the oscillatory OFT (curve 2 on Figure 4). For the case of the oscillatory OFT, the relevant complex Ginzburg-Landau equation was derived which describes the behavior of the system in the weakly nonlinear regime. The linear dispersion parameter in this equation turns out to be zero, while the nonlinear dispersion parameter is in such a range that one expects stable plane wave solutions, spirals and - in 1D - stable hole solutions [57]. The stability of the stationary distorted state above the OFT was also investigated. It was found that the secondary Hopf bifurcation which destabilizes it is homogeneous only if the Frank elastic constants are all equal ($K_1 = K_2 = K_3$), but it is not homogeneous otherwise. In fact, $\vec{k}_c \neq 0$ for any degree of anisotropy of the elastic constants (here reflection symmetry is broken by the primary transition, so $\vec{k}_c \neq 0$ is actually the generic case). This is contrary to the magnetic field induced transition, where there is a lower threshold to the ratio K_1/K_2 [54] or K_3/K_1 [55], below which $\vec{k}_c = 0$ and no stripes appear. Since the instability is nonstationary, a finite $\vec{k}_c$ means the appearance of traveling waves. The magnitude of $\vec{k}_c$ grows with the anisotropy of the elastic constants, and its direction (the direction of wave propagation) is roughly parallel to the in-plane component of the director $\vec{n}_\perp$.

Another simple model investigated in [11] is a model of the director dynamics induced by circularly polarized light using three variables. The stability of the basic state and the uniformly precessing state above the OFT was investigated, and it was found that both of these states remain stable against any finite k perturbation too. The phase diffusion equation for the precessing state, that describes the spatial evolution of phase disturbances [58] was also derived. The general form of this equation is $\partial_t \psi = a\nabla^2\psi + b(\nabla\psi)^2$. In our case, the spatially dependent phase perturbation is $\psi = B_0 - \Omega t$, and the real constants appearing in the equation turn out to be: $a = (K_1 + K_2)/K_3, b = 0$. This means that we have plane wave solutions ($B_0 \sim qx + py$) but without group velocity. These are the only attractors, and presumably all solutions decay to such states apart from topological point defects. Vortex-like topological defects, whose core however is not described by this equation, should also exist.

The linear stability analysis of the stationary distorted state above the OFT induced by obliquely incident, linearly polarized light was repeated in [12], this time without the numerous approximations used in earlier works, using numerical methods. In addition, dye-doped nematics were considered, for which the threshold for the OFT is much lower and thus permits the experimental realization of a light beam that is much wider than the cell width. For this case, however, absorption also has to be taken into account, which means that the OFT does not happen at $\rho = 1$ for perpendicular incidence (see Fig. 16). Taking all of this into account, the exact reorientation profiles for the director were calculated numerically first, and its stability was investigated with respect to spatially periodic perturbations (proportional to $\exp[i(qx + py)]$). Then the

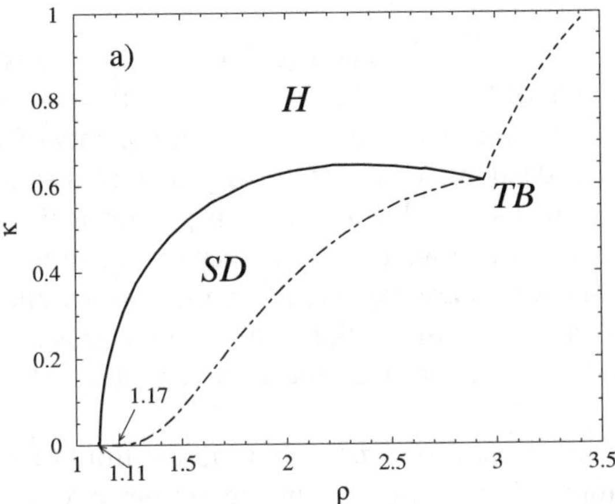

Figure 16. Bifurcation diagram of a dye-doped nematic excited by obliquely incident, linearly polarized light as a function of the intensity ρ and the dimensionless phase parameter κ, which is proportional to the $\sin^2(\alpha)$. H marks the domain where the homeotropic orientation is stable. The line of OFT consists of two parts: for small angles it is a stationary bifurcation (solid line) and for higher angles a Hopf bifurcation (dashed line). The domain of stationary distortion is marked SD. The dash-dotted line marks the secondary Hopf bifurcation which gives rise to traveling waves in the plane of the layer.

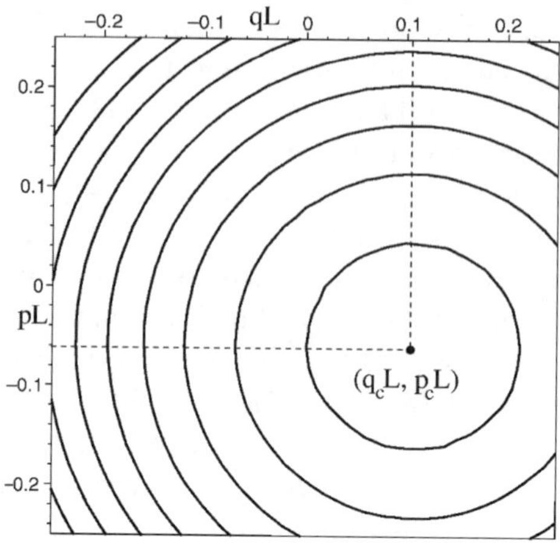

Figure 17.   Contour plot of the neutral surface as a function of the dimensionless wave vector components pL and qL calculated for $\alpha = 11°$. The minimum of the surface yields the critical wave vector components $q_cL = 0.11$ and $p_cL = -0.06$.

neutral surface $\rho(q, p)$ was calculated (defined by the vanishing real part of the linear growth rate of the perturbation $Re[\sigma(q, p)] = 0$ – see Fig. 17). The minimum of the neutral surface defines the components of the critical wave vector q_c, p_c, which destabilizes the stationary distorted state as the intensity is increased. The intensity value $\rho(q_c, p_c)$ is the critical intensity at which the transition occurs. Since we have a nonzero wave number and a nonzero frequency, traveling waves are expected to appear above the transition. This analysis also confirmed that the critical wave vector grows as the ratio of the elastic constants deviates from one, and it is zero if all elastic constants are equal.

An interesting situation also came to light in the limit of normal incidence. This case was impossible to analyze in the framework of the approximate model, as the modes become large quickly and violate the initial assumptions. It turned out that for $\alpha = 0$ (which is a peculiar case, since the external symmetry breaking in the x direction vanishes), another stationary instability precedes the secondary Hopf bifurcation that spontaneously breaks the reflection symmetry with respect to x. It is shown by point A in Fig. 18. It is also seen from this figure, that the secondary pitchfork bifurcation is destroyed in the case of oblique incidence, which can be interpreted as an imperfect bifurcation with respect to the angle α [43].

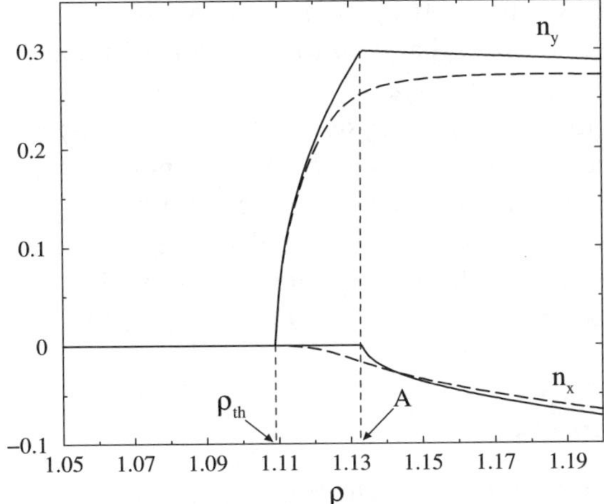

Figure 18. Profiles of the director components n_x, n_y versus ρ at some z inside the layer (not at the middle). Solid and dashed lines correspond to $\alpha = 0°$ and $\alpha = 0.5°$ respectively. ρ_{th} is the threshold intensity of the OFT. Point A is a pitchfork bifurcation to a stationary state with broken x-reflection symmetry ($\alpha = 0°$).

Conclusion and Outlook

As shown in the preceding sections, the behavior of nematics excited by light can be extremely complex. Theoretical models reveal relatively simple reasons behind the dynamics in only a few cases. More often, experiments and elaborate computer studies show that a number of factors govern the complex behavior together: director distortion induced by light, light propagation change due to reorientation, flow of the fluid and the effects of finite beam size all become factors to reckon with at some point or another.

There has also been a number of works devoted to some generalization of the simple system that was the subject of the present paper. One of these is the attempt to control the chaotic oscillations induced by an obliquely incident laser light with the use of additional laser beams [59]. Another one is the investigation of the response of nematics driven by circularly polarized light with a periodically modulated intensity near a Hopf bifurcation. Calculations showed that the $f_1/f = 2/1$ Arnold tongue (f [f_1] is the frequency of modulation [nutation]) has a large width, and there is a rather large region in the plane of amplitude - modulation frequency, where the director exhibits chaotic behavior followed by a cascade of period doubling bifurcations [60]. A third example is the case of a long-pitch cholesteric liquid crystal. There, the dynamics can be viewed as the result of the competition between the intrinsic unidimensional helical pattern (related with the chiral dopant) and the extrinsic one (related with the light). It was found [61] for the case of circular polarization, that the dynamics is more complex than that for a pure nematic and depends strongly on the amount of the chiral dopant.

The study of these complex systems is important, because they exhibit a large variety of nonlinear phenomena. While these are all known from the theory of nonlinear systems, many of these were investigated experimentally in only a few cases or sometimes not at all. Therefore, this relatively simple experimental system may be an important tool to realize and analyze various complex scenarios that appear in nature and may still have many surprises in store.

Acknowledgments

Financial support by the Deutsche Forschungsgemeinschaft under contracts Kr 690/16 and 436UNG113/151/1 is gratefully acknowledged.

References

[1] N. V. Tabiryan, A. V. Sukhov and B. Ya. Zel'dovich: Mol. Cryst. Liq. Cryst. **136**, 1-140 (1985).

[2] F. Simoni and O. Francescangeli, J. Phys.: Condens. Matter **11** R439 (1999).

[3] F. Simoni, *Nonlinear optical properties of liquid crystals and polymer dispersed liquid crystals* World Scientific, Singapore 1997.

[4] I. C. Khoo, *Liquid Crystals: Physical Properties and Nonlinear Optical Phenomena* Wiley Interscience, New York 1994; I. C. Khoo and S. T. Wu, *Optics and Nonlinear Optics of Liquid Crystals* World Scientific Singapore 1993.

[5] P. G. de Gennes and J. Prost, *The physics of liquid crystals* (Clarendon press, Oxford, 1993).

[6] F. M. Leslie, Quart. J. Mech. Appl. Math. **19**, 357 (1966).

[7] O. Parodi, J. Phys. (Paris) **31**, 581 (1970).

[8] L. Marrucci, G. Abbate, S. Ferraiuolo, P. Maddalena, and E. Santamato, Mol. Cryst. Liq. Cryst. **237**, 39 (1993).

[9] D. O. Krimer, G. Demeter, and L. Kramer, Phys. Rev E **71**, 051711 (2005).

[10] G. Demeter, D. O. Krimer and L. Kramer, Phys. Rev. submitted for publication.

[11] G. Demeter and L. Kramer, Mol. Cryst. Liq. Cryst. **366**, 2659 (2001).

[12] D. O. Krimer, G. Demeter and L. Kramer, Phys. Rev. E **66**, 031707 (2002).

[13] E. Brasselet, B. Doyon, T.V. Galstian and L. J. Dube, Phys. Rev. E **69**, 021701 (2004).

[14] B. L. Winkler, H. Richter, I. Rehberg, W. Zimmermann, L. Kramer and A. Buka, Phys. Rev. A **43**, 1940 (1991).

[15] D. W. Berreman, J. Opt. Soc. Am. **62**, 502 (1972).

[16] B. Ya. Zel'dovich, S. K. Merzlikin, N. F. Pilipetskii, A. V. Sukhov and N. V. Tabiryan, JETP Lett **37**, 677 (1983).

[17] A. S. Zolot'ko, V. F. Kitaeva, N. Kroo, N. N. Sobolev, A. P. Sukhorukov, V. A. Troshkin and L. Csillag, Sov. Phys. JETP **60**, 488 (1984); V. F. Kitaeva, N. Kroo, N. N. Sobolev, A. P. Sukhorukov, V. Yu. Fedorovich, and L. Csillag, Sov. Phys. JETP **62**, 520 (1985).

[18] A. S. Zolotko, V. F. Kitaeva, N. N. Sobolev, V. Yu. Fedorovich, A. P. Sukhorukov, N. Kroo and L. Csillag, Liq. Cryst. **15**, 787 (1993).

[19] G. Cipparrone, V. Carbone, C. Versace, C. Umeton, R. Bartolino and F. Simoni, Phys. Rev. E **47**, 3741 (1993).

[20] V. Carbone, G. Cipparrone, C. Versace, C. Umeton and R. Bartolino, Mol. Cryst. Liq. Cryst. Sci. Technol., Sect. A **251**, 167 (1994); V. Carbone, G. Cipparrone, C. Versace, C. Umeton and R. Bartolino, Phys. Rev. E **54**, 6948 (1996); C. Versace, V. Carbone, G. Cipparrone, C. Umeton and R. Bartolino, Mol. Cryst. Liq. Cryst. Sci. Technol., Sect. A**290**, 267 (1996).

[21] E. Santamato, P. Maddalena, L. Marrucci and B. Piccirillo, Liq. Cryst. **25**, 357 (1998); E. Santamato, G. Abbate, P. Maddalena, L. Marrucci, D. Paparo and B. Piccirillo, Mol. Cryst. Liq. Cryst. **328**, 479 (1999).

[22] N. V. Tabiryan, A. L. Tabiryan-Murazyan, V. Carbone, G. Cipparrone, C. Umeton and C. Versace, Optics Commun. **154**, 70 (1998).

[23] G. Demeter and L. Kramer, Phys. Rev. Lett. **83**, 4744 (1999); G. Demeter, Phys. Rev. E **61**, 6678 (2000).

[24] A. Arneodo, P. Coullet and C. Tresser, Phys. Lett. **81A**, 197 (1981); Y. Kuramoto and S. Koga, Phys. Lett. **92A**, 1 (1982).

[25] D. V. Lyubimov and M. A. Zaks, Physica D **9**, 52 (1983).

[26] G. Russo, V. Carbone and G. Cipparrone, Phys. Rev. E **62** 5036 (2000); V. Carbone, G. Cipparrone and G. Russo, Phys. Rev. E **63** 051701 (2001).

[27] G. Demeter and L. Kramer, Phys. Rev. E **64**, 020701 (2001); G. Demeter and L. Kramer: Mol. Cryst. Liq. Cryst. **375**, 745 (2002).

[28] E. Santamato, B. Daino, M. Romagnoli, M. Settembre, and Y.R. Shen, Phys. Rev. Lett. **57**, 2423 (1986).

[29] E. Santamato, M. Romagnoli, M. Settembre, B. Daino, and Y.R. Shen, Phys. Rev. Lett. **61**, 113 (1988).

[30] L. Marrucci, G. Abbate, S. Ferraiuolo, P. Maddalena, and E.Santamato, Phys. Rev. A **46**, 4859 (1992).

[31] R. A. Beth, Phys. Rev. **50**, 115 (1936).

[32] J. D. Jackson, *Classical electrodynamics* (Wiley, New York, 1962).

[33] L. Allen, M. W. Beijersbergen, R. J. C. Spreeuw, and J. P. Woerdmann, Phys. Rev. A **45**, 8185 (1992).

[34] A. S. Zolot'ko, and A. P. Sukhorukov, Sov. Phys. JETP Lett., **52**, 62 (1990).

[35] S. D. Durbin, S. M. Arakelian, and Y. R. Shen, Opt. Lett. **6**, 411 (1981).

[36] E. Brasselet and T. Galstian, Opt. Commun. **186**, 291 (2000).

[37] E. Brasselet, B. Doyon, T.V. Galstian and L. J. Dube, Phys. Lett. A **299**, 212 (2002).

[38] E. Brasselet, B. Doyon, T.V. Galstian and L. J. Dube, Phys. Rev. E **67**, 031706 (2003).

[39] D. O. Krimer, G. Demeter, and L. Kramer, Mol. Cryst. Liq. Cryst. **421**, 117 (2004).

[40] E. Brasselet, T.V. Galstian, L. J. Dube, D. O. Krimer and L. Kramer, J. Opt. Soc. Am. B, **22**, 1671 (2005).

[41] H. L. Ong, Phys. Rev. A **28**, 2393 (1983).

[42] P. Glendinning, *Stability, instability and chaos* (Cambridge University press, Cambridge, 1996).

[43] D. O. Krimer, PhD Thesis (2004).

[44] E. Brasselet, Phys. Lett. A **323**, 234 (2004).

[45] Istvan Janossy, J. Nonlin. Opt. Phys. Mat. **8**, 361 (1999).

[46] L. Marrucci, D. Paparo, P. Maddalena, E. Massera, E. Prudnikova, and E. Santamato, J. Chem. Phys **107**, 9783 (1997).

[47] B. Y. Zel'dovich and N. Tabiryan, Sov. Phys. JETP **55**, 656 (1982).

[48] E. Santamato, G. Abbate, P. Maddalena, L. Marrucci and Y. R. Shen, Phys. Rev. Lett. **64**, 1377 (1990).

[49] A. Vella, B. Piccirillo, and E. Santamato, Phys. Rev. E **65**, 031706 (2002).

[50] D. O. Krimer, L. Kramer, E. Brasselet, T.V. Galstian and L. J. Dube, J. Opt. Soc. Am. B, **22**, 1681 (2005).

[51] L. Csillag, J. Janossy, V. F. Kitaeva, N. Kroo and N. N. Sobolev, Mol. Cryst. Liq. Cryst. **84**, 125 (1982); E. Santamato and Y. R. Shen, Opt. lett. **9**, 564 (1984).

[52] I. C. Khoo, T. H. Liu and P. Y. Yan, J. Opt. Soc. Am. B **4**, 115 (1987).

[53] A. Vella, A. Setaro, B. Piccirillo, and E. Santamato, Phys. Rev. E **67**, 051704 (2003).

[54] F. Lonberg and R. B. Meyer, Phys. Rev. Lett. **55**, 718 (1985).

[55] D. W. Allender, R. M. Hornreich and D. L. Johnson, Phys. Rev. Lett. **59**, 2654 (1987).

[56] M. Argentina, P. Coullet, E. Risler: Phys. Rev. Lett. **86**, 807 (2001).

[57] S. Popp, O. Stiller, I. Aranson and L. Kramer, Physica D **84**, 398 (1995).

[58] Y. Kuramoto: *Chemical Oscillations, Waves and Turbulence*, Springer 1984.

[59] G. Russo, G. Cipparrone and V. Carbone, Europhys. Lett. **63** 180 (2003).

[60] D. O. Krimer, and E. Brasselet, to be published.

[61] E. Brasselet, D. O. Krimer, and L. Kramer, Europ. Phys. J. E **17** (2005).

SELF-ASSEMBLY OF QUANTUM DOTS FROM THIN SOLID FILMS

Alexander A. Golovin,[1] Peter W. Voorhees,[2] and Stephen H. Davis[1]

[1]*Department of Engineering Sciences and Applied Mathematics*

[2]*Department of Materials Science and Engineering*
Northwestern University
2145 Sheridan Rd, Evanston, IL 60201, USA

Abstract   Some aspects of self-assembly of quantum dots in thin solid films are considered. Nonlinear evolution equations describing the dynamics of the film instability that results in various surface nanostructures are analyzed. Two instability mechanisms are considered: the one associated with the epitaxial stress and the other caused by the surface-energy anisotropy. It is shown that wetting interactions between the film and the substrate transform the instability spectrum from the long- to the short-wave type, thus yielding the possibility of the formation of spatially-regular, stable arrays of quantum dots that do not coarsen in time. Pattern formation is analyzed by means of amplitude equations near the instability threshold and by numerical solution of the strongly nonlinear evolution equations in the small-slope approximation.

Keywords: Quantum dots, self-assembly, thin solid films, pattern formation, instabilities

1.   Introduction

In many applications in the electronics industry a thin film of a solid semiconductor material needs to be deposited on a solid semiconductor substrate. This deposition is usually made by means of a molecular beam, and during the growth process atoms of the film stick to the atoms of the substrate at its surface. Such a growth process is called epitaxial. If the material of the film differs from the material of the substrate the growth is called hetero-epitaxial; when the two materials are the same the growth is called homo-epitaxial. Very often the growing film does not remain planar during its growth and develops various kinds of surface structures. The types of these structures depend on physical characteristics of the materials as well as on the growth conditions. One particular type of surface structures has been attracting a great deal of attention during the last decade. It usually occurs during the hetero-epitaxial

123

A. A. Golovin and A. A. Nepomnyashchy (eds.),
Self-Assembly, Pattern Formation and Growth Phenomena in Nano-Systems, 123–158.
© 2006 *Springer. Printed in the Netherlands.*

growth and manifests itself as decomposition of a planar solid film into a system of islands that are divided from one another by an ultra-thin (a couple of atoms thick) layer, the so-called "wetting layer". A typical size of the islands is from a few to a few hundreds of nanometers which is of the same order of magnitude as the de Broglie wavelength of electrons in semiconductors. Due to this fact, such surface structures have quite interesting electronic properties, mainly associated with the possibility of electron localization. This is where the commonly used name for these nano-structures come from: they are called "quantum dots". Quantum dots that have formed on the surface of a solid film can be covered by another material and form localized bulk structures. They are considered very promising for creating new generation of electronic devices. Electronic properties of quantum dots, however, will not be discussed here; see [1] for a review. In these lecture notes we shall focus on the mechanisms and the nonlinear dynamics of quantum-dot formation and their evolution.

One of the typical features of quantum dots is that they can form spontaneously as the result of an instability of a thin solid film deposited on a solid substrate. Therefore, one can talk about *self-assembly* of quantum dots. Some examples of self-assembled quantum dots are shown in Fig.1. They can have various shapes: regular, as faceted pyramids; irregular, as small crystals with many facets in various orientations; rounded, as cones. They can form sparse or dense arrays, also regular or irregular, with broad or narrow size distribution. When an array of quantum dots formed on the surface of a solid film is kept (annealed) at a high, constant temperature, the dots can either exhibit coarsening (Ostwald ripening) or not. During coarsening the larger dots grow at the expense of the smaller ones so that the average dot size increases in time. In the absence of coarsening, the dot size distribution does not essentially evolve at all. Below we shall discuss the main mechanisms that govern the shape of quantum dots, the dynamics of their formation and the evolution of the quantum dot arrays.

2.   Mechanisms of morphological evolution of epitaxial films

Roughly speaking, the principal mechanisms that govern the formation, morphology and evolution of quantum dots are elastic stress, anisotropic surface energy and surface diffusion. The most common mechanism of instability that leads to decomposition of an initially planar solid film into a system of islands is associated with elastic stress. When a film of one material is deposited on a substrate made of another, the mismatch between the crystal-lattice spacings of the two materials (which is almost always present in hetero-epitaxial systems) results in elastic strain and stress in the film (epitaxial strain and epi-

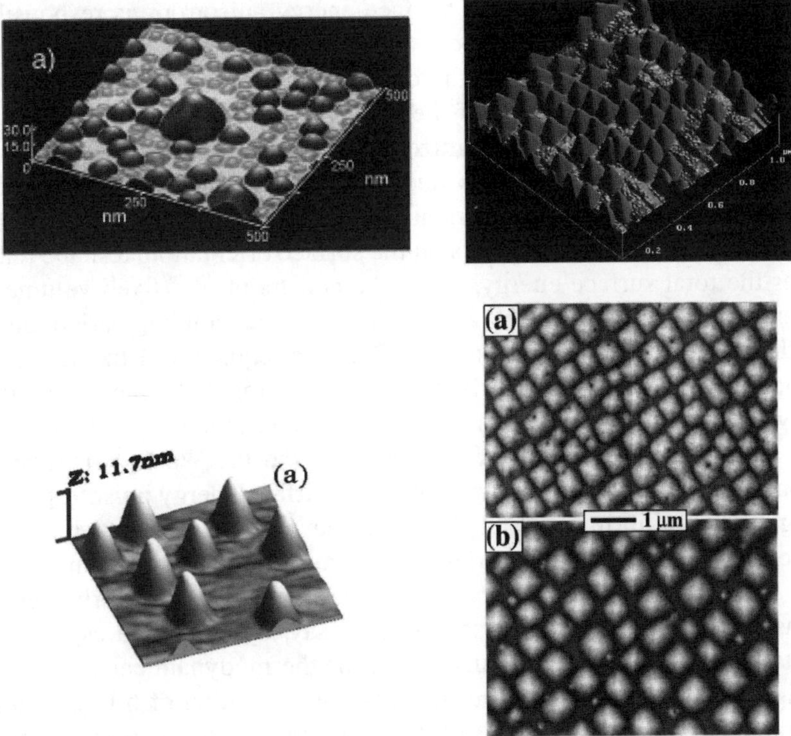

Figure 1. Examples of quantum dots (QDs): upper left: Ge QDs on Si(001) [2] (reprinted with permission from [2], ©1998 by the American Chemical Society); upper right: PbSe QDs on PbTe-on-Si(111) [3] (reprinted with permission from [3], ©2003 by the American Physical Society); lower left: InAs QDs on GaAs(001) [4] (reprinted with permission from [4], ©2003 by the American Institute of Physics); lower right: coarsening of SiGe QDs on Si(001) – figure b) corresponds to later time than a) [5] (reprinted with permission from [5], ©2000 by the American Physical Society).

taxial stress). As a result, the film has some excess elastic energy stored in it. When undulations on the film surface are formed, the epitaxial stress is released and the elastic energy is lowered. Therefore, the undulated surface is energetically more preferable than the planar one and this causes the instability of a planar film. This instability is called Asaro-Tiller-Grinfeld instability [6, 7]. A more detailed qualitative description of this instability can be found in [8].

Another mechanism is associated with the anisotropy of the film surface energy. Unlike liquids and amorphous materials, crystalline materials have internal spatially-regular structure, the crystal lattice. While surface energy (per unit area) of liquids and amorphous materials does not depend on the surface orientation, surface energy of a crystal usually does; thus the surface energy of a crystal is usually *anisotropic*. Surface-energy anisotropy is responsible for faceted equilibrium crystal shapes. The equilibrium shape of a given material with a fixed volume will be the one that minimizes the total energy of its surface. If the surface energy of the material is isotropic (that is a constant), the equilibrium shape must minimize the total surface area. When the volume is fixed this minimization is provided by a spherical shape, the one that a liquid drop has in the absence of gravity or other bulk forces. When the surface energy is anisotropic and depends on the surface orientation, a shape that minimizes the total surface energy, under the constraint of a fixed volume, is no longer spherical. It is given by a solution of a corresponding variational problem that leads to a nonlinear partial differential equation of the second order for the surface shape. Amazingly, the solution of this PDE can be obtained by a very elegant and simple geometrical construction which is called the Wulff plot. A detailed description of this problem and the Wulff construction can be found in [8]. It can be shown that if the surface energy anisotropy is large enough, the equilibrium crystal shape will have corners across which the surface orientation will jump. Orientations from the interval of this jump can never be seen in equilibrium crystal shapes; they are called "forbidden orientations". If one prepares a crystal surface (say, by a cut of a crystal) whose orientation is forbidden, such surface will be thermodynamically unstable and undergo "faceting instability" resulting in the formation of hill-and-valley or pyramidal structures. The slopes of these structures will correspond to thermodynamically preferable and stable orientations (i.e. with lower energy) so that, although the total surface area will increase, the total surface energy will decrease. Interestingly, such surface instability is completely analogous to spinodal decomposition that results in phase separation of binary systems. One can read more about this analogy in [9, 10].

Thus, the following mechanism of instability of a thin solid film deposited on a solid substrate is possible. The substrate can prescribe the film to grow in a specific crystallographic orientation that, in the absence of the substrate would have been forbidden. When the film becomes thick enough and does not "feel" the substrate any more it will undergo faceting instability and decompose into a system of faceted islands. This mechanism does not depend on the presence of epitaxial stress and can occur when the epitaxial stress is negligible or even in the course of a homo-epitaxial growth.

For any surface morphologies to appear, a surface reconstruction mechanism is required that would transform initially planar surface into a structured one. In epitaxially grown solid films such mechanism is surface diffusion in which atoms jump along the surface from one site to another, driven by the gradient of the surface chemical potential. The surface flux of atoms, $\mathbf{j}_s$, is given by the analog of the Fick's law: $\mathbf{j}_s = -M \nabla_s \mu_s$, where μ_s is the surface chemical potential, M is the atom mobility, and ∇_s is the surface gradient operator. A simple conservation of mass yields $\rho_s v_n + \nabla_s \cdot \mathbf{j}_s = 0$, where v_n is the normal velocity of the surface caused by the atoms redistribution and ρ_s is the surface density of atoms. The surface chemical potential is typically determined by the film elastic energy and surface energy, and as such it is a function of the film local thickness as well as its slope, curvature, and may be higher spatial derivatives (see below). For very thin films (a few atomic layers) *wetting interactions* between the film and the substrate can also become important. These interactions are somewhat similar to wetting interactions between a liquid film and a solid substrate. They are responsible for the presence of an ultra-thin *wetting layer* of the film material between the islands resulting from the film instability and depend on the film thickness and its slope. Naturally, this dependence decays rapidly with the increase of the film thickness.

Thus, a general evolution equation that describes the evolution of a thin solid film on a solid substrate has the following form:

$$\frac{h_t}{\sqrt{1+|\nabla h|^2}} = M\nabla_s^2 \mu_s(h, \nabla h, \nabla^2 h, ...), \qquad (1)$$

where the dependence of μ_s on the surface shape should be computed according to what type of interactions between the film and the substrate are important for a particular problem.

3.   Elastic effects and wetting interactions

In this section we consider evolution of a thin epitaxial film in the case when the film instability is caused by the epitaxial stress, the film surface energy is isotropic, and the film is thin enough so that wetting interactions between the film and the substrate are important. In more detail, the material discussed in this section can be found in [11]. (The figures are reprinted with permission from [11], ©2003 by the American Physical Society).

The general evolution equation for the shape of the film surface (1) can be written in this case as

$$\frac{h_t}{\sqrt{1+|\nabla h|^2}} = M\nabla_s^2[\mathcal{E}(h) + \gamma\mathcal{K} + \Phi], \qquad (2)$$

where $\mathcal{E}(h)$ is the elastic part of the surface chemical potential, γ is the film surface energy, $\mathcal{K}$ is the curvature of the surface, and $\Phi(h, |\nabla h|^2)$ is a wetting chemical potential that usually can be considered as a function of the film thickness and slope.

The functional $\mathcal{E}(h) = \frac{1}{2}[\sigma_{ij}\epsilon_{ij}]_{z=h}$, where σ_{ij} and ϵ_{ij} are stress and strain tensors, respectively. They should be found from the solution of the corresponding elasticity problem that would describe elastic equilibrium of the film and the substrate, with the boundary conditions of stress and displacement continuity at the film-substrate interface, as well as the stress-free condition at the surface of the film. This makes the functional $\mathcal{E}(h)$ very complicated and non-local. However, the problem simplifies considerably if one applies the small-slope approximation, assuming that the slopes of the emerging surface structures are small (which is true in many experimental systems). Also, the problem becomes simpler if one assumes that the substrate is much stiffer than the film and consider it as absolutely rigid, neglecting its elastic deformation.

Under these assumptions the evolution equation for the shape of an epitaxial film on a rigid substrate in the absence of wetting interactions was derived in [12]. Our problem differs from the one considered in [12] only by the presence of the wetting potential. As we shall see below, this difference becomes crucial for the possibility of self-assembly of spatially-regular arrays of quantum dots.

In the presence of wetting interactions, the evolution equation for the scaled film thickness H derived in [12] is generalized to be [11]

$$
\begin{aligned}
\partial_T H \;=\;& \nabla^2\{(H-1)\nabla^2 H + \frac{1}{2}(\nabla H)^2 \\
& + \frac{M\tau}{\alpha^2 L^3}\Phi(LH, \alpha^2|\nabla H|^2, \frac{\alpha^2}{L}\nabla^2 H)\} + O(\alpha^2).
\end{aligned}
\tag{3}
$$

Here, the spatial scale is $L = \gamma/(4\mu\varepsilon^2)$ and the time scale is $\tau = L^3/(4\mu\varepsilon^2 M)$, where $\varepsilon \ll 1$ is the lattice misfit of the epitaxially strained film (epitaxial strain in the horizontal direction), μ is the elastic shear modulus, M is the atom surface mobility, $\alpha \ll 1$ is the slope parameter, and dimensionless space and time coordinates are long-scale, i.e. $\nabla \sim \alpha$, $\partial_T \sim \alpha^2$.

We assume that $\Phi(h, |\nabla h|)$ satisfies the following scaling relations

$$
\frac{M\tau}{L^2}\frac{\partial\Phi(L,0)}{\partial h} = \alpha^4 a, \quad \frac{M\tau}{L}\frac{\partial^2\Phi(L,0)}{\partial h^2} = 2\alpha^2 b, \quad \frac{M\tau}{L^3}\frac{\partial\Phi(L,0)}{\partial|\nabla h|^2} = c,
\tag{4}
$$

where a, b and c are $O(1)$ constants. Then, taking $H = 1+\alpha^2[\eta+h(\mathbf{r}, t)]$, $\eta = const$, and introducing a new time scale, $t = \alpha^2 T$, we obtain, after appropriate rescaling, the following evolution equation for h:

$$
\partial_t h = g\nabla^2 h + \nabla^4 h + \nabla^6 h + \nabla^2[h\nabla^2 h + p(\nabla h)^2 + qh^2],
\tag{5}
$$

where

$$
g = a\beta\eta^{-2}, \quad p = \frac{1}{2} + c, \quad q = b\beta\eta^{-1}, \quad \beta = \frac{3+4\nu}{6(1-\nu)},
\tag{6}
$$

and ν is the Poisson ratio. The linear term with the sixth derivative comes from $O(\alpha^2)$ nonlinear terms in eq.(3) after the rescaling (see [12] for details).

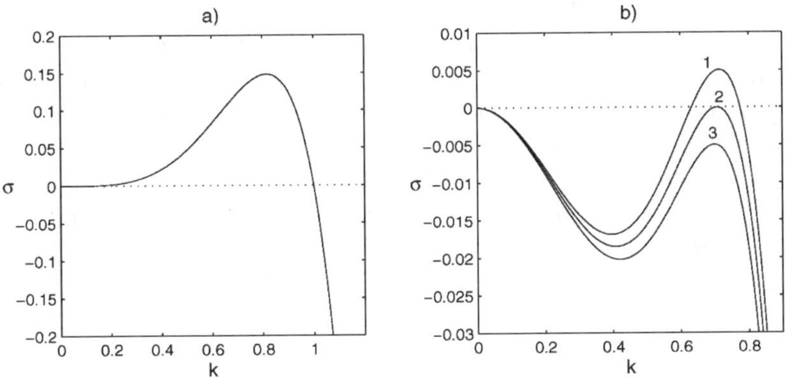

Figure 2. Dispersion curves $\sigma(k)$ for the ATG-instability of a solid film on a rigid substrate: (a) without wetting interactions; (b) with wetting interactions: (1) $g < g_c$, (2) $g = g_c$, (3) $g > g_c$.

Eq.(5) describes the nonlinear evolution of a thin, epitaxially-strained film in the presence of wetting interactions with the substrate. Without wetting interactions ($g = 0$), the dispersion relation for the growth rate of infinitesimal perturbations of the film surface $\tilde{h} \sim e^{\sigma t + i\mathbf{k} \cdot \mathbf{r}}$ is $\sigma = k^4 - k^6$, see Fig.2a (in the case of elastic substrate it changes to $\sigma = k^3 - k^4$ [7]). This is the so-called *long-wave* spectrum that leads to the growth of perturbations whose wavenumber is smaller than the cut-off value corresponding to $\sigma = 0$. Similar spectrum is described by the Cahn-Hilliard equation that describes spinodal decomposition and coarsening (Ostwald ripening) of phase separating systems. Typical nonlinear evolution in such systems could be either coarsening or spatio-temporal chaos, or a blow-up in a finite time [13].

When wetting interactions are present ($g > 0$) the dispersion relation changes to $\sigma = -gk^2 + k^4 - k^6$, see Fig.2b. One can see that wetting interactions suppress the elastically-driven instability. With wetting interactions, the instability occurs for $g < g_c = 1/4$ at a wavenumber $k_c = \sqrt{2}/2$. Thus, wetting interactions change the type of the instability spectrum from the long-wave to the *short-wave* type, in which the instability occurs at a fixed, non-zero wavenumber. In systems with short-wave instability formation of stationary, spatially periodic patterns becomes possible [13, 14].

Pattern formation in 1+1 system

First, we consider a 1+1 system (2D film with 1D surface) since some important features of the nonlinear evolution of the film instability can be studied in this case. In order to study the possibility of pattern formation one first performs a weakly nonlinear analysis of stationary solutions of eq.(5) near the instability threshold. A characteristic feature of the system described by eq.(5) is the presence of the zero mode, $\sigma = 0$, corresponding to $k = 0$ (see Fig.2) and associated with the conservation of mass. The nonlinear interaction between the zero mode and the unstable mode can substantially affect the system behavior near the instability threshold [15, 16] and must be taken into account in weakly nonlinear analysis.

Consider a 1D version of eq.(5). Take $g = g_c - 2\epsilon^2$ and

$$h \sim \epsilon A(X, T)e^{ik_c x} + c.c. + \epsilon^2 B(X, T) + \ldots,$$

where A is the complex amplitude of the unstable mode, B is the real amplitude of the zero mode, $X = \epsilon x$, $T = \epsilon^2 t$. Standard multiple-scale analysis near the bifurcation point yields the following system of equations for A and B:

$$\begin{aligned}
\partial_T A &= A + A_{XX} - \lambda_0 |A|^2 A + sAB, &\quad (7)\\
\partial_T B &= mB_{XX} + w(|A|^2)_{XX},
\end{aligned}$$

where

$$\lambda_0 = \frac{2}{9}(1 + p - 2q)(p + q - \frac{5}{4}), \qquad (8)$$

$$m = \frac{1}{4}, \; s = \frac{1}{4} - q, \; w = -1 + p + 2q.$$

System (7) can be considered as a generic system describing nonlinear evolution in a large class of unstable systems with a conserved quantity [16].

For $\lambda_0 > 0$ the periodic structure is supercritical and can be stable, while for $\lambda_0 < 0$ it is subcritical and blows up in a finite time. The conclusion about stability of the supercritical pattern, however, cannot be made unless the interaction with the zero mode is taken into account. For $\lambda_0 > 0$, stationary solution of the system (7),

$$A_0 = \lambda_0^{-1/2}, \; B_0 = 0, \qquad (9)$$

corresponds to a 1D periodic array of "islands". Consider perturbations $\tilde{A}$, $\tilde{B}$ of the solution (9) in the form

$$\tilde{A} = \tilde{a}e^{iQX + \omega T} + \tilde{b}e^{-iQX + \omega^* T}, \; \tilde{B} = \tilde{c}e^{iQX + \omega T} + c.c.,$$

where ω^* is the complex conjugate of ω. One easily obtains from (7) that the stationary solution becomes unstable with respect to monotonic perturbations, $\mathrm{Im}(\omega)=0$, for

$$\frac{sw}{m} + \lambda_0 < 0. \tag{10}$$

The condition (10) allows one to determine regions in the (q,p)-plane corresponding to different types of pattern excitation and stability near threshold as shown in Fig.3. The straight lines OCB and OGF correspond to $\lambda_0 = 0$ and the curves AB, CD, EF, GH are parts of the hyperbola $sw/m + \lambda_0 = 0$. It is interesting that at the intersection point O, $p = 1/2$, $q = 3/4$. Since $p = 1/2$ corresponds to $c = 0$, this means that unless the wetting potential depends on the film slope the periodic structure is always subcritical and therefore blows up. Weakly nonlinear analysis is not useful in this case.

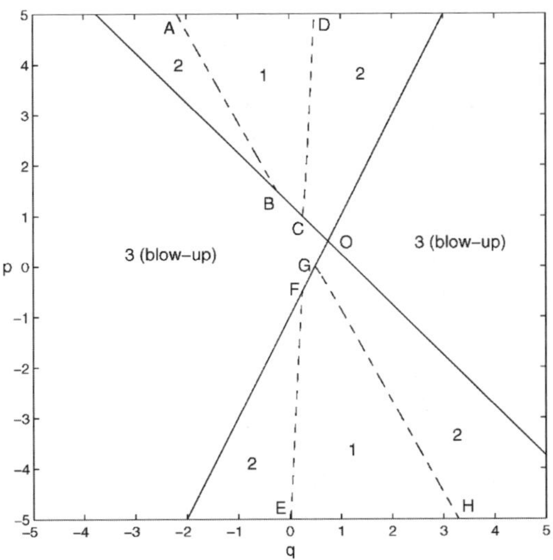

Figure 3. Regions corresponding to different types of excitation and stability of spatially periodic 1D solutions of eq.(5) near threshold: (1) – supercritical stable; (2) – supercritical unstable; (3) –subcritical. Coordinates of the points are: O(0.75,0.5), B(-0.25,1.5), C(0.25,1), F(0.25,-0.5), G(0.5,0).

Now we consider a strongly nonlinear evolution of 1D arrays of islands farther from the instability threshold studied by means of numerical simulations of eq.(5) using a pseudospectral method with periodic boundary conditions. For the parameter values corresponding to region 1 in Fig.3, near the instability threshold, one observes the formation of a sinusoidal surface profile. With the increase of the supercriticality (i.e. with the decrease of g from $g_c = 1/4$), the surface shape becomes significantly non-harmonic and exhibits two typical periodic patterns: periodic array of "cone"-type islands and periodic array of "cap"-type islands shown in Fig.4. "Cones" and "caps"are observed for $p > 0$ and $p < 0$, respectively.

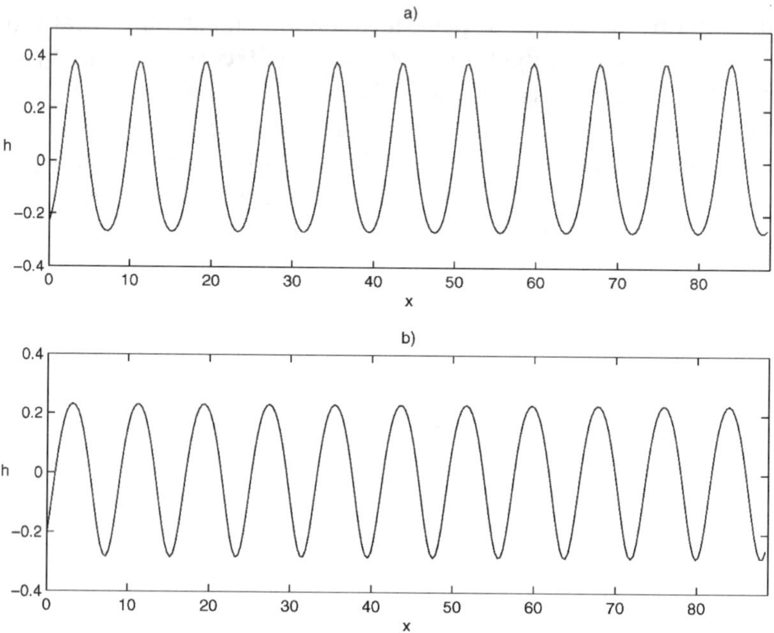

Figure 4.   Stationary numerical solutions of eq.(5) in 1D for $(g, q, p) = $ (a) $(0.1, -1.0, 4.0)$ ("cones") and (b) $(0.1, 1.0, -4.0)$ ("caps").

For the values of the parameters p and q from region 2 in Fig.3, where periodic arrays of islands near the instability threshold are unstable due to the presence of the zero mode, one observes the formation of localized (or strongly modulated) patches of islands shown in Fig.5. Depending on the sign of the parameter p, these can be either patches of "cones" or "caps".

Localized solutions shown in Fig.5 are found only near the instability threshold in the region 2 in Fig.3. With the increase of the supercriticality, one observes either the formation of periodic arrays of "cones" or "caps", or the blow-up. The latter can be of either "island"-type, shown in Fig.6a, or a "pit"-type, shown in Fig.6b. Spontaneous formation of nano-pits has been recently observed in experiments [17].

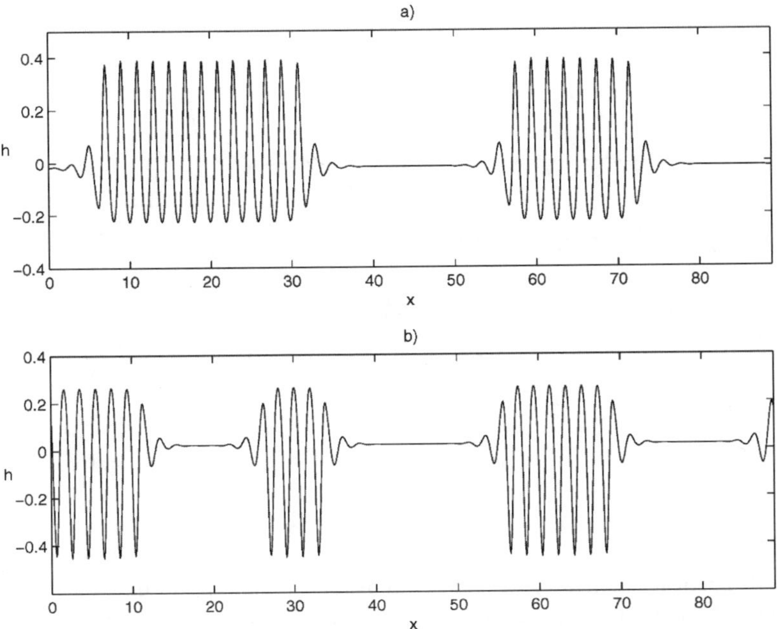

Figure 5.   Localized stationary solutions of eq.(5) in 1D: (a) patches of "cones", $g = 0.24$, $p = 4.0$, $q = -2.3$; (b) patches of "caps", $g = 0.248$, $p = -3.0$, $q = 3.0$.

Pattern formation in 2+1 system

Now we consider a more interesting, 2+1 case of a 3D film with a 2D surface whose evolution is described by eq.(5). Note that eq.(5) does not have the symmetry $h \rightarrow -h$. In this case, the instability whose threshold corresponds to a finite wavenumber (see Fig.2) usually results in a hexagonal pattern that occurs via a transcritical bifurcation [13, 14]. First we concentrate on the formation of surface structures with hexagonal symmetry.

Hexagonal arrays of dots and pits.   The weakly nonlinear evolution of a hexagonal surface structure, $h \sim \sum_{j=1}^{3} A_j e^{i\mathbf{k}_j \cdot r} + c.c.$, where the wavevectors $\mathbf{k}_j$ form an equilateral triangle with the side $k_j = k_c \equiv \sqrt{2}/2$, is usually described by three Ginzburg-Landau-type equations for the complex amplitudes A_j [13, 14]. However, in our case the interaction with the zero mode must also be taken into account. Thus, consider $g = g_c - 2g_1\epsilon^2$, and

$$h = \epsilon \sum_{j=1}^{3} A_j(\mathbf{R}, T)e^{i\mathbf{k}_j \cdot \mathbf{r}} + \epsilon^2 B(\mathbf{R}, T) + c.c. + \ldots, \qquad (11)$$

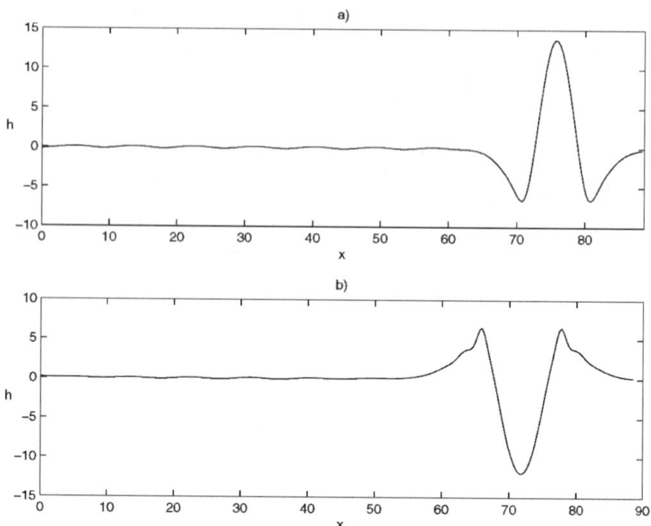

Figure 6.   Numerical solutions of eq.(5) in 1D at particular moments of time showing intermediate stages of the two types of the blow-up: (a) "island"-type, $g = 0.23$, $p = -4.0$, $q = -1.0$, and (b) "pit"-type, $g = 0.23$, $p = 4.0$, $q = 1.0$.

where A_j, $j = 1, 2, 3$ are three complex amplitudes of the unstable modes with wavevectors $\mathbf{k}_j$, respectively, B is the real amplitude of the zero mode, $\mathbf{R} = \epsilon \mathbf{r}$, $T = \epsilon^2 t$. Use the multiple-scale analysis near the bifurcation point to obtain the following system of coupled equations for A_j and B:

$$\partial_T A_j = g_1 A_j + 2(\mathbf{k}_j \cdot \nabla)^2 A_j + r_0 A_l^* A_n^* \tag{12}$$

$$+ i \sum_{l \neq n \neq j} [A_l^*(r_1 \mathbf{k}_l + r_2 \mathbf{k}_n) \cdot \nabla A_n^*]$$

$$- [\lambda_0 |A_j|^2 + \lambda_1(|A_l|^2 + |A_n|^2)]A_j + s A_j B,$$

$$\partial_T B = m \nabla^2 B + w \nabla^2(|A_1|^2 + |A_2|^2 + |A_3|^2), \tag{13}$$

where

$$\lambda_1 = (-1 + \frac{3}{4}p + q)(1 + \frac{1}{2}p - 2q), \tag{14}$$

$$r_0 = \frac{1}{2}(1 - \frac{1}{2}p - 2q), \ r_1 = -p - 2r_0, \ r_2 = 1 - 2r_0,$$

and other parameters are defined in (8). The indices j, l, n run from 1 to 3.

Eqs. (12), (13) are similar to those derived previously for several other pattern forming systems with conserved quantities [16]. Note that since a hexagonal pattern occurs via a transcritical bifurcation, eqs.(12) are valid, strictly speaking, only for $r_0 = O(\epsilon)$. Otherwise, these equations should be considered as model equations describing weakly nonlinear evolution of a hexagonal pattern (see [18]). System (12),(13) has the following stationary solutions

$$A_j \;=\; A_0 = \pm\frac{|r_0| + \sqrt{r_0^2 + 4g_1(\lambda_0 + 2\lambda_1)}}{2(\lambda_0 + 2\lambda_1)}, \qquad (15)$$
$$B \;=\; 0,$$

corresponding to a spatially regular pattern of equilateral hexagons, with the signs $\pm$ corresponding to $r_0 > 0$ and $r_0 < 0$, respectively. Thus, depending on the sign of the resonant-interaction coefficient, r_0, the system can exhibit formation of hexagonal arrays of either mounds (dots) or pits. Dots occur for $r_0 > 0$ $(A_0 > 0)$ and pits occur for $r_0 < 0$ $(A_0 < 0)$. In both cases, the hexagonal pattern can be stable only for $\lambda_0 > 0$ and $\lambda_0 + 2\lambda_1 > 0$ [14]. Also, in our case, the presence of the zero mode strongly affects the stability of the pattern. A detailed stability analysis of hexagonal patterns interacting with the zero mode, within the framework of the system (12) and (13), has been recently carried out in the long-wave approximation [16], and it has been shown that if

$$2ws + m(\lambda_0 + \lambda_1) < 0, \qquad (16)$$

a hexagonal pattern is unstable at any supercriticality g_1 [16].

The described stability conditions determine regions in the (q, p)-plane where self-organization of hexagonal arrangements of dots or pits can be observed. These regions are shown in Fig.7. The lines BC and FG correspond to $\lambda_0 = 0$ and the curves AB, CD, EF and GH are parts of the hyperbola $\lambda_0 + 2\lambda_1 = 0$. Condition (16) holds outside the region bounded by the dashed lines. Inside this region surface structures with hexagonal symmetry can be stable in a certain range of the supercriticality g_1 and the pattern wavenumber (a "Busse balloon") (see [13, 16] and references therein). The lines BK and FL correspond to $r_0 = 0$ and divide the regions where hexagonal arrays of dots or pits can be observed. Note that the full stability analysis must include the consideration of finite-wavelength instabilities [18].

One can see from Fig.7 that, unlike the 1+1 case, in the 2+1 system there is an interval, $0.4 < p < 0.6$ (or $|c| < 0.1$, see eq.(6)) in which hexagonal arrays of dots or pits are always subcritical and therefore unstable. Thus, the wetting interaction between the film and the substrate can lead to the self-organization of dots or pits with the almost uniform sizes only if the wetting potential has strong-enough dependence on the surface slope (e.g. sufficient anisotropy), namely, for $|c| > 0.1$ or (see eq.(4))

$$\left|\frac{\partial \Phi(L,0)}{\partial |\nabla h|^2}\right| > 0.4\mu\varepsilon^2. \qquad (17)$$

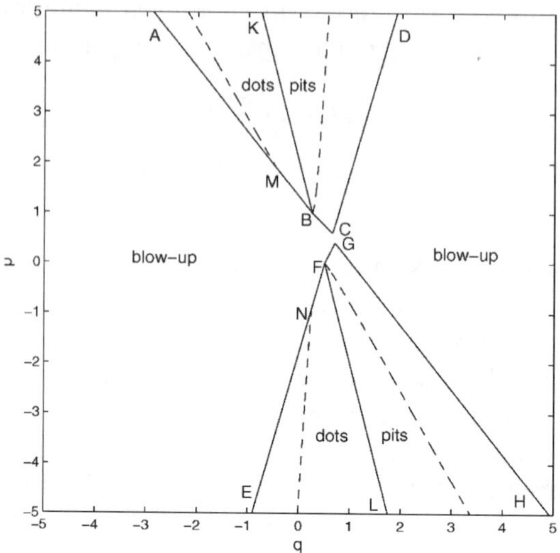

Figure 7.   Regions in (q, p)-plane where self-organization of hexagonal arrays of dots or pits can be observed in solutions of eq.(5) in 2D. The points B, C, F, G, M and N are (1/4, 1), (13/20, 3/5), (1/2, 0), (7/10, 2/5), (-0.46, 1.91), (0.24,-0.91), respectively.

This conclusion is related to the formation of only those regular structures whose characteristic scale is much larger than the critical film thickness.

Numerical solutions of eq.(5) in 2D by means of a pseudospectral method with periodic boundary conditions are shown in Fig.8. One observes the formation of hexagonal arrays of dots or pits in the parameter regions predicted by the weakly nonlinear analysis. It is interesting that, similar to the 1+1 case, the formation of two types of dots is possible: "cone"-like and "cap"-like (Fig.8a,b). With the increase of the supercriticality "cones" transform into "caps". Similarly, the formation of two types of pits is observed: "anticones" at small supercriticality and "anticaps" at larger ones (Fig.8c,d).

Wires, rings and other surface structures.   It is known that hexagonal patterns can become unstable with respect to patterns with other symmetries [13, 14]. This instability is determined by the Landau coefficient, $\lambda(\phi)$, governing the nonlinear interaction between two modes characterized by two wavevectors, $\mathbf{k}_1$ and $\mathbf{k}_2$, $k_1 = k_2 = k_c$, $\mathbf{k}_1 \cdot \mathbf{k}_2 = k_c^2 \cos \phi$. The Landau coefficient is a function of the angle ϕ between the two wavevectors. Obviously, $\lambda(\phi)$ is defined for $0 < \phi < \pi$ and $\lambda(\phi) = \lambda(\pi - \phi)$. Standard multiple-scale analysis

Figure 8.   Numerical solutions of eq.(5) for various parameter values: (a) hexagonal array of "cone"-type dots, $p = 2.0$, $q = -0.5$, $g = 0.2$; (b) hexagonal array of "cap"-type dots, $p = -4.0$, $q = 1.0$, $g = 0.01$; (c) hexagonal array of "anticone"-type pits, $p = -4.0$, $q = 2.0$, $g = 0.24$; (d) hexagonal array of "anticap"-type pits, $p = -4.0$, $q = 2.0$, $g = 0.2$.

of eq.(5) near the bifurcation threshold $g = g_c$ yields

$$\lambda(\phi) = (-\frac{3}{2} + p + 2q)[\alpha_+(\phi) + \alpha_-(\phi)]$$
$$+(-1 + p)\cos\phi[\alpha_+(\phi) - \alpha_-(\phi)], \tag{18}$$
$$\alpha_\pm(\phi) = \frac{1 \pm p\cos\phi - 2q}{(\cos\phi \pm 1/2)^2}.$$

Note that the function $\lambda(\phi)$ is singular at $\phi = \pi/3$, $2\pi/3$ due to the resonant quadratic interaction between these modes which is responsible for the formation of a hexagonal pattern and is not taken into account in the computation of $\lambda(\phi)$. The Landau (cubic) interaction coefficient in this case is equal to λ_1 in (14). Note also that $\lambda(0) = \lambda_0 + sw/m$, which explains the stability condition (10).

Stability of a hexagonal pattern with respect to patterns with other symmetries was investigated in a number of works (see, e.g. [13, 14] and references therein). It was shown that, with the increase of the supercriticality, a hexagonal pattern can become unstable with respect to a stripe pattern if $\lambda_1/\lambda_0 > 1$.

The presence of the zero mode, as well as the presence of the quadratic non-linear terms with the coefficients r_1 and r_2 in eqs. (12) (that characterize the dependence of the resonant quadratic interaction coefficient on the mode wavevectors [18]), can promote the instability [18, 16]. Results of the numerical simulations of eq.(5) that show the transition from hexagonal arrays of dots or pits to stripe patterns ("wires") with the increase of the supercriticality are shown in Fig.9. Transition from dots to wires in epitaxially strained films has been observed in experiments [19].

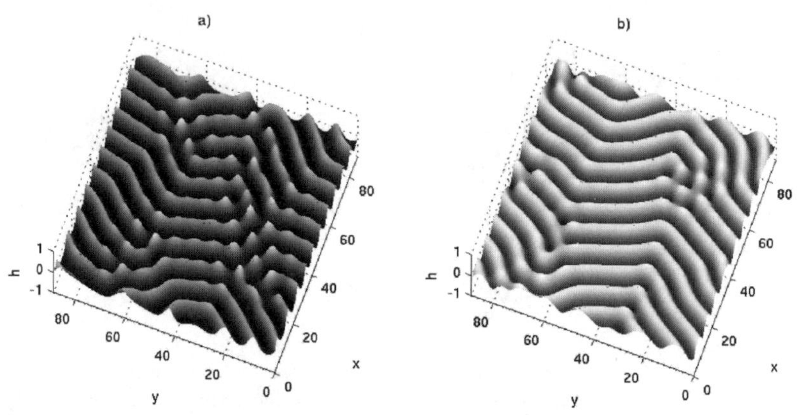

Figure 9.   Formation of "wire" patterns via transition from (a) dots, $p = 4.0$, $q = -1.0$, $g = 0.1$; (b) pits, $p = -4.0$, $q = 2.0$, $g = 0.1$.

An interesting surface structure can develop in the parameter regions where $\lambda_0 > 0$, $\lambda_1 > 0$ and $\lambda(\pi/2) < 0$. In this case, there is a strong interaction between the modes with orthogonal wavevectors that can lead to a resonant coupling between two system of hexagons whose basic wavevector sets, $(\mathbf{k}_1, \mathbf{k}_2, \mathbf{k}_3)$ and $(\mathbf{k}_4, \mathbf{k}_5, \mathbf{k}_6)$, each forming an equilateral triangle, are mutually orthogonal. This can lead to the formation of a dodecagonal quasiperiodic pattern [20, 21]. The numerical solution of eq.(5) show that such quasiperiodic dodecagonal arrangement of dots can indeed form; it is shown in Fig.10a. However, this dodecagonal structure occurs only at the beginning of pattern formation; later in time it either gets replaced by a hexagonal structure, or grows further and ultimately blows up. The intermediate stage of the blow-up is shown in Fig.10b.

In the parameter regions where the hexagonal patterns are subcritical, or unstable according to the condition (16), solutions of eq.(5) blow up in a finite time. Depending on the resonant interaction coefficient r_0, the blow-up occurs through the formation of high, spatially localized mounds, or deep pits.

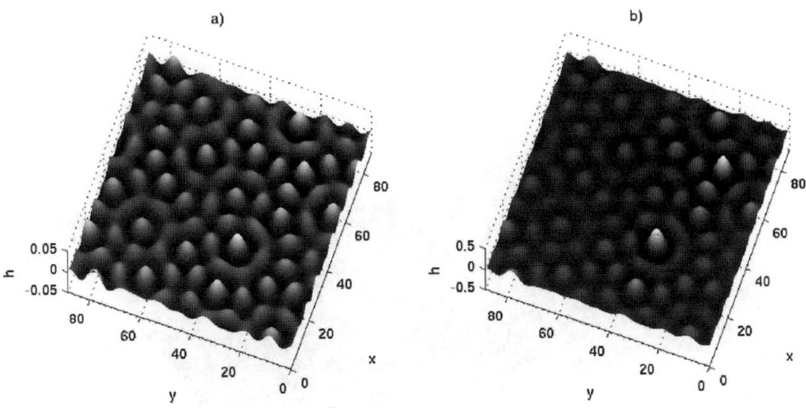

Figure 10. Formation of a quasiperiodic array of dots with dodecagonal symmetry, $p = -3.0$, $q = -0.1$, $g = 0.249$; (a) initial stage of formation; (b) intermediate stage of the blow-up.

The latter develop into the structures, shown in Fig.11, that strongly resemble "quantum rings" [22] or "quantum fortresses" [17] observed in experiments.

To conclude, wetting interactions between an epitaxial solid film and a solid substrate change the instability spectrum from the long-wave type to a short-wave type and make the formation of spatially-regular arrays of quantum dots possible. In the next section we shall consider self-assembly of quantum dots driven by a different mechanism: anisotropy of the film surface energy. We shall show that the presence of wetting interactions between the film and the substrate can in this case also lead to self-organization of regular arrays of anisotropic islands.

4.   Surface-energy anisotropy and wetting interactions

Now let us discuss another mechanism that can lead to the formation of quantum dots, the one associated with the anisotropy of the film surface energy. The material of this section is discussed in more detail in [23]. (The figures are reprinted with permission from [23], ©2004 by the American Physical Society).

Consider a thin, solid film grown on a solid substrate where the lattice mismatch between the two materials is negligible, the surface energy γ of the film is strongly anisotropic, the film wets the substrate and it is thin enough for the wetting interaction energy to affect the chemical potential of the film.

Let us assume that the substrate determines the initial crystallographic orientation of the free surface of a growing film. Let us also assume that *in the*

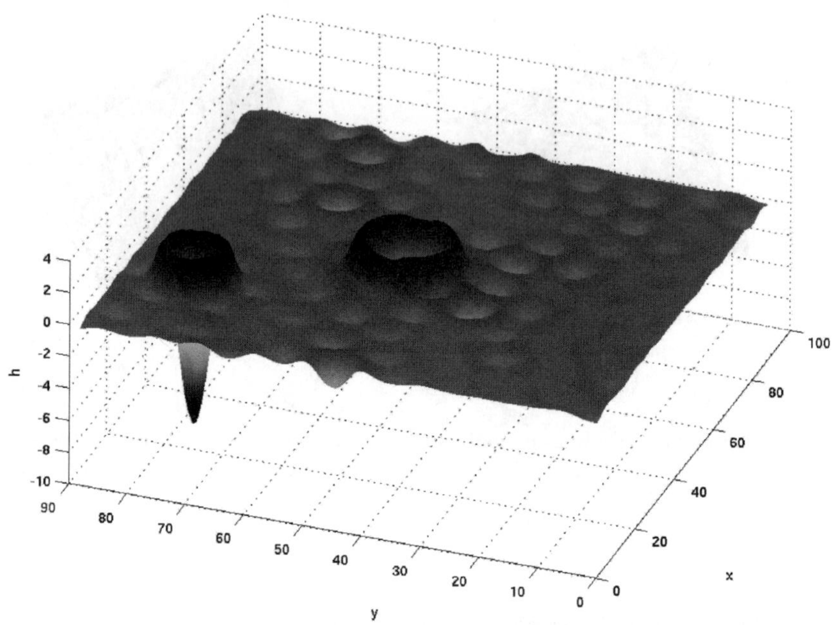

Figure 11.   Intermediate stage of the blow-up solution of eq.(5) in the form of a "quantum ring"; $p = 2.0$, $q = 1.0$, $g = 0.2$.

absence of the substrate, or when the film is thick enough so it does not "feel" the substrate, this orientation would be in the range of "forbidden orientations."

Here we consider only high-symmetry orientations, such as [001] and [111]. In this case, the forbidden orientation implies that the surface-stiffness tensor [24, 8],

$$\tilde{\gamma}_{\alpha\beta} = \gamma\delta_{\alpha\beta} + \frac{\partial^2\gamma}{\partial\theta_\alpha\partial\theta_\beta}, \tag{19}$$

is diagonal for this orientation and has two equal negative components,

$$\tilde{\gamma}_{11} = \tilde{\gamma}_{22} \equiv -\sigma < 0. \tag{20}$$

(Here $\theta_{\alpha,\beta}$ are the surface angular coordinates and $\delta_{\alpha\beta}$ is the Kroneker delta). In the absence of wetting interactions between the film and the substrate, such a surface is thermodynamically unstable and exhibits spontaneous formation of pyramidal "faceted" structures that coarsen in time [10]. However, as we show below, the presence of wetting interactions can *suppress* this instability, or qualitatively change it, so that it would lead to the formation of spatially-regular arrays of islands.

The continuum evolution of the film free surface can be described by evolution equation (1) where the chemical potential

$$\mu_s = \frac{\delta \mathcal{F}}{\delta h},$$

(21)

and $\mathcal{F}$ is the free energy functional. In the *absence* of elastic stresses and wetting interactions between the film and the substrate,

$$\mathcal{F} = \int [\mu_0 h + I(h_x, h_y) + \frac{1}{2}\nu(\Delta h)^2]\, dx dy,$$

(22)

where μ_0 is the volume part of the free energy (μ_0 is the constant chemical potential of a planar film), $I = \gamma(h_x, h_y)\sqrt{1 + (\nabla h)^2}$ is the weighted anisotropic surface energy that depends on the local surface slope, and ν is the regularization coefficient that measures the energy of edges and corners [9, 10, 25] (for simplicity, we write this term in the small-slope approximation that will be further employed here). The free energy (22) gives the chemical potential

$$\mu_s = \mu_0 + \mu_\gamma \equiv \mu_0 + \tilde{\gamma}_{\alpha\beta} C_{\alpha\beta} + \nu \Delta^2 h,$$

(23)

where $C_{\alpha\beta}$ is the surface curvature tensor.

In the presence of wetting interactions between the film and the substrate, the film chemical potential strongly depends on the film thickness h for $h \sim \delta_w$, where δ_w is the characteristic wetting length, and $\mu \to \mu_0$ for $h \gg \delta_w$. In this case the film free energy can be written as

$$\mathcal{F} = \int [f(h, h_x, h_y) + \frac{1}{2}\nu(\Delta h)^2]\, dx dy,$$

(24)

where $f(h, h_x, h_y) \to \mu_0 h + I(h_x, h_y)$ for $h \gg \delta_w$. The *wetting* part of the free energy can be then defined as

$$\mathcal{F}_w = \int [f(h, h_x, h_y) - \mu_0 h - I(h_x, h_y)]\, dx dy.$$

(25)

Here we consider the following two models for wetting interactions between the film and the substrate.

A *two-layer wetting model*, according to which the wetting interactions between the film and the substrate are described as a thickness-dependent surface energy of the film, $\gamma(h)$. This dependence is usually taken to be [26]

$$\gamma(h) = \gamma_f + (\gamma_s - \gamma_f)\exp(-h/\delta),$$

(26)

where $\gamma_s = const$ is the surface energy of the substrate in the absence of the film, γ_f is the energy of the film free surface far from the substrate, and

δ is the characteristic wetting length. This model is consistent with *ab initio* calculations [27]. For anisotropic surface energy of the film,

$$\gamma_f = \gamma_f^0 \left[1 + \varepsilon(h_x, h_y)\right], \tag{27}$$

where $\gamma_f^0 = const$ and $\varepsilon(h_x, h_y)$ is the anisotropy function that depends on the orientation of the film surface. Thus, in this model the free energy density in (24) is $f(h, h_x, h_y) = \gamma(h, h_x, h_y)\sqrt{1 + |\nabla h|^2}$, and the chemical potential is computed as $\mu_s = \mu_\gamma + \mu_w$, where μ_γ is defined by (23) and

$$\mu_w = \frac{\frac{\partial \gamma}{\partial h} - [\frac{\partial^2 \gamma}{\partial h \partial h_x} h_x + \frac{\partial^2 \gamma}{\partial h \partial h_y} h_y](1 + |\nabla h|^2)}{\sqrt{1 + |\nabla h|^2}} \tag{28}$$

Note that in this case $\tilde{\gamma}_{\alpha\beta}$ in μ_γ depends on h.

A *glued wetting-layer model*, that considers isotropic wetting free energy, additive to the anisotropic surface energy, yielding $\mu_s = \mu_\gamma + \mu_w$, with μ_γ defined by (23) and μ_w being an exponentially decaying function of h that has a singularity at $h \to 0$:

$$\mu_w = -w \, (h/\delta)^{-\alpha_w} \exp\left(-h/\delta\right). \tag{29}$$

Here $w > 0$ characterizes the "strength" of the wetting interactions, and $\alpha_w > 0$ characterizes the singularity of the wetting potential at $h \to 0$. This singularity is a simple continuum phenomenological model of a very large potential barrier for removal of an ultra-thin (possibly monolayer) wetting layer that persists between islands.

Thus, in the small-slope approximation and for high-symmetry orientations, the surface chemical potential in both of these models have the same form,

$$\mu_s = \mu_\gamma^0 + \mu_w, \tag{30}$$

where $\mu_\gamma^0 = \mu_\gamma(h_0)$ is defined by (23) and evaluated at the initial film thickness h_0, and the part of chemical potential due to wetting can be expanded as

$$\mu_w = W_0(h) + W_2(h)\,(\nabla h)^2 + W_3(h)\nabla^2 h + \ldots, \tag{31}$$

where $W_{0,2,3}(h)$ are smooth functions, rapidly (exponentially) decaying with the increase of h, $W_3(h_0) = 0$, and $2W_2 = dW_3/dh$ (due to (21)).

In the small-slope approximation, and in the particular cases of high-symmetry orientations ([001] or [111]) of a crystal with cubic symmetry, the evolution equation (1) for the film thickness can be written in the following form:

$$\begin{aligned} \partial_t h \;=\; & M\Delta[\sigma\Delta h + \nu\Delta^2 h - \Gamma_{ijk}[h] \\ & + W_0(h) + W_2(h)\,(\nabla h)^2 + W_3(h)\Delta h], \end{aligned} \tag{32}$$

where for the orientations [001] and [111] the nonlinear differential operator $\Gamma_{ijk}[h]$ has the following forms [10], respectively,

$$
\begin{aligned}
\Gamma_{001} &= (a\,h_x^2 + b\,h_y^2)\,h_{xx} + (b\,h_x^2 + a\,h_y^2)\,h_{yy} \\
&\quad + 4b\,h_x h_y h_{xy},
\end{aligned}
\tag{33}
$$

$$
\begin{aligned}
\Gamma_{111} &= a\,[h_x^2 h_{xx} + h_y^2 h_{yy} + 2h_x h_y h_{xy}] \\
&\quad + \frac{a}{3}\,[h_y^2 h_{xx} + h_x^2 h_{yy} - 2h_x h_y h_{xy}] \\
&\quad + b\,[(h_{xx} - h_{yy})h_y + 2h_{xy}h_x].
\end{aligned}
\tag{34}
$$

Here the coefficients a and b characterize the surface-energy anisotropy and can be computed from the surface-energy dependence on the surface orientation. Naturally, the nonlinear operator Γ_{001} is invariant with respect to rotations by $\pi/2$, as well as any of the transformations $x \to -x,\ y \to -y,\ x \to y$, while Γ_{111} is invariant with respect to rotations by $2\pi/3$ as well as the transformation $y \to -y,\ b \to -b$. The functions $W_{0,2,3}(h)$ are determined by the type of a wetting interaction model and can also differ for different orientations of the film surface.

Note that eq.(32) with the nonlinear operators Γ_{ijk} defined by (33), (34) can be written in a variational form

$$
\partial_t h = M \nabla^2 \frac{\delta \mathcal{F}}{\delta h}
\tag{35}
$$

where $\mathcal{F} = \int F\,dx dy$, and the free energy density

$$
\begin{aligned}
F &= -\frac{\sigma}{2}(\nabla h)^2 + \frac{\nu}{2}(\Delta h)^2 + \mathcal{G}_{ijk} \\
&\quad + \int W_0(h)\,dh - \frac{1}{2}W_2(h)(\nabla h)^2,
\end{aligned}
\tag{36}
$$

with

$$
\mathcal{G}_{001} = \frac{a}{12}\,(h_x^4 + h_y^4) + \frac{b}{2}\,h_x^2 h_y^2,
\tag{37}
$$

$$
\mathcal{G}_{111} = \frac{a}{12}\,(\nabla h)^4 + \frac{b}{6}\,(3h_x^2 h_y - h_y^3).
\tag{38}
$$

Below we investigate the stability and nonlinear dynamics of the solid-film surface governed by eq.(32).

Faceting instability in the presence of wetting interactions.

Consider infinitesimal perturbations of a planar film surface, $h = h_0 + \tilde{h}\,e^{i\,\mathbf{k}\cdot\mathbf{x} + \omega t}$, and linearize eq.(32) to obtain the following dispersion relation between the perturbation growth rate, ω, and the wavevector $\mathbf{k}$:

$$
\omega = M(-W_{01}k^2 + \sigma k^4 - \nu k^6),
\tag{39}
$$

where $k = |\mathbf{k}|$ and

$$W_{01} = \left(\frac{\partial W_0}{\partial h}\right)_{h=h_0}. \tag{40}$$

One can see that if the film wets the substrate, i.e. when $W_{01} > 0$, the wetting interactions suppress the long-wave faceting instability caused by the surface-energy anisotropy. The instability occurs only for

$$\frac{\sigma^2}{4\nu W_{01}} > 1, \tag{41}$$

i.e. if either the wetting interaction is less than the threshold value, $W_{01} < W_{01}^c = \sigma^2/(4\nu)$, or the surface stiffness is larger than the threshold value, $\sigma > \sigma_c = 2\sqrt{W_{01}\nu}$. At the instability threshold, the wavelength of the unstable perturbations λ is *finite*, $\lambda = \lambda_c = 2\pi/k_c$, where

$$k_c = \sqrt{\frac{\sigma}{2\nu}}. \tag{42}$$

Typical dispersion curves defined by (39) are qualitatively the same as shown in Fig.2b. Note that the critical wavenumber at the threshold does not depend on the wetting potential and is determined only by the surface stiffness and the energy of edges and corners. For the parameter values typical of semiconductors like Si or Ge, with the surface energy $\gamma \sim 2.0\,\mathrm{Jm}^{-2}$, surface stiffness $\sigma \sim 0.2\,\mathrm{Jm}^{-2}$, the lattice spacing $a_0 \sim 0.5\,\mathrm{nm}$ and the regularization parameter $\nu \sim \gamma a_0^2 \sim 5.0 \times 10^{-19}\,\mathrm{J}$, the wavelength of the structure at the onset of instability is $14.0\,\mathrm{nm}$.

Thus, in the presence of the wetting interactions with the substrate, the faceting instability becomes *short-wave*. This is qualitatively different from the case of the faceting instability in the absence of the wetting interactions when the instability is *long-wave*, i.e. when all perturbations whose wavelengths are larger than a certain threshold are unstable. In other words, wetting interactions with the substrate change the faceting instability from the spinodal decomposition type [9] (*long-wave*) to the Turing type [14] (*short-wave*), thus leading to the possibility of changing the system evolution from Ostwald ripening (coarsening) to the formation of spatially regular patterns.

Formation of surface structures: 1+1 case

Here we discuss the nonlinear evolution of surface structures resulting from the faceting instability in the presence of wetting interactions with the substrate in a 1+1 case of a two-dimensional film with a one-dimensional surface. In this case, the evolution equation (32) for the shape of the film surface, after the rescaling $x \to (\nu/\sigma)^{1/2}\, x$, $t \to [\nu^2/(M\sigma^3)]\, t$, $h \to (\nu/a)^{1/2}\, h$, becomes

$$\partial_t h = [h_{xx} + h_{xxxx} - h_x^2 h_{xx}$$
$$+ w_0(h) + w_2(h)\, h_x^2 + w_3(h)\, h_{xx}]_{xx}, \tag{43}$$

where $w_{0,2,3}(h)$ are the rescaled functions $W_{0,2,3}(h)$, respectively ($w_3(h_0) = 0$, $2w_2 = dw_3/dh$). In this scaling, the instability occurs for $(\partial w_0/\partial h)_{h=h_0} \equiv w_{01} < 1/4$ at the wavenumber $k_c = \sqrt{2}/2$.

First, let us discuss the evolution near the instability threshold by means of weakly nonlinear analysis. Consider $w_{01} = 1/4 - 2\epsilon^2$, $\epsilon \ll 1$, introduce the long-scale coordinate $X = \epsilon x$ and the slow time $T = \epsilon^2 t$, and expand

$$\tilde{h} = h - h_0 = \epsilon[A(X,T)e^{ik_c x} + c.c.]$$
$$+ \epsilon^2[A_2(X,T)e^{2ik_c x} + B(X,T) + c.c.] + \dots, \tag{44}$$
$$w_0(h) = w_{00} + w_{01}\tilde{h} + w_{02}\tilde{h}^2 + w_{03}\tilde{h}^3 + \dots, \tag{45}$$
$$w_2(h) = w_{20} + w_{21}\tilde{h} + \dots, \tag{46}$$
$$w_3(h) = w_{31}\tilde{h} + w_{32}\tilde{h}^2 + \dots, \tag{47}$$

where $w_{31} = 2w_{20}$, $w_{32} = w_{21}$. Substitute (44)-(47) into eq.(43) to obtain the corresponding problems in the successive orders of ϵ. From the problem at second order one finds

$$A_2 = \frac{2}{9}(3w_{20} - 2w_{02})\, A^2. \tag{48}$$

As the solvability condition at the third order, one obtains the evolution equation for the complex amplitude of the unstable, spatially-periodic mode, $A(X,T)$. The solvability condition at the fourth order yields the evolution equation for the real amplitude $B(X,T)$ of the zero mode associated with the conservation of mass. Together, the two equations form the system of coupled equations (7) with

$$\lambda_0 = \frac{1}{8} - \frac{1}{9}(3w_{20} - 2w_{02})^2 - \frac{1}{2}w_{21} + \frac{3}{2}w_{03}, \tag{49}$$

$$s = \frac{1}{2}w_{20} - w_{02}, \quad m = \frac{1}{4}, \quad w = -2s. \tag{50}$$

The system of amplitude equations (7) has a stable, stationary solution, $A = \lambda_0^{-1/2}$, $B = 0$, corresponding to spatially periodic pattern (array of dots), if [16, 11]

$$\lambda_0 > 8s^2 = 2\,(w_{20} - 2w_{02})^2. \tag{51}$$

Condition (51) defines a region in the parameter space in which one can observe the formation of stable periodic arrays of dots. First, consider a *glued-layer* wetting potential defined by (29). From (41) one obtains that the planar film surface becomes unstable with respect to periodic structure for

$$-\frac{\sigma^2\delta}{w\nu} > 4(\alpha_w + \zeta)\,\zeta^{-(\alpha_w+1)}e^{-\zeta}, \tag{52}$$

where $\zeta = h_0/\delta$. Since for the wetting potential (29) $w_2(h) = w_3(h) \equiv 0$, one obtains from (51) and (52) that a near-threshold periodic surface structure is stable if

$$\frac{a\delta^2}{\nu} > f(\zeta, \alpha_w), \tag{53}$$

where

$$f(\zeta, \alpha_w) = [18\,\zeta^2(\zeta + \alpha_w)^2]^{-1}[10\zeta^4 + 40\alpha_w\zeta^3$$
$$+\alpha_w(11 + 60\alpha_w)\zeta^2 + 2\alpha_w(20\alpha_w^2 + 11\alpha_w - 9)\zeta$$
$$+\alpha_w^2(10\alpha_w^2 + 11\alpha_w + 1)]. \tag{54}$$

Conditions (52) and (53) are shown in Fig.12.

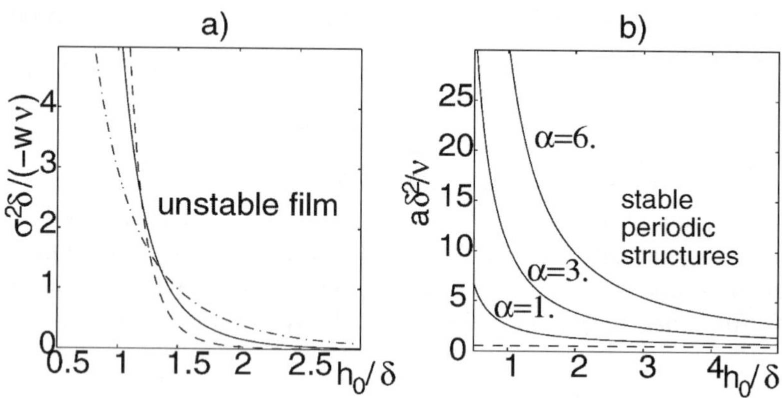

Figure 12.　　(a) Parameter regions where a planar film surface is unstable (above the corresponding curves) for $\alpha_w = 6.0$ (dashed line), $\alpha_w = 3.0$ (solid line) and $\alpha_w = 1.0$ (dashed-dotted line). (b) Parameter regions where weakly nonlinear periodic surface structures are stable (above the corresponding curves) for different values of α_w.

Now consider a *two-layer* wetting potential defined by (26) with

$$\gamma_f = \gamma_f^0[1 + \varepsilon\cos 4(\theta_0 + \theta)], \tag{55}$$

where $\theta = \arctan(h_x)$ and θ_0 corresponds to the orientation of the planar surface of the film, parallel to the substrate; for the high-symmetry orientations

[01], [11] $\theta_0 = 0$, $\pi/4$, respectively. It is convenient to introduce the following dimensionless parameters:

$$\Gamma = \frac{\gamma_{f0}}{\gamma_s}(15\varepsilon - 1), \ \zeta = \frac{h_0}{\delta}, \ \tilde{\varepsilon}_1 = \frac{\varepsilon + 1}{15\varepsilon - 1}, \ \tilde{\varepsilon}_2 = \frac{95\varepsilon - 1}{15\varepsilon - 1}.$$

The film wets the substrate if $\Gamma < \tilde{\varepsilon}_1^{-1}$, or $\gamma_s/\gamma_{f0} > \varepsilon + 1$. In this case, the nonlinear anisotropy coefficient a in eq.(32) is always positive. The faceting instability requires a negative surface stiffness that can be achieved only if $15\varepsilon - 1 > 0$, and

$$\zeta > \ln(1 + \Gamma^{-1}). \tag{56}$$

The instability threshold condition (41) gives

$$\frac{\gamma_s \delta^2}{\nu} \geq \frac{4\,e^\zeta\,[1 - \Gamma\tilde{\varepsilon}_1]}{[\Gamma(e^\zeta - 1) - 1]^2}. \tag{57}$$

The analysis of the conditions (56) and (57) shows that the short-wave instability of the film surface that can lead to pattern formation can occur only if the film thickness is above a threshold value determined only by the surface-energy anisotropy and the wetting length, namely, for

$$h_0 > \delta \ln\left[\frac{16\varepsilon}{15\varepsilon - 1}\right]. \tag{58}$$

Using (51) one can show that the weakly-nonlinear periodic structure is stable if

$$\frac{\gamma_s \delta^2}{\nu} > f(\zeta, \Gamma, \varepsilon), \tag{59}$$

where

$$f = \frac{2}{27} \frac{\Gamma^2(5e^{2\zeta} + 14e^\zeta + 35) + 14\Gamma(e^\zeta + 5) + 35}{e^{-\zeta}[\Gamma(e^\zeta - 1) - 1]^2\,[\tilde{\varepsilon}_2\Gamma(e^\zeta - 1) - 1]} \tag{60}$$

The conditions (56)-(60) allow one to determine regions in the (Γ, ζ) parameter plane where spatially-regular surface structures can occur as a result of thermodynamic instability of the film surface caused by strongly-anisotropic surface-tension in the presence of wetting interaction described by the two-layer model (26). Examples of these regions for different values of the anisotropy parameter ε are shown in Fig.13. Solid lines correspond to the condition (57), the dashed lines correspond to the condition (60). The film is unstable in the regions above the solid lines, and the stable periodic structures can form in the region near the solid line which lies above the dashed curve. One can see that for given values of the surface-energy anisotropy, ε, and the value of $\gamma_s\delta/\nu$, the formation of stable periodic structures occurs if the ratio of the initial film thickness to the wetting length is within a certain interval.

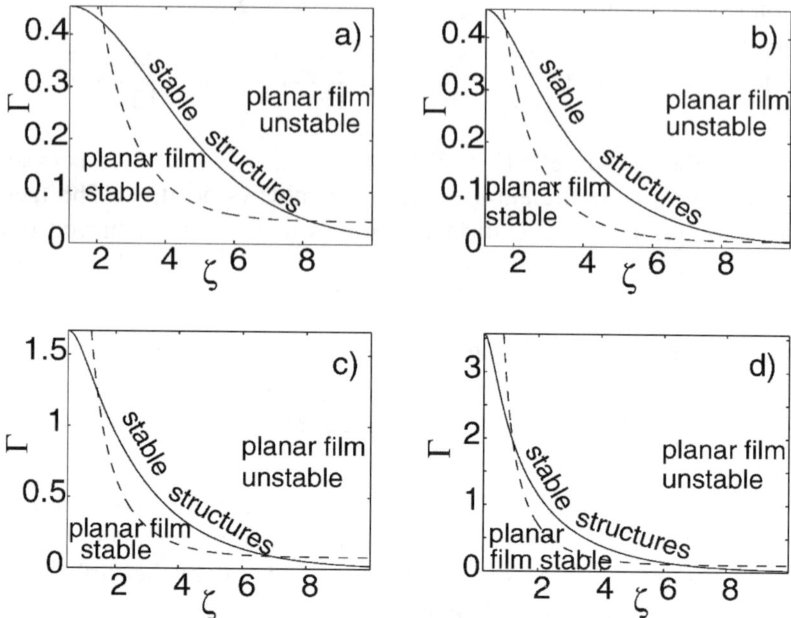

Figure 13.   Parameter regions where a planar film surface is unstable (above the solid line) and where stable periodic structures can form near the instability threshold (near the solidline, above the dashed line): a) $\varepsilon = 0.1$, $\gamma_s\delta^2/\nu = 0.5$; b) $\varepsilon = 0.1$, $\gamma_s\delta^2/\nu = 2.0$; c) $\varepsilon = 0.2$, $\gamma_s\delta^2/\nu = 0.5$; d) $\varepsilon = 0.4$, $\gamma_s\delta^2/\nu = 0.5$;

Numerical solutions of eq.(43) for the case of the glued wetting layer model are shown in Figs.14,15. The shown length scales correspond to the following parameter values, typical of semiconductors like Si or Ge: $\gamma \sim 2.0\,\mathrm{Jm}^{-2}$, $\sigma \sim 0.2\,\mathrm{Jm}^{-2}$, $\delta \sim 1.5\,\mathrm{nm}$, and the estimates of $\alpha = 3.0$ and $\nu \sim \gamma a_0^2 \sim 5.0 \times 10^{-19}\,\mathrm{J}$, where $a_0 \sim 0.5\,\mathrm{nm}$ is the crystal lattice spacing; the nonlinear coefficient of the surface-energy anisotropy, a, the initial film thickness, h_0, and the wetting interaction strength parameter, w, are varied. In experiment, the film thickness is the main parameter that controls the film instability. For example, for $w \sim \Delta\gamma/\delta \sim 6.7 \times 10^7\,\mathrm{Jm}^{-3}$, where $\Delta\gamma \sim 0.1\,\mathrm{Jm}^{-2}$ is the surface-tension difference between the substrate and the film, one finds that if the initial film thickness, h_0, ranges from 1.2 to 5.3 nm, the parameter w_{01} changes from 2.3 to 7.3×10^{-4}, respectively. The instability occurs in this case for a film thicker than $h_0 \approx 1.9\,\mathrm{nm}$, and the wavelength of the structure at the onset of instability is 14.0 nm.

Fig.14 presents the stationary solutions of eq.(43) for different values of the dimensionless wetting parameter, w_{01}, corresponding to different values of the wetting interaction strength, w, and the initial film thickness, h_0. One can see

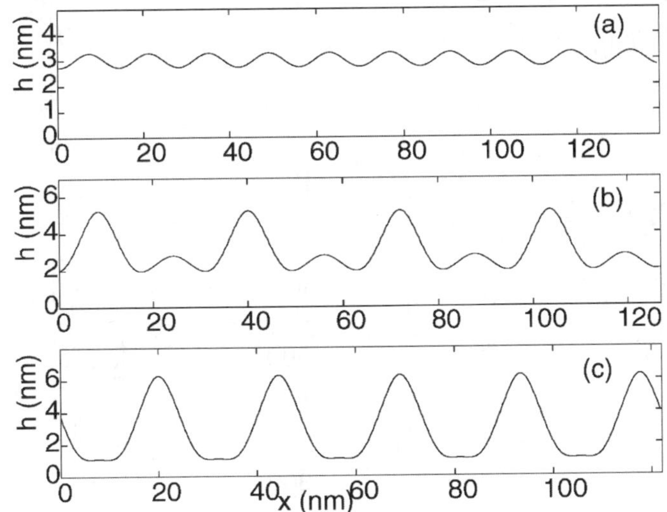

Figure 14. Stationary numerical solutions of eq.(43) with the wetting potential (29) showing stationary surface structures for $h_0 = 3.0\,\mathrm{nm}$, $(\zeta = 2.0)$, $a = 0.17\,\mathrm{Jm}^{-2}$, and (a) $w = 6.8 \times 10^8\,\mathrm{Jm}^{-3}$ ($w_{01} = 0.24$); (b) $w = 2.8 \times 10^8\,\mathrm{Jm}^{-3}$ ($w_{01} = 0.1$); (c) $w = 2.8 \times 10^7\,\mathrm{Jm}^{-3}$ ($w_{01} = 0.01$).

that, near the instability threshold, an almost harmonic small-amplitude periodic structure is formed. Farther from the threshold, the formation of periodic structures with larger amplitude and larger wavelengths can be observed. In the parameter regions where a near-threshold periodic structure with the wavelength corresponding to the most rapidly growing linear mode is unstable, it undergoes coarsening right after formation, and finally evolves into spatially *localized* dots. Different stages of this coarsening process and the formation of localized islands are shown in Fig.15. At the last stage, shown in Fig.15, the coarsening either completely stops or becomes logarithmically slow.

Thus, the presence of wetting interactions between the film and the substrate can suppress the faceting instability of the film surface, that is thermodynamically unstable due to strong anisotropy of the surface energy, and lead to the formation of spatially-regular surface structures, or to the formation of spatially-localized dots divided by a thin wetting layer. Below we consider evolution of a more realistic, 2+1 system.

Formation of surface structures: 2+1 case

Here we discuss the nonlinear evolution of surface structures resulting from the faceting instability of a three-dimensional film with a two-dimensional

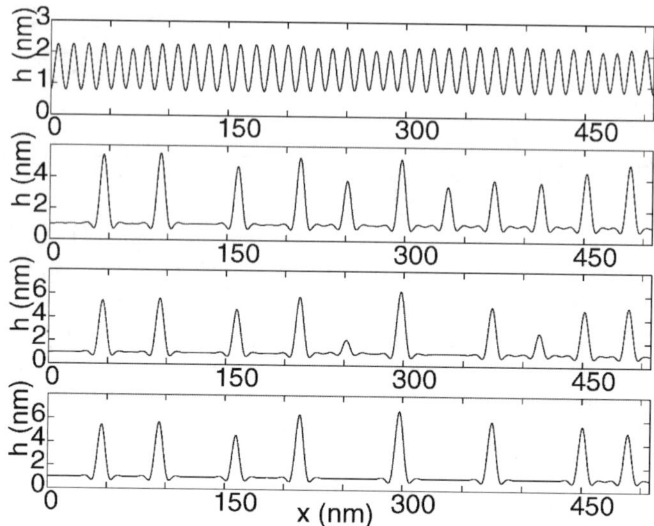

Figure 15.   Different stages of coarsening of the initial periodic structure, yielding the formation of localized dots divided by a thin wetting layer – numerical solution of eq.(43) with the wetting potential (29); $h_0 = 1.5\,\mathrm{nm}$, $w = 8.2 \times 10^6\,\mathrm{Jm}^{-3}$, ($\zeta = 1.0$, $w_{01} = 0.1$), $a = 0.22\,\mathrm{Jm}^{-2}$.

surface (2+1 case) in the presence of wetting interactions with the substrate. We consider high-symmetry orientations only, [001] and [111], described by eq.(32).

After the appropriate rescaling, eq.(32) can be written as

$$\partial_t h \;=\; \Delta\{\Delta h + \Delta^2 h - g[h]$$
$$+\,w_0(h) + w_2(h)\,(\nabla h)^2 + w_3(h)\,\Delta h\}, \tag{61}$$

where the nonlinear differential operator $g[h]$ for [001] orientation is

$$g_{001} = (h_x^2 + p\,h_y^2)\,h_{xx} + (h_y^2 + p\,h_x^2)\,h_{yy} + 4p\,h_x h_y h_{xy}, \tag{62}$$

and for [111] orientation it is

$$g_{111} \;=\; (h_x^2 + \frac{1}{3}\,h_y^2)\,h_{xx} + (h_y^2 + \frac{1}{3}\,h_x^2)\,h_{yy} + \frac{4}{3}\,h_x h_y h_{xy}$$
$$+\,q\,[(h_{xx} - h_{yy})h_y + 2h_{xy}h_x]. \tag{63}$$

Eq.(61) has a special structure in that the linear operator is isotropic, while the nonlinear operator is anisotropic. The linear growth rate near the instability threshold, thus, does not depend on the wavevector orientation and the resulting dispersion relation is the same as in the 1+1 case, $\omega = -w_{01}k^2 + k^4 - k^6$, with the instability threshold $w_{01} = 1/4$ at $k = k_c = \sqrt{2}/2$. It is the non-linear interaction between the modes that will determine the symmetry of the emerging pattern. This situation is similar to the one considered in [28] where the effect of surface-energy anisotropy on the formation of cellular patterns with different symmetries at a crystal-melt interface caused by morphological instability during directional solidification was studied. First we consider the weakly nonlinear analysis near the instability threshold.

Since the linear operator of eq.(61) is isotropic and the nonlinear operator of eq.(61) has a quadratic nonlinearity that breaks $h \to -h$ symmetry, the preferred pattern near the instability threshold will have a *hexagonal* symmetry, caused by the quadratic resonant interaction between three different modes oriented at $120°$ with respect to one another and having the same linear growth rate. The specific type of the pattern in this case is determined by the phase locking of the three resonant modes that depends on the quadratic resonant interaction coefficient. In order to compute this coefficient, take $w_{01} = 1/4 - 2\gamma\epsilon$, $\epsilon \ll 1$, introduce the slow time $\tau = \epsilon t$, and use the expansions (45)-(47), as well as the expansion

$$
\begin{aligned}
h \;=\; & \epsilon \sum_{n=1}^{3} A_n(\tau) e^{\mathbf{k}_n \cdot \mathbf{r}} + \epsilon^2 \sum_{n=1}^{3} B_n(\tau) e^{\mathbf{k}_n \cdot \mathbf{r}} \\
& + \epsilon^2 \sum_{n=1}^{3} [B_{n,n}(\tau) e^{2\mathbf{k}_n \cdot \mathbf{r}} + B_{n,n-1}(\tau) e^{(\mathbf{k}_n - \mathbf{k}_{n-1}) \cdot \mathbf{r}}] \\
& + c.c. + O(\epsilon^3)
\end{aligned}
\tag{64}
$$

where $A_n(\tau)$, $B_n(\tau)$, $B_{n,n}(\tau)$, $B_{n,n-1}(\tau)$ are complex amplitudes (the spatially-uniform mode $B_{n,-n}$ is missing due to the conservation of mass), $\mathbf{r}$ is a vector in the (x, y)-plane, $k_n = 1/\sqrt{2}$ and $\mathbf{k}_1 + \mathbf{k}_2 + \mathbf{k}_3 = 0$ ($n = 0$ and $n = 3$ correspond to the same mode with the wavevector $\mathbf{k}_3$). We neglect here spatial modulations of the pattern near the instability threshold. The solvability condition for the problem for B_n in the order ϵ^2 yields the following three evolution equations for the amplitudes $A_{1,2,3}$:

$$
\partial_\tau A_1 = \gamma A_1 + \alpha A_2^* A_3^*,
\tag{65}
$$

where the other two equations are obtained by the cyclic permutation of the indices in eq.(65). The resonant quadratic interaction coefficient is different for different surface orientations:

$$
\alpha^{001} \;=\; \frac{3}{4} w_{20} - w_{02},
\tag{66}
$$

$$
\alpha^{111} \;=\; \alpha^{001} - i\frac{q}{4\sqrt{2}} \sin 3\phi_0,
\tag{67}
$$

where the angle ϕ_0 characterizes the orientation of the resonant triad $\{\mathbf{k}_1, \mathbf{k}_2, \mathbf{k}_3\}$ in the surface plane, $\mathbf{k}_1 = (\cos\phi_0, \sin\phi_0)$. Thus, in the case of the [001] surface, the quadratic mode interaction is isotropic, while in the case of [111] surface it depends on the pattern orientation within the [111] plane.

For equilateral patterns $A_k = \rho e^{i\theta_k}$ and using $\alpha = |\alpha| e^{i\delta}$ one obtains from (65) the following system of equations for ρ and $\Theta = \theta_1 + \theta_2 + \theta_3$:

$$\partial_\tau \rho = \gamma\rho + |\alpha|\rho^2 \cos(\Theta - \delta), \tag{68}$$
$$\partial_\tau \Theta = -3\rho|\alpha| \sin(\Theta - \delta). \tag{69}$$

Eq.(69) has two critical points: stable, $\Theta = \delta$, and unstable, $\Theta = \pi + \delta$. Thus, the system (68)-(69) describes an unbounded growth of a pattern given by a function

$$\begin{aligned} h &= \rho\left[\cos(\mathbf{k}_1 \cdot \mathbf{x} + \theta_1) + \cos(\mathbf{k}_2 \cdot \mathbf{x} + \theta_2)\right. \\ &+ \left. \cos(\mathbf{k}_3 \cdot \mathbf{x} + \theta_3)\right], \end{aligned} \tag{70}$$

in which the phases are locked: $\theta_1 + \theta_2 + \theta_3 = \delta$. If the resonant interaction coefficient is real then $\delta = 0$ ($\alpha > 0$) or $\delta = \pi$ ($\alpha < 0$), and the function (70) describes a spatially regular array of hexagons with $h > 0$ ($h < 0$) in the centers of the hexagons for $\alpha > 0$ ($\alpha < 0$). Therefore, in the case of the [001] surface when α^{001} is real, one could observe the growth of regular hexagonal arrays of dots for $\alpha^{001} > 0$ or pits for $\alpha^{001} < 0$. Note that for [001] orientation, the pattern type is determined purely by the details of the wetting potential (the coefficients w_{02} and w_{20}) since in this case the anisotropic surface energy enters only through the quartic terms in the free energy functional yielding cubic nonlinear terms in the evolution equation for the surface shape. For example, for a glued wetting potential of the type (29), $w_{20} = 0$, $\alpha^{001} = -w_{02} > 0$ and therefore the formation of only hexagonal arrays of *dots* is possible, an array of pits cannot form.

The situation is different for [111] orientation when the free-energy functional has anisotropic cubic terms leading to anisotropic quadratic terms in the evolution equation for the surface shape and the complex quadratic resonant interaction coefficient. In this case, the imaginary part of the resonant interaction coefficient depends on the surface-energy anisotropy coefficient, q, and the pattern orientation within the [111] plane (angle ϕ_0). As one can see from (68), the most rapidly growing pattern corresponds to the maximum of $|\alpha|$ that is achieved for $\phi_0 = \pi/6$. Thus, one would observe in this case the growth of a pattern described by the function (70) with the phases locked at

$$\Theta = \arctan\left[\frac{q/\sqrt{2}}{4w_{02} - 3w_{20}}\right]. \tag{71}$$

Example of patterns corresponding to different values of Θ are shown in Fig.16 (see also [28]). One can see that for intermediate values of Θ the growing pattern consists of a regular hexagonal array of triangular pyramids. Similar hexagonal arrays of triangular pyramids were observed in experiments reported in [29]. Although the physical mechanism of the formation of ordered arrays of triangular pyramids observed in [29] was different (elastic interaction of multiple epitaxial layers), the *nonlinear mechanism* based on the resonant quadratic interaction of unstable modes in the presence of the anisotropy of the [111] orientation is universal and may well be the same in the system studied in [29].

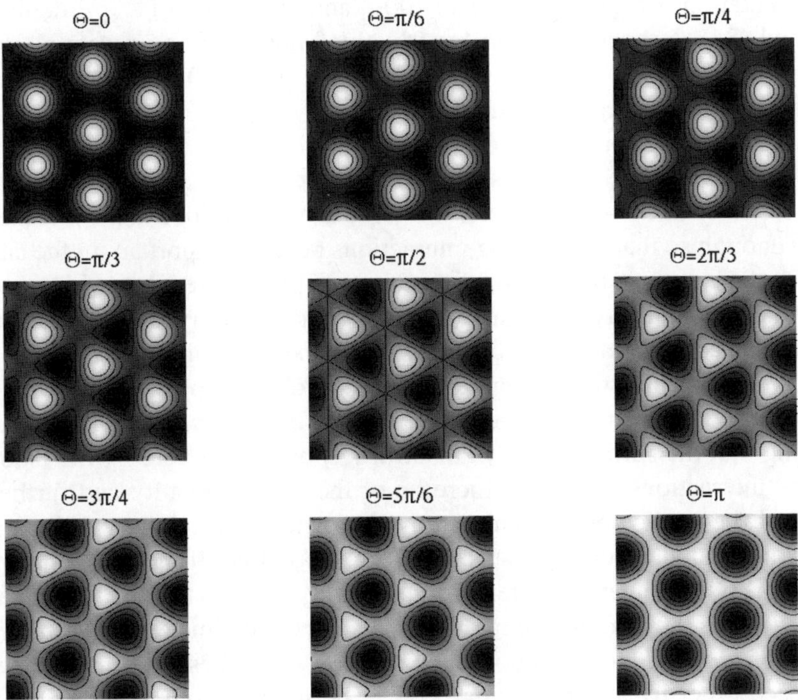

Figure 16. Spatial patterns described by (70) with different values of $\Theta = \theta_1 + \theta_2 + \theta_3$.

The amplitude equations (65) cannot describe the nonlinear stabilization of a growing surface structure and cannot provide conditions for the formation of stable, spatially regular structures near the instability threshold. In order to obtain such conditions higher order (usually cubic) nonlinear terms in the amplitude equations need to be taken into account. However, in the presence of the resonant quadratic interaction, the addition of cubic terms in the amplitude equations near the instability threshold is asymptotically rigorous only if the quadratic interaction coefficient is small, $|\alpha| \sim \epsilon$, which restricts the validity of the weakly nonlinear analysis to a narrow range of physical parameters. The Landau cubic interaction coefficients will be anisotropic and depend on the pattern orientation in the surface plane [28]. Besides, if one allows for long-scale spatial modulations of the patterns, the interaction between the unstable periodic modes and the zero mode will strongly affect the pattern stability (see above) and must be taken into account. The resulting system of amplitude equations with cubic terms, coupled to an equation for the zero mode, will be similar to the system (12)-(13), but will have anisotropic Landau coefficients and anisotropic and complex quadratic coefficients.

Now we shall discuss the results of numerical simulations of eq.(32) for the two orientations of the film surface: [001] and [111], with $\Gamma_{ijk}[h]$ defined by (33) and (34), respectively, and for the glued-layer wetting potential $W_0(h)$ defined by (29) (so that $W_2(h) = W_3(h) = 0$ in (32)). If the initial film thickness h_0 is so large that the film does not "feel" the substrate, $h_0 \gg \delta$, the numerical solutions of eq.(61) exhibit the formation of "faceted" pyramidal structures (square pyramids for [001] surface and triangular pyramids for [111] surface) that coarsen in time, similar to those described in [10]. If the film is thin enough so that the wetting interactions become important, in the case of [001] surface one can observe the formation of spatially-regular (with some defects), hexagonal arrays of rounded dots. We have observed that these arrays of equal-sized dots can be stable for small supercriticality and large enough surface-energy anisotropy (large enough coefficient a together with the ratio a/b in eq.(32)). Example of such stable array is shown in Fig.17a. It is interesting that the surface-energy anisotropy is overcome here by the isotropic wetting interactions. With the increase of the supercriticality and further increase of the surface-energy anisotropy, formation of spatially-regular square arrays of equal-sized dots shown in Fig.17b is possible. If the anisotropy coefficient a is not sufficiently large, hexagonal arrays of dots that are formed at the initial stage of the film instability (Fig.18a) coarsen in time resulting in the formation of rounded localized dots shown in Fig.18b. These dots are connected with each other by a thin wetting layer. The mound slope remains constant during the coarsening. At the late stages, when the dots become localized, the coarsening rate decreases sharply and the coarsening apparently stops.

In the case of the [111] orientation of the film surface we have not observed the formation of regular arrays of dots even near the instability threshold and for large surface-tension anisotropy coefficients. For all studied parameter values we have observed the initial formation of a hexagonal array of triangular pyramids that further coarsened and evolved towards localized triangular pyramidal structures; different stages of the coarsening process are shown in Fig.19. As the localized dots shown in Fig.18b, the localized pyramids here are divided from one another by a thin wetting layer and at the late stages the coarsening apparently stops. It is interesting that, unlike [001] surface orientation, the

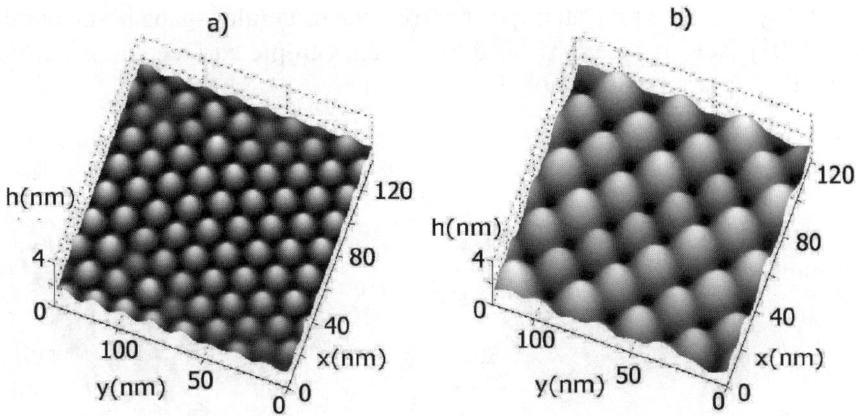

Figure 17. Formation of spatially-regular arrays of dots: numerical solutions of eq.(32) for [001] surface orientation. (a) Stationary hexagonal array of equal-size dots, $h_0 = 1.5\,\mathrm{nm}$, $w = 1.89 \times 10^7\,\mathrm{Jm}^{-3}$, $a = 6.6\,\mathrm{Jm}^{-2}$, ($w_{01} = 0.23$, $\zeta = 1.0$). (b) Stationary square array of equal-size dots, $h_0 = 1.5\,\mathrm{nm}$, $w = 8.2 \times 10^5\,\mathrm{Jm}^{-3}$, $a = 11.1\,\mathrm{Jm}^{-2}$, ($w_{01} = 0.01$, $\zeta = 1.0$); Other parameters are the same as in Figs.14,15, and $b = 0$.

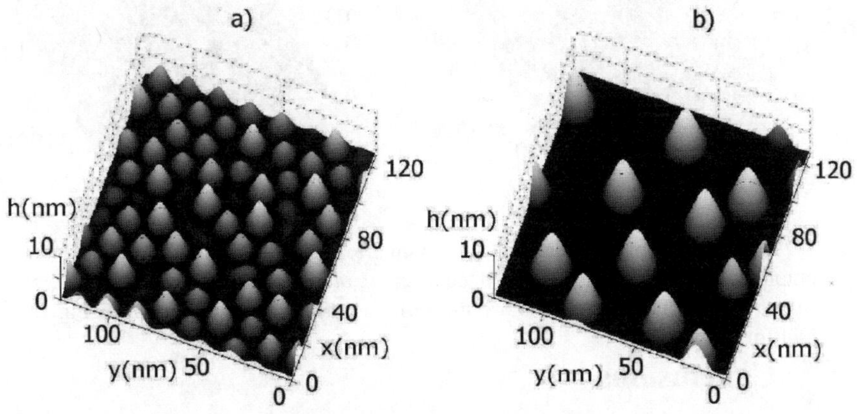

Figure 18. Formation of localized dots via coarsening: numerical solutions of eq.(32) for [001] surface orientation. (a) Nearly hexagonal array of dots (initial stage); (b) spatially-localized dots divided by a thin wetting layer (late stage); $h_0 = 1.5\,\mathrm{nm}$, $w = 8.2 \times 10^6\,\mathrm{Jm}^{-3}$, $a = 0.22\,\mathrm{Jm}^{-2}$, ($w_{01} = 0.1$, $\zeta = 1.0$). Other parameters are the same as in Figs.14,15 and $b = 0$.

localized dots grown on [111] surface are strongly anisotropic. This is due to anisotropic cubic terms in the free energy functional (anisotropic quadratic terms in the evolution equation) that, in the small slope approximation, become dominant. Note that self-organization of quantum dots in the form of localized triangular pyramids on a [111] surface was observed in [30]. Although in experiments described in [30] an elastic mechanism of the solid film instability seems to play an important role, the triangular shape of the pyramids with [001] faces is clearly caused by the anisotropic surface energy, which is correctly captured by the model discussed here.

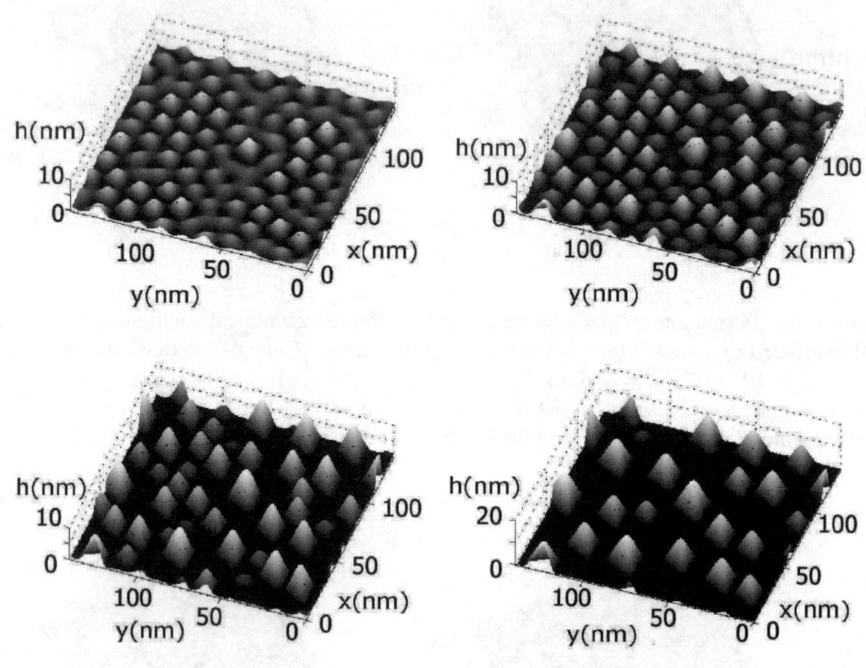

Figure 19. Formation of localized dots: numerical solutions of eq.(32) for [111] surface orientation showing different stages of coarsening of initial regular hexagonal array of triangular pyramids. The parameters are the same as in for the case shown in Fig.18, except $b = 0.44\,\mathrm{Jm}^{-2}$.

5. Conclusions

We have discussed certain aspects of self-assembly of quantum dots from thin solid films epitaxially grown on solid substrates. We have considered two principle mechanisms of instability of a planar film that lead to the formation of quantum dots: the one associated with epitaxial stress and the one associated with the anisotropy of the film surface energy. We have focused on the case of particularly thin films when wetting interactions between the film and the substrate are important and derived nonlinear evolution equations for the film surface shape in the small-slope approximation. We have shown that wetting interactions between the film and the substrate damp long-wave modes of instability and yield the short-wave instability spectrum that can result in the formation of spatially-regular arrays of islands. We have discussed the nonlinear evolution of such arrays analytically, by means of weakly nonlinear analysis, and numerically, far from the instability threshold and have shown that for certain values of physical parameters such arrays can be observed in experiments.

We note that the area of modeling the self-assembly of quantum dots is very active. Various numerical methods have been developed for the solution of the evolution equation for the film surface shape that include non-local elastic effects and anisotropy, such as phase-field methods, finite element methods and others [31]. Also, numerous investigations are devoted to atomistic modeling of self-assembly of quantum dots, as well as to the combination of modeling at small and large scales. These investigations are reviewed in [32].

Acknowledgment

This work was supported by the NATO ASI grant 980684 and by the US National Science Foundation NIRT Program grant DMR-0102794.

References

[1] V. A. Shchukin and D. Bimberg, Rev. Mod. Phys. **71**, 1125 (1999).

[2] R.S. Williams, G. Medeiros-Ribeiro, T.I.Kamins and D.A.A. Ohlberg, J. Phys. Chem. B **102**, 9605 (1998).

[3] K. Alchalabi, D. Zimin, G. Kostorz and H. Zogg, Phys. Rev. Lett. **90**, 026104 (2003).

[4] D. Granados and J.M. Garcia, Appl. Phys. Lett. **82**, 2401 (2003).

[5] J.A. Floro, M.B. Sinclair, E. Chason, L.B. Freund, R.D. Twesten, R.Q. Hwang and G.A. Lucadamo, Phys. Rev. Lett. **84**, 701 (2000).

[6] R.J. Asaro and W.A. Tiller, Metall. Trans. **3**, 1789 (1972); M.Ya. Grinfeld, Sov. Phys. Dokl. **31**, 831 (1986).

[7] D.J. Srolovitz, Acta Metall. **17**, 621 (1989); B.J. Spencer, P.W. Voorhees, and S.H. Davis, Phys. Rev. Lett. **67**, 3696 (1991). B.J. Spencer, P.W. Voorhees, and S.H. Davis, J. Appl. Phys. **73**, 4955 (1993).

[8] A. Pimpinelli and J. Villain, *Physics of Crystal Growth*. Cambridge Univ. Press, Cambridge, 1998.

[9] J. Stewart and N. Goldenfeld, Phys. Rev. A **46**, 6505 (1992); F. Liu and H. Metiu, Phys. Rev. B **48**, 5808 (1993).

[10] A.A. Golovin, S.H. Davis and A.A. Nepomnyashchy, Phys. Rev. E **59**, 803 (1999); T.V. Savina, A.A. Golovin, S.H. Davis, A.A.Nepomnyashchy, and P.W. Voorhees, Phys. Rev. E. **67**, 021606 (2003).

[11] A.A.Golovin, S.H.Davis and P.W.Voorhees, "Self-Organization of Quantum Dots in Epitaxially-Strained Solid Films", Phys. Rev. E **68**, #056203 (2003).

[12] B.J. Spencer, S.H. Davis and P.W. Voorhees, Phys. Rev. B **47**, 9760 (1993).

[13] M.C. Cross and P.C. Hohenberg, *Rev. Mod. Phys.* **65**, 851 (1993).

[14] D. Walgraef, *Spatio-Temporal Pattern Formation*, Springer, 1997.

[15] M.I. Tribelskii, Usp. Fiz. Nauk **167**, 167 (1997).

[16] P.C. Matthews and S.M. Cox, Nonlinearity **13**, 1293 (2000); S.M. Cox and P.C. Matthews, Physica D **175**, 196 (2003).

[17] J.L. Gray, R. Hull and J.A. Floro, Appl. Phys. Lett. **81**, 2445 (2002).

[18] A.E. Nuz, A.A. Nepomnyashchy, A.A. Golovin, A. Hari and L.M. Pismen Physica D **135**, 233 (2000).

[19] T. Mano, R. Nøtzel, G.J. Hamhuis, T.J. Eijkemans, and J.H. Wolter, J. Appl. Phys. **92**, 4043 (2002).

[20] B.A. Malomed, A.A. Nepomnyashchy and M.I. Tribelsky, Sov. Phys. JETP **69**, 388 (1989).

[21] A.A. Golovin, A.A.Nepomnyashchy and L.M.Pismen, Physica D **81**, 117 (1995).

[22] D. Granados and J.M. Garcia, Appl. Phys. Lett. **82**, 2401 (2003).

[23] A.A. Golovin, M.S. Levine, T.V. Savina and S.H. Davis, Phys. Rev. B **70**, #235342 (2004).

[24] P. Nozieres, in: *Solids far from Equilibrium*, ed.: C. Godrȩche, Cambridge Univ. Press, Cambridge, 1992.

[25] M.E. Gurtin, *Thermomechanics of Evolving Phase Boundaries in the Plane*. Clarendon Press, Oxford, 1993.

[26] C.H. Chiu and H. Gao in *Thin Films: Stresses and Mechanical Properties V*, ed.: S.P. Baker e.a., MRS Symposia Proc. #356 (Mater. Res. Soc., Pittsburgh 1995), p.33.

[27] M.J. Beck, A. Van de Walle, and M. Asta, Phys. Rev. B **70**, 205337 (2004).

[28] R.B. Hoyle, G.B. McFadden and S.H. Davis, Phil. Trans. Roy. Soc. Lond. A **354**, 2915 (1996).

[29] G. Springholtz, V. Holy, M. Pinczolits and G. Bauer, Science **282**, 734 (1998); M. Pinczolits, G. Springholtz and G. Bauer, Phys. Rev. B **60**, 11524 (1999).

[30] K. Alchalabi, D. Zimin, G. Kostorz, and H. Zogg, Phys. Rev. Lett. **90**, 026104 (2003)

[31] S.M. Wise, J.S. Lowengrub, J.S. Kim, and W.C. Johnson, Superlattices and Microstructures **36**, 293 (2004); Y.W. Zhang, Phys. Rev. B **61**, 10388 (2000); P. Liu, Y.W. Zhang and C. Lu, Phys. Rev. B **68**, 195314 (2003); C. Chiu, Phys. Rev. B **69**, 165413 (2004).

[32] D.D. Vvedensky, J. Phys.: Cond. Matter **16**, R1537 (2004).

MACROSCOPIC AND MESOPHYSICS TOGETHER: THE MOVING CONTACT LINE PROBLEM REVISITED

Yves Pomeau

Laboratoire de Physique Statistique de l'Ecole normale supérieure, 24 Rue Lhomond, 75231 Paris Cedex 05, France
and
Department of Mathematics, University of Arizona, Tucson, AZ 85720, USA

Abstract   This communication reexamines a famous riddle in fluid mechanics: the moving contact line problem. I show that there is a solution for the wedge of a viscous fluid sliding on a solid surface without dropping any basic tenet of the fluid mechanics of viscous fluids. The solution satisfies in particular the balance of normal forces. It can be taken as a starting point for a numerical simulation of 'large' scale motion like a droplet sliding on an incline, without introducing any new scaling parameter.

Keywords: Moving contact line, fluid mechanics, viscous fluid, mesoscopic effects

The moving contact line problem can be summarized as follows: we know that, at equilibrium, the contact angle between a liquid/vapor (l/v) interface and a flat solid is determined by the Young-Laplace condition relating this angle to a combination of various equilibrium surface energies of the thermodynamic phases under consideration. Suppose now that, by some external forcing one manages to slide the contact line with respect to the solid surface. This leads immediately to a major problem, since the usual boundary condition for fluid is the continuity of velocity between the solid and the fluid in contact. Therefore, if one insists that the l/v interface is a material surface carried by the fluid motion, there is no way for the contact line to move with respect to the solid. In this situation mesoscopic and macroscopic physics do mix together in a way that is very hard to disentangle. Below I present some new ideas on this topic by assuming that the solid is perfectly smooth, something that is very hard to achieve concretely in real life experiments.

This moving contact line problem has received a lot of attention over the years [2]. Recently there has been significant progress in this problem, both in experiments and theory. The experiments have clearly shown what could be called a dynamical wetting transition. This was already in the work of Blake

159

A. A. Golovin and A. A. Nepomnyashchy (eds.),
Self-Assembly, Pattern Formation and Growth Phenomena in Nano-Systems, 159–166.
© 2006 *Springer. Printed in the Netherlands.*

and collaborators [3] and was upheld in more recent experiments on droplets sliding down an incline [4]: at low speed the shape of the droplet remains similar to that of a droplet resting on a flat horizontal plane. Beyond a critical receding speed, the contact line shows a cusp becoming more and more acute as the speed of descent increases. This confirms the idea that the relationship between the speed of sliding of the triple line and the contact angle depends on microscopic process going on near the moving line. Such a relationship has to be derived from microscopic models of interaction between the solid and the two fluids (vapor and liquid) merging along the same line. This is not surprising, since the Laplace theory of the equilibrium contact angle relies on the balance of molecular forces.

The difficulty with this idea is that one does not know how the contact angle/velocity relation should enter the equations of fluid mechanics, which is the question addressed below. Actually, the very problem of the motion of a wedge of fluid sliding along a solid surface has no known satisfactory solution, except when the solid is not wet at all by the liquid [5]. However, this problem can be bypassed in the often realistic limit of a small capillary number [6]. In this limit, the main contribution to the stress on the interface comes from the capillary forces. Therefore a quasistatic approximation yields the right shape of the interface, given the curve followed by the contact line and the volume trapped inside. This yields also the value of the contact angle, which, in general, is not the Young-Laplace equilibrium angle. The contact angle/velocity relation accounts for this difference. It gives a well defined and successful schema to describe various physical phenomena where the contact line motion is crucial [9], [7]. However, it leaves out other cases where the capillary number is not small (although this is not so common, except with very viscous fluids). I deal below with some aspects of this problem.

Specifically, I show that some of the problems that are present in the wedge-like solution of Huh and Scriven disappear if one drops the condition that the liquid-vapor interface is a material surface. This is important, because, besides a logarithmic divergence of the dissipation, it is not really a solution to the problem at hand. It does not satisfy the balance of normal forces across the liquid-vapor interface. I propose a way to get around this difficulty. The net result is a set of formula that satisfy all the requirements coming from the mechanical balance. I shall discuss afterwards the consequence of this formulation.

The early paper [1] solves the Stokes problem in a wedge-like flow, with boundary conditions that I shall discuss later on (see below for the details). Let a liquid/vapor (l/v) flat interface cross an equally flat solid surface at a certain prescribed angle denoted Φ and in between 0 and π. The polar coordinates are such that $r = 0$ is at the intersection of the solid and l/v interface. The fluid of index B is on the right side and corresponds to the values of the polar angle

$0 < \theta < \Pi$, and the fluid of index A is on the left side and corresponds to $\Phi < \theta < \pi$. The solid is moving at a speed U parallel to itself. The calculation is done in the frame of reference moving with the A/B interface, in such a way that every quantity is time independent in this frame of reference. The equations to be solved are, therefore, the Stokes equations in a wedge, with various boundary conditions.

In order to make comparison of results easier, the notations used here are close to the ones of Huh and Scriven. The flow is assumed to be incompressible in each fluid phase. The incompressibility conditions are satisfied there by means of two stream functions $\Psi_A(r,\theta)$ and $\Psi_B(r,\theta)$, defined in such a way that the radial and azimuthal components of the fluid velocity are

$$u_r = -\frac{1}{r}\frac{\partial\Psi}{\partial\theta},$$

and

$$u_\theta = \frac{\partial\Psi}{\partial r}$$

. In such a two dimensional geometry, the Stokes equations are reduced to the biharmonic equation for the stream function

$$\Delta^2\Psi(r,\theta) = 0, \tag{1}$$

where the Laplacian Δ in polar coordinates is:

$$\Delta = \frac{\partial^2}{\partial r^2} + \frac{1}{r}\frac{\partial}{\partial r} + \frac{1}{r^2}\frac{\partial^2}{\partial\theta^2}.$$

The relevant solution of the equation (1) is

$$\Psi(r,\theta) = r\left[a\sin(\theta) + b\cos(\theta) + c\theta\sin(\theta) + d\theta\cos(\theta)\right], \tag{2}$$

where (a,b,c,d) make up two sets of four coefficients corresponding to each phase (that is, a set (a_A, b_A, c_A, d_A) and a set (a_B, b_B, c_B, d_B)). They are constrained by two types of boundary conditions: some for the no-slip condition on the solid, and the others for the continuity across the l/v interface. The boundary conditions on the surface of the solid are $u_\theta = 0$ both for $\theta = 0$ and $\theta = \pi$, $u_r = U$ for $\theta = 0$ and $u_r = -U$ for $\theta = \pi$. The normal speed of liquid A on the solid vanishes, which yields:

$$b_A + \pi d_A = 0. \tag{3}$$

The same condition for liquid B yields:

$$b_B = 0. \tag{4}$$

The condition for the tangential component on the solid on the A side reads:

$$a_A + d_A + \pi c_A = -U. \tag{5}$$

On the B side, it is:

$$a_B + d_B = U. \tag{6}$$

The other set of boundary conditions expresses the continuity of tangential speed and tangential stress along the l/v interface (namely, for $\theta = \Phi$). The tangential speed on the l/v interface is just the u_r component. Therefore, the continuity of this velocity component reads:

$$\sin(\Phi)\left[(a_A - a_B) + \Phi(c_A - c_B)\right] + \cos(\Phi)\left[b_A + \Phi(d_A - d_B)\right] = 0. \tag{7}$$

The stress is a rank two tensor. The relevant components here are the normal and tangential stress, namely, the components $\Sigma_{\theta\theta}$ and $\Sigma_{r\theta}$. They both enter the expression of forces exerted on the element of surface parallel to the l/v interface. Let us first look at $\Sigma_{r\theta}$ part. Its only contribution comes from the viscous stress. Its general expression in cylindrical coordinates is:

$$\Sigma_{r\theta} = -\eta \left(\frac{\partial u_\theta}{\partial r} + \frac{1}{r}\frac{\partial u_r}{\partial \theta} \right),$$

where η is the shear viscosity. Since u_θ is independent on r, one finds that, with the stream function given in equation (2), the tangential stress reads:

$$\begin{aligned}
\Sigma_{r\theta} &= -\frac{\eta}{r}\left[-a\sin(\theta) - b\sin(\theta) + c(2\cos(\theta) - \theta\sin(\theta))\right. \\
&\quad \left. -d(\theta\cos(\theta) + 2\sin(\theta))\right]
\end{aligned}$$

So the continuity of tangential stress gives:

$$\begin{aligned}
\eta_B &\left[-a_B\sin(\Phi) + c_B(2\cos(\Phi) - \Phi\sin(\Phi))\right. \\
&\quad \left. -d_B(\Phi\cos(\Phi) + 2\sin(\Phi))\right] \\
= \eta_A &\left[-a_A\sin(\Phi) - b_A\sin(\Phi) + c_A(2\cos(\Phi) - \Phi\sin(\Phi))\right. \\
&\quad \left. -d_A\Phi\cos(\Phi) + 2\sin(\Phi))\right]
\end{aligned} \tag{8}$$

Until now, our analysis followed the steps of Huh and Scriven. Before explaining what I propose, I shall discuss what they do next.

Huh and Scriven add, to all constraints already taken into account, that the l/v interface is a material surface, that is not crossed by any matter flux. This adds two conditions, namely, the vanishing of the normal velocity on both sides of the interface. This results in four additional boundary conditions along the l/v interface: continuity of tangential stress, of tangential speed and one condition for the normal speed on each side. Now there are eight conditions in all, exactly the number of free parameters in the solution of the biharmonic equation in the two wedges. However, a major difficulty is looming, namely, that the continuity of normal stress is not satisfied. Because one condition cannot be satisfied, one could choose as well to drop another one, like the continuity of tangential stress or the continuity of tangential velocity for instance, instead of the balance of normal forces. In this respect this freedom is unphysical and means that this solution, however interesting it is, cannot be seen as physically sound. The core difficulty is that, by imposing the shape of the l/v interface, one has lost the freedom existing in fully time-dependent problems when the interface is taken as a material surface: the normal speed of this interface is left free. The freedom added in this way permits at the end to satisfy all conditions.

Many attempts to resolve this difficulty have been made over the last thirty years or so. A solution to this riddle is provided by the phase field approach of Seppecher [8]. It is *a priori* free of any divergence related to discontinuities on the l/v interface, assumed to have a finite thickness. This does not mean, however, that Seppecher's solution is a fully satisfactory representation of the physics of the moving contact line, including at distances far from the triple point. The phase field model has, I believe, a weakness : it allows the fluid to evaporate or condense from one fluid phase to the other too easily. I tried in previous work [2] to get around this by imposing a slow relaxation toward equilibrium on the van der Waals equation of the phase field model. This was based upon the idea that this transformation from one thermodynamic phase (the liquid, for instance) to the other (vapor) requires the molecules to jump above some barrier, with the jumps being governed by an Arrhenius exponential that can be very large. This is likely relevant for non-volatile fluids, where this barrier is very high. Here I consider another situation, namely, a volatile fluid.

In a volatile fluid, the evaporation is not dominated by the activation energy but rather by the release or intake of latent heat. That the exchange of latent heat in the evaporation/condensation process is relevant can be understood as follows. Suppose that evaporation and condensation are only relevant to account for the mass exchange across the interface. Then, the velocity of this interface becomes arbitrary. In classical fluid mechanics, one always assumes that the interface is a material surface so that the Stokes or Navier-Stokes equations have the right number of boundary conditions at the interface. If one omits one boundary condition at the interface, that is if one replaces the two constraints of zero normal speed across the interface on the two sides by the single condition of continuity of the normal mass flux, one degree of freedom is added to the system and the velocity of the interface becomes arbitrary. This freedom is lost if one takes into account the 'energetics' of the phase changes. Actually, the motion of the interface is completely determined (in the absence of flow) in a temperature field by the conservation of energy via the Stefan relation between the jump of the normal heat flux and the actual velocity of the interface, related, in turn, to the growth of one thermodynamic phase at the expense of the other. Therefore, the right number of boundary conditions at the l/v interface is recovered, in principle, by incorporating the energy condition into eventual evaporation/condensation effects. Indeed, the most common situation by far in fluid mechanics is the one where evaporation/condensation is negligible and where the boundary condition is the one of a material surface. This is the limit in which the heat flux is small enough to make the velocity related to the evaporation/condensation process negligible compared to the fluid velocity. In this limit, one recovers the usual no-cross flow condition. However, in the contact line problem there is no parameter making a priori this cross flow negligible. Therefore, it should be included together with the Stefan condition and the Fourier heat equation for the temperature, in order to make the problem fully consistent. I plan to come back to this general issue later on. Below I assume that, because of the presence of the solid and because the latent heat of transformation is small (that is consistent with the assumption of "easy evaporation" of a volatile liquid) the energy transfer does not play any role. Moreover, I show that a little evaporation or condensation can be significant according to the Maxwell's conclusion that the shear viscosity of a dilute gas is independent on its density.

Returning to the analysis of the contact line motion, I shall write the last two conditions of continuity along the l/v interface. The boundary conditions on the surface of the solid are in equations (3), (4), (5) and (6) and the continuity of tangential speed and stress along the l/v interface is in equations (7) and (8).

The remaining boundary conditions come from the continuity of normal stress across the l/v interface and of the matter flux. The continuity of the matter flux expresses that the product ρu_θ is the same on both sides of the l/v interface. This yields in general:

$$\rho_A \left[a_A \cos(\Phi) + b_A \sin(\Phi) + c_A \Phi \cos(\Phi) + d_A \Phi \sin(\Phi) \right]$$
$$= \rho_B \left[a_B \cos(\Phi) + c_B \Phi \cos(\Phi) + d_B \Phi \sin(\Phi) \right]. \tag{9}$$

I shall assume now that one of the phase is actually much more dilute than the other, namely, that fluid A is a vapor. Practically speaking, this means that the dimensionless ratio $\frac{\rho_A}{\rho_B}$ is very small. This makes one of the two dimensionless numbers in the problem disappear, the only one remaining being the ratio of the shear viscosities, $\frac{\eta_A}{\eta_B}$. One may think that, if the ratio $\frac{\rho_A}{\rho_B}$ is small, the same is true for the ratio of shear viscosities $\frac{\eta_A}{\eta_B}$. This is not so because, as was shown long ago by Maxwell, the shear viscosity of a dilute gas is independent of its density, and I assume realistically that the vapor is a dilute gas. This sharpens the physical image of the moving contact line: because the shear viscosity of the vapor is of the same order as that of the liquid, a little bit of evaporation or condensation is enough to generate stresses on the vapor side that may balance those on the liquid side. Therefore, the evaporation speed on the liquid side can be neglected while it yields the required stress on the vapor side. In this limit, the continuity of flux across the l/v interface becomes:

$$a_B \cos(\Phi) + c_B \Phi \cos(\Phi) + d_B \Phi \sin(\Phi) = 0. \tag{10}$$

The continuity of normal stress remains to be worked out. The normal stress is the sum of the hydrostatic pressure p and of the viscous stress $-\frac{2\eta}{r}\frac{\partial u_\theta}{\partial \theta}$, and it must be continuous across the l/v interface. Note that, because I assume a flat interface, there is no contribution from the Laplace capillary pressure. In the Stokes approximation, p is a harmonic function. It scales like $U\frac{f(\theta)}{r}$, where $f(\theta)$ is a dimensionless function of θ to be determined. Because p is harmonic $f(\theta)$ has to be of the form $f(\theta) = \alpha \sin(\theta) + \beta \cos(\theta)$, where α and β are some numbers. The equation for momentum balance in the fluid when written in polar coordinates is, for a velocity field that depends only on θ:

$$r^2 \frac{\partial p}{\partial r} = \eta \left[\frac{\partial^2 u_r}{\partial \theta^2} - u_r - 2\frac{\partial u_\theta}{\partial \theta} \right]. \tag{11}$$

Substituting the explicit expressions for u_r and u_θ given above into this equation, one obtains after tedious algebra

$$p = -\frac{2\eta}{r} \left(c \sin(\theta) + d \cos(\theta) \right) \tag{12}$$

Now we can write the continuity of the normal stress $p - \frac{2\eta}{r}\frac{\partial u_\theta}{\partial \theta}$:

$$\begin{aligned}
\eta_A \left[a_A \cos(\Phi) - b_A \sin(\Phi) + c_A(2\sin(\Phi) + \Phi\cos(\Phi)) \right. \\
\left. + d_A(2\cos(\Phi) - \Phi\sin(\Phi)) \right] \\
= \eta_B \left[a_B \cos(\Phi) + c_B(2\sin(\Phi) + \Phi\cos(\Phi)) \right. \\
\left. + d_B(2\cos(\Phi) - \Phi\sin(\Phi)) \right].
\end{aligned} \tag{13}$$

There are now eight linear relations between the eight coefficients $(a_A, b_A, c_A,$ $, d_A)$ (a_B, b_B, c_B, d_B). Unless the determinant of the eight by eight matrix of coefficients is zero, which may happen for some peculiar values of the angle Φ, for instance, this problem has a solution.

This set of conditions is the one that is satisfied at large distances from the triple point in the Seppecher's phase field equations for the contact line motion: as in the phase field equations, our model assumes an unhindered interface motion, without caring for the heat transfer due to the latent heat. It also provides a theory consistent with the balance of normal stress across the fluid/fluid interface. The idea presented here opens the way to model fluid mechanical problems with contact line motion. This is something that is impossible with the Huh-Scriven solution: without balancing the normal stress one cannot match a large-scale solution (like the one relevant for a sliding drop, for instance) with the local relation between the contact angle and the speed of the moving contact line. Indeed, there is still a logarithmic divergence of the total force on the contact line, because the pressure diverges like $\frac{1}{r}$ at the distance r from the moving line. Such a divergence could be taken care of by merging the large-scale solution (at the scale of the whole droplet, for instance) with a Seppecher-like solution near the triple line, the solution presented here being the one valid in the intermediate "matching" domain. Indeed, such a "large scale" solution cannot keep the freedom coming from the unhindered evaporation/condensation. With such a freedom, the shape of a sliding droplet would be arbitrary, because, as in the wedge problem, it leaves enough free parameters to satisfy the boundary conditions for the Stokes equations. This is where the latent heat should be taken into account: the cross flow is a function of the local temperature field, that is itself defined by various boundary conditions of the problem (including, of course, the value of the equilibrium temperature at the l/v interface). This fully constrains the problem to bring a unique and divergenceless solution. I plan to come back to this issue in future publications.

References

[1] C. Huh and Scriven L.E., J. of Colloid and Interface science, **35**, 85-101 (1971).

[2] Y. Pomeau, C.R. MECANIQUE **330** 207(2002).

[3] T.D. Blake and Ruschak K.J., Nature (London) **282**, 489 (1979).

[4] T. Podgorski, Flesseles J.M. and Limat L., Phys. Rev. Lett., **87**, 036102 (2001).

[5] L. Mahadevan and Pomeau Y., Phys. of Fluids **11** 2449 (1999).

[6] Y. Pomeau, CRAS Serie II **328** 411 (2000).

[7] C. Andrieu, Beysens D., Nikolayev V. and Pomeau Y., J. of Fluid Mech. **453** 427 (2002).

[8] P. Seppecher, Int. J. Eng. Sci. **34**, 977 (1996); L. Pismen and Pomeau Y., Phys. Rev. **E 62**, 2480 (2000); U. Thiele, Neuffer K, Bestehorn M., Velarde M.G. and Y. Pomeau, Coll. and Surf.A -Phys.and Eng.**206** 87 (2002).

[9] M. Ben Amar, Cummings L.J. and Pomeau Y., Phys.of Fluids **15** 2949 (2003).

NANOSCALE EFFECTS IN MESOSCOPIC FILMS

L. M. Pismen
Department of Chemical Engineering and Minerva Center for Nonlinear Physics of Complex Systems, Technion – Israel Institute of Technology, 32000 Haifa, Israel

Abstract   This chapter reviews the ways to incorporate nanoscale molecular interactions into hydrodynamic theory of mesoscopic films and moving contact lines exploring wide separation of different characteristic scales. The approach uses the lubrication approximation, and successfully reduces the description from density functional theory to equations of Cahn-Hilliard type incorporating effects of interaction with the substrate and interfacial curvature. These equations are in turn analyzed to obtain mobility relations for droplets on a precursor film.

Keywords: Molecular interactions, thin films, contact line, disjoining potential, mobility.

Introduction

Many important technological processes involving flow, transport and chemical reactions take place on or near fluid-solid or fluid-fluid interfaces. Both equilibrium properties of a fluid and transport coefficients are modified in the vicinity of interfaces. The effect of these changes is crucial in the behavior of ultra-thin fluid films, fluid motion in microchannels, etc. It is no less important in macroscopic phenomena involving interfacial singularities, such as rupture, coalescence and motion of three-phase contact lines.

Any interphase boundary is essentially a *mesoscopic* structure. While the material properties vary smoothly at macroscopic distances along the interface, the gradients in the normal direction are steep, approaching a molecular scale in the vicinity of the interface. Under these conditions, a nonlocal character of long-range intermolecular interactions becomes important. This problem becomes even more acute in the vicinity of solid surfaces, where the fluid is subject to van der Waals, polar and electrical double layer forces. Such factors as chemical and geometric inhomogeneities, induced anisotropy, steric constraints and partial ordering of the fluid may come into play there as well.

On the one hand, interfacial processes are governed by a variety of microscopic factors that lie outside the scope of conventional fluid dynamics and

167

A. A. Golovin and A. A. Nepomnyashchy (eds.),
Self-Assembly, Pattern Formation and Growth Phenomena in Nano-Systems, 167–193.
© 2006 *Springer. Printed in the Netherlands.*

theories of macroscopic transport. On the other hand, both the number of particles involved and the characteristic scale of the spatial structures are too large to be treated by molecular dynamics or similar methods. This implies a contradiction between the need in macroscopic description and a necessity to take into consideration microscopic factors, which come to influence the fluid motion and transport on incommensurately larger scales. This contradiction can be only resolved by matching the processes taking place on widely separated scales and described by models of dissimilar kind.

A middle ground between classical hydrodynamics (inapplicable at molecular scales) and molecular dynamics (inapplicable to macroscopic volumes) is taken by mesoscopic continuum theories introducing intermolecular forces into coarse-grained equations of motion. Theories of this kind account for special properties of an interfaces in a natural way by considering it as a *diffuse* region interpolating between the two phases. The origin of this approach is in the diffuse interface model going back to van der Waals [1]. Much later, it became prominent in the phase field models [2], used mostly in phenomenological theory of solidification where a fictitious phase field, rather than density, plays the role of a continuous variable changing across the interphase boundary. The theory of van der Waals was widely used for description of equilibrium fluid properties, including surface tension and line tension in three-phase fluid systems [3]. Significant progress has been achieved in application of more sophisticated methods of this type, in particular, density functional theory, to the study of equilibrium properties of thin films [4].

Applications of this theory to dynamical processes in fluids is much more difficult, as it requires coupling to hydrodynamics. So far, non-equilibrium continuum theory has been largely based on classical fluid mechanics [5–7]. Intermolecular interactions are taken into account in this theory through disjoining potential, computed in lubrication approximation, while slip and interface relaxation are taken into account phenomenologically, if at all. Coupling the van der Waals–Cahn–Hilliard diffuse interface theory of to hydrodynamics involves incorporating the free energy of an inhomogeneous fluid in hydrodynamic equations [8]. Several computational studies based on this set of equations have been published in the last decade [9–12].

Extension of more advanced methods, in particular, density functional theory, to non-equilibrium phenomena is the principal aim of this survey. We shall consider a simple one-component fluid with van der Waals interactions as a suitable medium for exploration of basic theoretical problems of interfacial dynamics. In the case when intermolecular interactions are long-range, in particular, in the most important case of Lennard–Jones potential, the transformation from the nonlocal (density functional) to local (van der Waals–Landau–Cahn) equations fails due to divergences appearing in the commonly used expansion of the interaction term in the expression for free energy. Setting boundary conditions on the fluid-solid interface necessary for a local theory is also contradictory, since it neglects a gradual change of fluid properties near this boundary.

Formulation of a full dynamic nonlocal theory is not practical due to intrinsic difficulties in separation of advective and diffusional fluxes, aggravated in nonlocal theory by impossibility to reduce external forces to boundary integrals. Even in the framework of local theory, computational difficulties of a straightforward approach make it so far impossible to span the entire range from nanoscopic scale of molecular interactions to observable macroscopic scales.

Our strategy is therefore to use natural scale separation in weakly distorted thin films and other systems with almost parallel interacting interfaces to separate molecular interactions and diffusion in the direction normal to the interfaces from hydrodynamic motion in spanwise directions. The former is treated nonlocally, resulting in computation of quasiequilibrium disjoining potential. This input is further used in long-scale hydrodynamic computations.

The disparity of scales relieves the formidable task of solving coupled kinetic and hydrodynamic equations. It allows one to separate the inner interfacial region where macroscopic flow velocity is constant and the interface is close to equilibrium, and the outer region where flow is incompressible, while weak gradients of chemical potential are relaxed by diffusion. Macroscopic flow driven by external sources will remain unaffected by gradients of chemical potential almost everywhere. Nevertheless, coupling to the inner region influences the flow through boundary conditions sensitive to changes of surface tension, interfacial curvature, and interphase transport. The corrections, even minute, may become essential when the classical hydrodynamic solution is singular.

1.   Hydrodynamic Equations

Modified Stokes Equation

Classical hydrodynamic equations should be modified at mesoscopic distances from interfaces by including thermodynamic driving forces arising in a non-equilibrium fluid. Local mesoscopic thermo-hydrodynamic theory assuming linear coupling between fluxes and thermodynamic forces in the spirit of Onsager's non-equilibrium thermodynamics [8] modifies hydrodynamic equations by including in the stress balance a reversible part of the stress tensor, called *capillary tensor*, which is derived from an applicable free energy functional. This tensor is related to the momentum conservation law, which follows, by Noether theorem, from the translational invariance of the system. The capillary tensor complements the usual (irreversible) viscous stress tensor in the hydrodynamic equations. This leads, after some transformations, to a modified Stokes equation in the form

$$-\rho\nabla\mu + \nabla\cdot(\eta\nabla\boldsymbol{v}) = 0, \tag{1}$$

where $\boldsymbol{v}$ is the velocity field and η is the dynamic viscosity (generally, dependent on the particle density ρ). The local chemical potential μ, dependent on proximity of fluid-solid and liquid-vapor interfaces and interfacial curvature (and, possibly, also on external forces) serves as a driving force of fluid motion.

The capillary tensor, being a local quantity, cannot be unequivocally defined in a nonlocal framework; nevertheless, we expect the modified Stokes equation (1) to remain applicable also in this case. It can be introduced directly by including molecular interactions into the common driving term (enthalpy gradient) in the Stokes equation. Since molecular interactions are essential only in a thin interfacial layer, standard interfacial boundary conditions follow from the sharp interface limit of Eq. (1), valid when the relevant macroscopic distances far exceed the range of intermolecular forces. As usual, the Stokes equation is complemented by the continuity equation

$$\rho_t + \nabla\cdot(\rho\boldsymbol{v}) = 0. \tag{2}$$

The system of equations is closed by thermodynamic equations defining μ, to be derived in Sections 2 and 3, which replace an equation of state used in standard hydrodynamic theory. Combined thermo-hydrodynamic theory based on Eq. (1) and including nonlocal interactions is formidable, but the problem can simplified using a natural scale separation in thin films [13, 14], which is also used in the standard hydrodynamic lubrication approximation [7].

Lubrication Scaling

The lubrication approximation assumes the characteristic length scale in the "vertical" direction z (normal to the substrate) to be much smaller than that in the "horizontal" (parallel) directions spanned by the 2D vector $\boldsymbol{x}$. The approximation is applicable to liquid films with a large aspect ratio, when the interface is weakly inclined and curved. The scaling is consistent if one assumes $\partial_z = O(1)$, $\nabla = O(\sqrt{\epsilon}) \ll 1$, where ∇ is now the 2D gradient in the plane of the solid support. Then the continuity equation requires that the vertical velocity v should be much smaller than the horizontal velocity (denoted by the 2D vector $\boldsymbol{u}$): $\boldsymbol{u} = O(\sqrt{\epsilon})$, $v = O(\epsilon)$.

The lubrication equations are derived by expanding both equations and boundary conditions in powers of ϵ and retaining the lowest-order terms. First, we deduce from the vertical component of the Stokes equation, reduced in the leading order to $\partial\mu/\partial z = 0$, that chemical potential $\mu(\boldsymbol{x}, t)$ is constant across the layer. The horizontal component of the Stokes equation takes now the form

$$-\rho\nabla\mu + \partial_z(\eta\boldsymbol{u}_z) = 0. \tag{3}$$

The continuity equation is rewritten as

$$\rho_t + \nabla \cdot (\rho \boldsymbol{u}) + \partial_z(\rho v) = 0. \tag{4}$$

Matching the lubrication equation to thermodynamic theory requires some caution, since thermodynamic theory yielding an expression for μ should be applied to the entire system including dense (liquid) and dilute (vapor) phases in equilibrium, whereas only the dense phase may have a suitable aspect ratio. To make the approximation applicable, one has to assume that the interface dividing the dense and the dilute phase is only weakly inclined relative to the substrate and weakly curved, so that its position can be expressed by a function $h(\boldsymbol{x}, t)$ with derivatives obeying the above lubrication scaling. Thermodynamic theory, either local or nonlocal, can be used to compute an equilibrium density profile across the interface (in the "vertical" direction), $\rho_0(z - h(\boldsymbol{x}, t))$, which is weakly dependent on the "horizontal" 2D position and time only through its dependence on h, e.g.

$$\rho_t = -\rho_0'(z - h)h_t, \quad \nabla^2 \rho = -\rho_0'(z - h)\nabla^2 h + \rho_0''(z - h)|\nabla h|^2. \tag{5}$$

As the interface is diffuse, h can be only defined as a *nominal* interface position, which may be identified with the location of a particular isodensity level. The most natural choice is the Gibbs equimolar surface, which satisfies the relation

$$\int_{-\infty}^{h} (\rho^+ - \rho)\,\mathrm{d}z = \int_{h}^{\infty} (\rho - \rho^-)\,\mathrm{d}z, \tag{6}$$

where $\rho^\pm$ are densities of the dense and dilute phases. This means that the total mass of an unbounded fluid would not change when the actual profile is replaced by a sharp boundary located at $z = h$. The interfacial thickness is commonly of molecular dimensions, so that on the macroscopic scale $\rho'(z)$ approaches a delta function, $\rho'(z) = -(\rho^+ - \rho^-)\delta(0)$. In the absence of interphase mass transport, the velocity normal to any isodensity level should vanish, $\boldsymbol{n} \cdot \boldsymbol{v} = v - \boldsymbol{u} \cdot \nabla h = 0$. Integrating Eq. (4) and replacing v with the help of this condition yields the material balance equation averaged across the liquid layer:

$$h_t + \nabla \cdot \boldsymbol{j} = 0, \quad \boldsymbol{j} = \int_0^h \boldsymbol{u}(z)\,\mathrm{d}z. \tag{7}$$

Mobility Coefficient

Solving Eq. (3) with appropriate boundary conditions gives, generally, a linear relationship between the total flux through the film $\boldsymbol{j}$ and the 2D gradient of chemical potential:

$$\boldsymbol{j} = -k(h)\rho\nabla\mu, \tag{8}$$

where $k(h)$ is a *mobility coefficient* .

With the density, as well as viscosity of the dilute phase neglected, the boundary condition at the free interface for macroscopic flow in the dense layer is the condition of vanishing tangential stress, which reduces in the leading order of the lubrication approximation to $u_z(h) = 0$. The classical no-slip condition on a solid substrate is $u(0) = 0$. Assuming η and $\rho = \rho^+$ to be constant, this yields

$$u = -\eta^{-1}z\left(h - \tfrac{1}{2}z\right)\rho^+\nabla\mu, \quad k(h) = \tfrac{1}{3}\eta^{-1}h^3. \tag{9}$$

The no-slip condition becomes, however, inapplicable on molecular scales. It is known, in particular, that it generates a multivalued velocity and, hence, an infinite stress in the vicinity of a contact line, leading formally to an infinite drag force [15, 16].

The physics of motion in a layer adjacent to the solid surface is quite different from the bulk motion described by the Stokes equation. This generates effective slip at a microscopic scale comparable with intermolecular distances. The presence of a slip in dense fluids it is confirmed by molecular dynamics simulations [17, 18] as well as experiment [19]. The two alternatives are slip conditions of "hydrodynamic" and "kinetic" type. The version of the slip condition most commonly used in fluid-mechanical theory is a linear relation between the velocity component along the solid surface and the shear stress [20], leading to the boundary condition $u = bu_z$ at $z = 0$ where b is a phenomenological parameter – *slip length.* interaction between the fluid and the substrate. In liquids this length should be of molecular dimensions.

The kinetic slip condition assumes diffusive motion in a thin layer of molecular thickness d adjacent to the solid substrate driven by the chemical potential gradient. The total flux through the molecular layer can be expressed as

$$j = -(dD/T)\nabla\mu, \tag{10}$$

where D is the surface diffusivity and temperature T is measured in energy units (with the Boltzmann constant rescaled to unity). This relation, first introduced to describe motion in a molecularly thin precursor layer [21, 22], is extended to mesoscopic films [23] when the usual no-slip boundary condition on the solid support $u(0) = 0$ is replaced by the slip condition at the molecular cut-off distance d with $u(d)$ given by Eq. (10). The solution in the bulk layer $d < z < h$ is

$$u = -\eta^{-1}\rho^+\left[\lambda^2 + h(z - d) - \tfrac{1}{2}(z^2 - d^2)\right]\rho^+\nabla\mu, \tag{11}$$

where $\lambda = \sqrt{D\eta/(\rho^+T)} = O(d)$ is the effective slip length. The mobility coefficient is computed as

$$k(h) = \eta^{-1}\left[\lambda^2 h + \tfrac{1}{3}(h - d)^3\right], \tag{12}$$

which reduces to $k(h)$ in Eq. (9) in the limit $\lambda, d \to 0$.

2. Thermodynamic Equations

Free Energy Functional

The starting point for computation of chemical potential is the free energy functional written in the local density functional approximation [24] as

$$\mathcal{F} = \int \rho(\boldsymbol{x}) f[\rho(\boldsymbol{x})]\, \mathrm{d}^3\boldsymbol{x} + \frac{1}{2} \int \rho(\boldsymbol{x})\, \mathrm{d}^3\boldsymbol{x} \int_{r>d} U(r)[\rho(\boldsymbol{x}+\boldsymbol{r}) - \rho(\boldsymbol{x})]\, \mathrm{d}^3\boldsymbol{r},$$

(13)

where $f(\rho)$ is free energy per particle of a homogeneous fluid and $U(r)$ is an isotropic pair interaction kernel with a short-scale cut-off d. The contribution of density inhomogeneities is expressed by the last term vanishing in a homogeneous fluid.

The chemical potential $\mu = \delta\mathcal{F}/\delta\rho$ enters the respective Euler–Lagrange equation obtained by minimizing the grand ensemble thermodynamic potential $\Phi = \mathcal{F} - \mu \int \rho\, \mathrm{d}^3\boldsymbol{x}$, which defines the equilibrium particle density distribution $\rho(\boldsymbol{x})$:

$$g(\rho) - \mu + \int_{r>d} U(r)[\rho(\boldsymbol{x}+\boldsymbol{r}) - \rho(\boldsymbol{x})]\, \mathrm{d}^3\boldsymbol{r} = 0,$$

(14)

where $g(\rho) = \mathrm{d}[\rho f(\rho)]/\mathrm{d}\rho$. The function $F(\rho) = \rho[f(\rho) - \mu]$ should have two minima $\rho^{\pm}$ corresponding to two stable uniform equilibrium states of higher and lower density (liquid and vapor).

A simple example of long-range potential is the modified Lennard–Jones potential with hard-core repulsion:

$$U = \begin{cases} -Ar^{-6} & \text{at} \quad r > d \\ \infty & \text{at} \quad r < d \end{cases},$$

(15)

where d is the nominal hard-core molecular diameter. The interaction kernel $U(r)$ gives the free energy density of a homogeneous van der Waals fluid

$$f(\rho, T) = T \ln \frac{\rho}{1 - b\rho} - a\rho,$$

(16)

where T is temperature, $b = \frac{2}{3}\pi d^3$ is the excluded volume and

$$a = -2\pi \int_d^{\infty} U(r) r^2\, \mathrm{d}r = \frac{2\pi A}{3d^3}.$$

(17)

Equilibrium between the two homogeneous states, $\rho = \rho^{\pm}$ is fixed by the Maxwell condition

$$\mu_0 = \frac{\rho^+ f(\rho^+) - \rho^- f(\rho^-)}{\rho^+ - \rho^-},$$

(18)

which defines, together with $\mu_0 = g(\rho^{\pm})$, the equilibrium chemical potential $\mu = \mu_0$ and both equilibrium densities.

The integral equation (14) can be converted to a much simpler differential equation assuming that density is changing only slightly over distances comparable with the characteristic interaction length. Then one can expand

$$\rho(\boldsymbol{x}+\boldsymbol{r}) = \rho(\boldsymbol{x}) + \boldsymbol{r}\cdot\nabla\rho(\boldsymbol{x}) + \tfrac{1}{2}\boldsymbol{rr}:\nabla\nabla\rho(\boldsymbol{x}) + \ldots \tag{19}$$

Using this in Eq. (13) we see that the contribution of the linear term to the nonlocal integral vanishes when the system is isotropic and, as a consequence, the interaction term is spherically symmetrical, and the lowest order contribution is due to the quadratic term:

$$F_2(\boldsymbol{x}) = -\tfrac{1}{2}K\int \rho(\boldsymbol{x})\nabla^2\rho(\boldsymbol{x})\,\mathrm{d}^3\boldsymbol{x} = \tfrac{1}{2}K\int |\nabla\rho(\boldsymbol{x})|^2\,\mathrm{d}^3\boldsymbol{x}, \tag{20}$$

where

$$K = -\frac{2\pi}{3}\int_d^\infty U(r)\,r^4\,\mathrm{d}r = -\int_0^\infty Q(z)z^2\,\mathrm{d}z = \frac{2\pi A}{3d}. \tag{21}$$

Thus, Eq. (13) is replaced by

$$\mathcal{F} = \int \left[\rho f(\rho) - \mu\rho + \tfrac{1}{2}K|\nabla\rho(\boldsymbol{x})|^2\right]\mathrm{d}^3\boldsymbol{x}. \tag{22}$$

This derivation, going back to van der Waals [1], has a disturbing flaw. If the expansion is continued to the next non-vanishing order (fourth), the expression for the respective coefficient, computed analogous to Eq. (21), diverges when the common Lennard–Jones potential is used.

Surface Tension and Density Profile

The equation for density distribution near a flat boundary normal to the z axis is obtained by assuming ρ to be constant in each lateral plane and integrating Eq. (13) in the lateral directions. This yields the free energy per unit area, or surface tension

$$\gamma = \int_{-\infty}^\infty \rho(z)[f(\rho)-\mu]\mathrm{d}z + \frac{1}{2}\int_{-\infty}^\infty \rho(z)\,\mathrm{d}z \int_{-\infty}^\infty Q(\zeta)[\rho(z+\zeta)-\rho(z)]\,\mathrm{d}\zeta. \tag{23}$$

The interfacial energy is contributed both by deviations from the equilibrium density levels in the transitional region and by the distortion energy localized there. The 1D interaction kernel $Q(z)$ lumps intermolecular interaction between the layers $z = $ const. It is computed by lateral integration using as an integration variable the squared distance $q = r^2 = \xi^2 + z^2$, where ξ is radial distance in the lateral plane. Taking note that the lower integration limit for q is $q_0 = z^2$ at $|z| > d$, $q_0 = d^2$ at $|z| \leq d$, we compute

$$Q(z) = -\pi A \int_{q_0}^{\infty} q^{-3}\, \mathrm{d}q = \begin{cases} -\frac{1}{2}\pi A z^{-4} & \text{at} \quad |z| > d \\ \\ -\frac{1}{2}\pi A d^{-4} & \text{at} \quad |z| \leq d. \end{cases} \tag{24}$$

The respective 1D Euler–Lagrange equation, replacing Eq. (14), is

$$g\left[\rho(z)\right] - \mu + \int_{-\infty}^{\infty} Q(\zeta)[\rho(z+\zeta) - \rho(z)]\, \mathrm{d}\zeta = 0. \tag{25}$$

This equation can be rewritten in a dimensionless form

$$g(\rho) - \mu + \frac{3}{4}\beta \int_{-\infty}^{\infty} Q(\zeta)[\rho(z+\zeta) - \rho(z)]\, \mathrm{d}\zeta = 0, \tag{26}$$

where

$$g(\rho) = \frac{1}{1-\rho} - \ln\left(\frac{1}{\rho} - 1\right) - 2\beta\rho. \tag{27}$$

Here the length is scaled by the nominal molecular diameter d, the density by b^{-1}, and the chemical potential by T; the interaction kernel is $Q(z) = -z^{-4}$ at $|z| > 1$, $Q(z) = -1$ at $|z| \leq 1$, and the only remaining dimensionless parameter is the rescaled inverse temperature $\beta = a/(bT) = (A/T)d^{-6}$.

An example of a density profile obtained by solving numerically Eq. (26) is shown in Fig. 1. The density tail asymptotics can be estimated by considering a location far removed from the interface placed at the origin $[|z| \gg 1$ in the dimensionless units of Eq. (26)] where a sharp interface limit can be implemented. The density is presented as $\rho = \rho^{\pm} + \tilde{\rho}$, where $\tilde{\rho}/\rho \sim 1/|z|^3 \ll 1$. Inserting this in (26) and linearizing around $\rho = \rho^{\pm}$, we see that the densities inside the integral are well approximated in the leading order by the two limiting constants, which is equivalent to the sharp interface limit. For example, for the vapor tail at $z > 0$, $|z| \gg 1$ we have $\rho(z) = \rho^-$ and $\rho(z+\zeta) = \rho^+$ for $\zeta > |z|$, $\rho(z+\zeta) = \rho^-$ for $\zeta < |z|$. Thus, we obtain

$$\rho = \rho^{\pm} + \frac{\beta(\rho^+ - \rho^-)}{4g'(\rho^{\pm})} \frac{1}{z^3}. \tag{28}$$

This is in good agreement to the numerical solution, as seen in the inset of Fig. 1. One can check *a posteriori* using this expression that the contribution to the integral of neighboring locations with $|\zeta| = O(1)$ is of a higher order $\propto |z|^{-5}$ and therefore can be neglected.

For comparizon, the 1D version of the local Euler–Lagrange equation derived from Eq. (22),

$$K\rho''(z) - g(\rho) + \mu = 0. \tag{29}$$

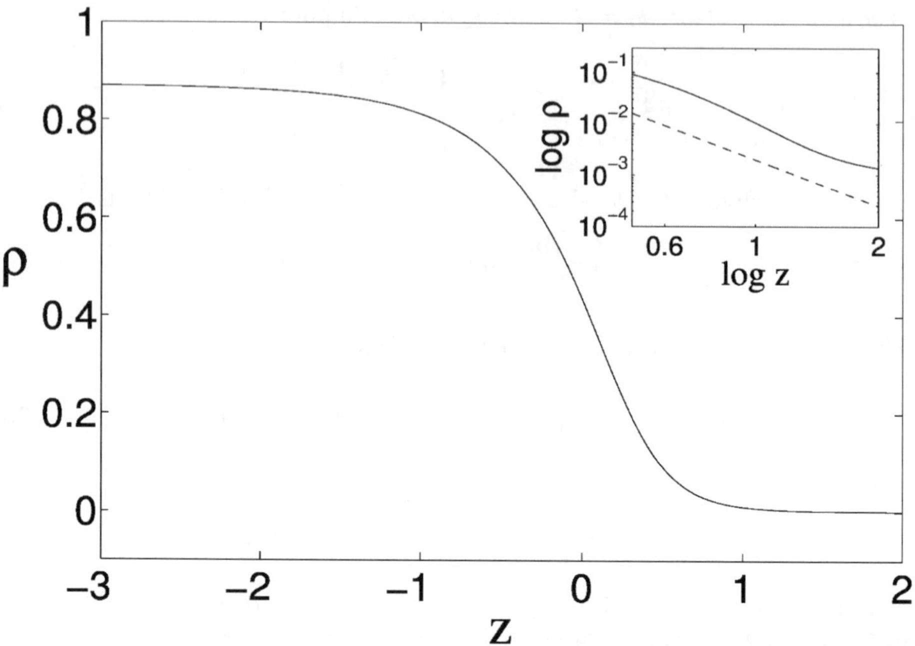

Figure 1.   The density profile of the liquid-vapor interface obtained by numerical solution of (26) for $\beta = 9$. The inset shows the vapor-phase tail of the numerical solution (solid line) compared to the asymptotic form (28) depicted by the dashed line.

has exponential asymptotics, as the deviation from either homogeneous state decays at $|z| \to \infty$ as $\exp(-\sqrt{g'(\rho^{\pm})}|z|)$. This qualitative difference in asympotics is a consequence of incorrect truncation involving divergent higher-order terms.

Interfacial Curvature

Suppose now that the interface is weakly curved, so that isodensity levels do not coincide anymore with planes $z = \text{const}$. The nominal location of a curved diffuse interface (e.g. the Gibbs equimolar surface) can be used to describe it in the language of differential geometry commonly applied to sharp interfaces. Its spatial position can be defined in a most general way as a vector function $X(\xi)$ of surface coordinates ξ. A curved interface can be approximated locally by an ellipsoid with the half-axes equal to the principal curvature radii. If both radii far exceed the characteristic interface thickness, all isodensity levels are approximated by ellipsoidal segments equidistant from the interface. The density changes along the direction z normal to isodensity surfaces, and the density profile along each normal is defined in the zero order by the function $\rho(z)$ computed above.

A more precise approach is to introduce a coordinate frame aligned with a weakly deformed interface. Given the interface $X(\xi)$, one can find unit tangent vectors $t_\alpha = \partial X/\partial \xi_\alpha$ along the surface coordinates ξ_α, the surface metric tensor $g_{\alpha\beta} = t_\alpha \cdot t_\beta$, and the normal vector $n = \frac{1}{2}\epsilon^{\alpha\beta} t_\alpha \times t_\beta$, where $\times$ is the 3D cross product and $\epsilon_{\alpha\beta}$ is the antisymmetric tensor; the Greek indices taking the values (1,2) are lowered and raised with the help of the metric tensor $g_{\alpha\beta}$ and its inverse $g^{\alpha\beta}$. The curvature tensor $\kappa_{\alpha\beta}$ is defined through the covariant derivatives of the tangent or normal vectors with respect to the surface coordinates:

$$\nabla_\beta t_\alpha = \kappa_{\alpha\beta} n, \quad \nabla_\alpha n = \kappa_{\alpha\beta} g^{\beta\gamma} t^\gamma. \tag{30}$$

Next, we define the coordinate axis z directed along n with the origin on the Gibbs surface. To fix the signs, we assume that the dense phase prevails at $z < 0$. The coordinate surfaces $z = \text{const}$ are obtained by shifting the interface along the normal by a constant increment. Evidently, this shift causes the length to increase on convex, and to decrease on concave sections. The aligned frame is not well defined far from the interface due to a singularity developing on the concave side at distance about the smallest value of the local curvature radius, i.e. the smallest inverse eigenvalue of the curvature tensor. Since the aligned frame is well defined only sufficiently close to the interface, we have to assume that the curvature is of $O(\epsilon) \ll 1$ when measured on the characteristic scale of intermolecular interactions that defines the effective interface thickness.

The metric tensor of the aligned coordinate system extends the surface metric to the neighboring layers, so that the infinitesimal interval is computed as

$$dr^2 = dz^2 + (g_{\alpha\beta} + \epsilon z \kappa_{\alpha\beta}) \xi^\alpha \xi^\beta + O(\epsilon^2). \tag{31}$$

The free energy integral (32) is rewritten in the aligned frame as a 2D integral along the Gibbs surface. The free energy per unit area γ is a functional of the density profile $\rho(z)$, and is computed in the leading order using the standard two-phase solution on the infinite line $\rho_0(z)$. This approximation can be used whenever the density changes between the two extreme values ρ^- and ρ^+ within a thin layer where the aligned frame remains well defined. The interfacial curvature induces, however, an $O(\epsilon)$ correction to the local chemical potential, denoted as $\widetilde{\mu}(\xi)$. Thus, we write

$$\mathcal{F} = \int \left\{ \gamma[\rho_0(z)] + \epsilon\widetilde{\mu} \int_{-\infty}^{\infty} \rho_0(z) dz \right\} \sqrt{g}\, d^2\xi, \tag{32}$$

where g is the determinant of the surface metric tensor.

The equation defining $\widetilde{\mu}(\boldsymbol{\xi})$ (or, after imposing the condition $\widetilde{\mu} = \text{const}$, the equilibrium shape of the interface) is obtained by varying Eq. (32) with respect to normal displacements $\delta \boldsymbol{X}^{(n)} = \boldsymbol{n}\,\delta z$ reshaping the Gibbs surface. The variation of the area element is expressed through the mean Gaussian curvature $\kappa = g^{\alpha\beta}\kappa_{\alpha\beta}$:

$$\begin{aligned}
\delta\sqrt{g} &= \tfrac{1}{2}\sqrt{g}\,g^{\alpha\beta}\delta g_{\alpha\beta} = \sqrt{g}\,g^{\alpha\beta}\boldsymbol{t}_\alpha \cdot \delta\boldsymbol{t}_\beta \\
&= -\left(\sqrt{g}\,g^{\alpha\beta}\boldsymbol{t}_\alpha\right)_{,\beta} \cdot \boldsymbol{n}\,\delta z = -\sqrt{g}\,g^{\alpha\beta}\kappa_{\alpha\beta}\,\delta z.
\end{aligned} \tag{33}$$

The variation of the other term in Eq. (32) is

$$\widetilde{\mu}\,\delta z \int_{-\infty}^{\infty} \rho_0'(z)\mathrm{d}z = \widetilde{\mu}(\rho^+ - \rho^-). \tag{34}$$

Assuming, in the leading order, $\gamma = \text{const}$, the first-order variation is computed as

$$(\rho^+ - \rho^-)\widetilde{\mu} + \gamma\kappa = 0. \tag{35}$$

First-order terms are added as well when Eq. (32) is varied with respect to ρ; the respective Euler–Lagrange equation can be used to compute first-order correction to the density profile, which we shall not need.

Equation (35) is equivalent to the Gibbs–Thomson law relating the equilibrium chemical potential with interfacial curvature. This relation is valid only when the surface tension γ is independent of curvature, but curvature-dependent corrections to γ, stemming from corrections to the 1D interaction kernel (24) due to lateral integration along curved isodensity levels, are of $O(\epsilon)$ and do not affect Eq. (35).

3.   Fluid-Substrate Interactions

Disjoining Potential

In the proximity of a substrate, the additional term in the free energy integral (13) is

$$\mathcal{F}_s = \int \rho(\boldsymbol{x})\,\mathrm{d}^3x \int_s U_s(|\boldsymbol{x}-\boldsymbol{r}|)\rho_s(\boldsymbol{r})\,\mathrm{d}^3r\,, \tag{36}$$

where U_s is the attractive part of the fluid-substrate interaction potential, ρ_s is the substrate density, and $\int_s$ means that the integration is carried over the volume occupied by the substrate; all other integrals in Eq. (13) are now restricted to the volume occupied by the fluid.

We shall consider a flat interface parallel to the substrate surface $z = 0$, and suppose that liquid-substrate interactions are also of the van der Waals type with a modified constant $A_s = \alpha_s A$. Then the free energy per unit area is expressed, after some rearrangements, as

$$\gamma = \int_0^\infty \rho(z) \left\{ f(\rho) + \psi(z) \left[\alpha_s \rho_s - \frac{1}{2}\rho(z) \right] \right\} dz$$

$$+ \frac{1}{2} \int_0^\infty \rho(z) \, dz \int_0^\infty Q(z - \zeta)[\rho(\zeta) - \rho(z)] \, d\zeta. \qquad (37)$$

The first term contains the same local part as in Eq. (23) complemented by the liquid-substrate interaction energy. The latter is computed by integrating the attracting part of the fluid-fluid and fluid-substrate interaction energy laterally as in Eq. (24) and represents the shift of energy compared to the unbounded fluid. The term $\rho(z)/2$ compensates lost fluid-fluid interactions in the substrate domain which are included in the homogeneous part $f(\rho)$. The function $\psi(z)$ is computed as

$$\psi(z) = -\pi A \int_0^\infty d\zeta \int_{q0}^\infty q^{-3} \, dq = \int_0^\infty Q(\zeta - z) d\zeta, \qquad (38)$$

where the integration limit is $q_0 = (z - \zeta)^2$ at $|z - \zeta| > d$, $q_0 = d^2$ at $|z - \zeta| \leq d$. The result is

$$\psi(z) = \begin{cases} -\frac{1}{6}\pi A z^{-3} & \text{at} \quad |z| > d \\ -\pi A d^{-3} \left(\frac{2}{3} - \frac{z}{2d} \right) & \text{at} \quad |z| < d. \end{cases} \qquad (39)$$

This expression, however, does not take into account steric effects causing liquid layering in the vicinity of the substrate.

The last term in Eq. (37) expresses, as before, the distortion energy, now restricted to the half-space $z > 0$. The Euler–Lagrange equation derived from Eq. (37) is the familiar Eq. (25) with an additional z-dependent term $\psi(z)[\alpha_s \rho_s - \rho(z)]$.

In a situation compatible with the lubrication approximation, perturbations due to the proximity of a solid surface are weak. In this case, the translational invariance of an unbounded two-phase system is weakly broken, and both the shift of the equilibrium chemical potential due to interactions with the solid surface and the deviation from the zero-order density profile are small. Since molecular interactions have a power decay with a nanoscopic characteristic length, this should be certainly true in layers exceeding several molecular diameters. A necessary condition for the perturbation to remain weak even as the liquid-vapor and liquid-solid interfaces are drawn together still closer, as it should happen in the vicinity of a contact line, is smallness of the dimensionless Hamaker constant $\chi = \alpha_s \rho_s / \rho^+ - 1$. Even under these conditions, the perturbation, however, ceases to be weak when the density in the layer adjacent to the solid deviates considerably from ρ^+. This means that low densities near the solid surface are strongly discouraged thermodynamically, and a dense *precursor layer* should form on a solid surface in equilibrium with bulk liquid even when the liquid is weakly nonwetting. Since $\rho^+ - \rho(z) \propto |z|^{-3}$ at $z \to -\infty$ in an unbounded fluid, the thickness of the precursor layer is estimated as $h_0 \propto \chi^{-1/3}$ at $\chi \ll 1$.

Sharp Interface Limit

The equilibrium chemical potential is shifted from the Maxwell construction, $\mu = \mu_0$ in the proximity of the solid surface. In the sharp interface theory, this shift, called *disjoining potential* [25], is defined as

$$\mu_s = \frac{1}{\rho^+ - \rho^-} \frac{\partial \gamma}{\partial h}, \tag{40}$$

where h is the distance between vapor-liquid interface, which should be identified here with the Gibbs equimolar surface (6), and the substrate.

Returning to Eq. (37), one can observe that only the non-autonomous (z-dependent) part of the first term is responsible for the disjoining potential proper, caused by replacing liquid molecules by the solid in the half-space $z < 0$. The other terms express the energy of the liquid-vapor interface, which is modified when the fluid is restricted to the half-space $z > 0$. The shift of the chemical potential can be computed, in the leading order, by using in Eq. (37) the zero-order density profile centered at the nominal interface $\rho = \rho_0(z - h)$.

We shall separate several constituent parts of this shift. The derivative of the non-autonomous term is

$$F_h^{(1)} = - \int_0^\infty \psi(z) \left[\alpha_s \rho_s - \rho_0(z - h) \right] \rho_0'(z - h) \mathrm{d}z. \tag{41}$$

Before differentiating the remaining terms in Eq. (37), it is convenient to transfer the h-dependence to the integration limits by using a shifted integration variable $z' = z - h$. The derivative of the local algebraic part is

$$F_h^{(2)} = \rho_0(-h)\overline{f}(\rho_0(-h)). \tag{42}$$

The nonlocal term in Eq. (37) is transformed after differentiating with respect to h using the symmetry of the interaction kernel $Q(z)$, and, after shifting the variable back, integrated by parts with the help of Eq. (38). The result is

$$\begin{aligned}
F_h^{(3)} &= -\frac{1}{2} \int_0^\infty Q(z)[\rho_0(-h) - \rho_0(z - h)]^2 \, \mathrm{d}z \\
&= \int_0^\infty \psi(z)[\rho_0(-h) - \rho_0(z - h)]\rho_0'(z - h) \, \mathrm{d}z. \tag{43}
\end{aligned}$$

The latter expression partly cancels with Eq. (41) when all contributions to μ_s are summed up. It is further convenient to separate the term proportional to the dimensionless Hamaker constant $\chi = \alpha_s \rho_s / \rho^+ - 1$. The resulting expression for the disjoining potential is

$$
\begin{aligned}
\mu_s \;=\; & \frac{1}{\rho^+ - \rho^-} \left\{ \rho_0(-h)\overline{f}(\rho_0(-h)) - \chi\rho^+ \int_0^\infty \psi(z)\rho_0'(z-h)\mathrm{d}z \right. \\
& \left. - \int_0^\infty \psi(z)[\rho^+ - \rho_0(-h)]\rho_0'(z-h)\mathrm{d}z \right\}.
\end{aligned}
\tag{44}
$$

In the limit $h \gg d$, when $\rho_0'(z-h)$ can be replaced by the delta function $-(\rho^+ - \rho^-)\delta(z-h)$, the second term yields the standard disjoining potential of a liquid layer with sharp interface and uniform density ρ^+:

$$
\mu_s^{\mathrm{st}} = -\frac{\pi\chi\rho^+ A}{6h^3} = -\frac{H}{6\pi h^3 \rho^+},
\tag{45}
$$

where $H = \pi^2 \rho^+ A(\alpha_s \rho_s - \rho^+)$ is the Hamaker constant defined in the standard way. The remaining terms vanish in the sharp interface limit when $\rho_0(-h) = \rho^+$.

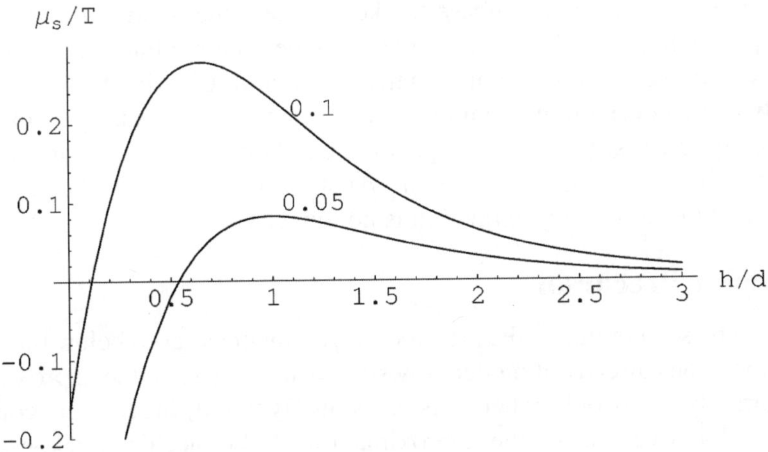

Figure 2. The dependence of the disjoining potential μ_s on the nominal layer thickness h defined by Eq. (44) for $\beta = 9$ and $\chi = 0.1$ and 0.05 (as indicated at the respective curves).

Thick Precursor

A formal first correction to the sharp interface limit valid at $h \gg 1$ can be obtained using the asymptotics of $\rho_0(z)$ at $z \to -\infty$ given by Eq. (28). The first term in Eq. (44), proportional to $[\rho^+ - \rho_0(-h)]^2$, decays asymptotically as h^{-6}. The same asymptotics is obtained for the last term in Eq. (44) when the integral is computed in the sharp interface limit. Moreover, both terms differ by a factor $\frac{1}{2}$ only in this approximation. When $|\chi| \ll 1$, all terms in Eq. (44) are of the same order of magnitude χ^2 when $h = O(|\chi|^{1/3})$. Neglecting the vapor density, we compute in the limit $h \gg 1$

$$\mu_s = -\frac{\pi A \chi \rho^+}{6h^3} - \frac{\rho^+}{2g'(\rho^+)} \left(\frac{\pi A}{6h^3}\right)^2.$$ (46)

The dependence $\mu_s(h)$ defined by this asymptotic formula is non-monotonic at $\chi < 0$, passing a maximum at $h_m = (\pi A/6|\chi|g'(\rho^+))^{1/3}$ and crossing zero at $h_0 = 2^{-1/3}h_m$, as seen in Fig. 2). The maximum $h = h_m$ corresponds to a minimal thickness of a liquid nucleus condensing on the solid surface. At h larger than the critical thickness h_m, the density profiles are non-monotonic. Such a solution describes a liquid layer sandwiched between the vapor and the solid, with a weakly depleted density near the solid surface. Non-monotonic density profiles are unstable with respect to perturbations with a sufficiently long wavelength. This instability is inherent to any nonwetting liquid, but the dynamics is practically frozen whenever the layer has a macroscopic thickness.

At smaller values of h, the maximum disappears, and the solution can be interpreted as a pure vapor phase thickening near the solid wall. The value $h = h_0$ such that $\mu_s(h_0) = 0$ corresponds to the nominal interface position on the "dry" surface in equilibrium with the bulk liquid. Clearly, the surface is not literally dry, as even on the nominally "dry" patches the density must be close to bulk liquid density under the specified conditions. At still smaller values of h (which may be also negative) $\mu_s(h)$ sharply decreases to large negative values, and the above approximation is no longer valid.

Molecular Precursor

As can be seen in Fig. 2, the precursor layer thickness goes below the molecular dimensions already at moderately small values of χ, so the approximation (46) formally holds only when χ is exceedingly small; indeed, the condition $\chi^{1/3} \ll 1$ is relevant for the approximation. Submolecular thicknesses are clearly non-physical. Even though density functional computations, in spite of their continuous formulation, can be continued to these distances, simple model potentials cannot account for the various steric constraints, fluctuations and multiple correlations relevant at this distances. Rather than introducing excessive physical detail of dubious veracity, we shall modify the model in a simple way to avoid a divergency at $h \to 0$ arising in a sharp interface model and to allow for a precursor in equilibrium with bulk fluid.

For this purpose, we assume the precursor film to be of a *constant* molecular thickness d, but have variable liquid density. The respective free energy functional can be obtained then by singling out the $z = d$ section in Eq. (37)

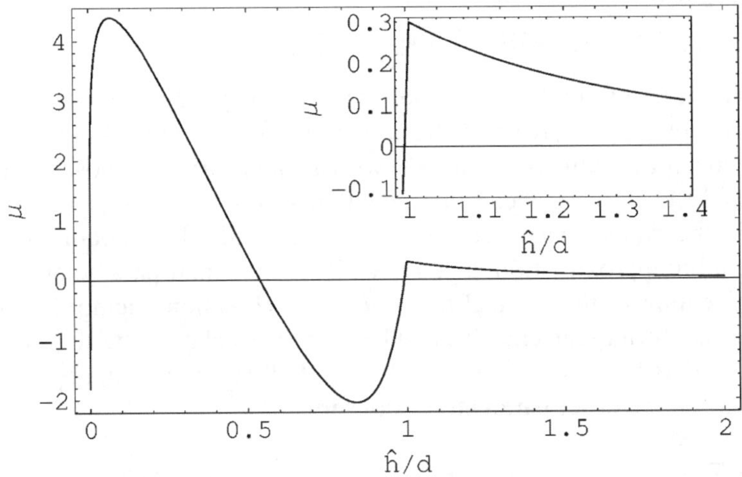

Figure 3. A typical dependence of the chemical potential μ on the effective layer thickness (see the definition in the text). The inset is the enlarged high-density part of the picture.

and computing the change of chemical potential needed to restore equilibrium shifted under the influence of fluid-substrate interactions. The Euler–Lagrange equation, replacing Eq. (25), obtained by minimizing the grand ensemble thermodynamic potential $\Phi = \gamma - \mu\rho$ is expressed now as

$$g(\rho) - \mu + \psi(d)[\rho^+(\chi + 1) - \rho] = 0, \quad \psi(d) = -\frac{\pi A}{6d^3}. \qquad (47)$$

A combined dependence of chemical potential on effective layer thickness $\widehat{h}$, defined as $\widehat{h} = h$ at $h \geq d$ and $\widehat{h} = d\rho/\rho^+$ at $h \leq d$, is shown in Fig. 3. The dependence merges the sharp-interface formula (45) at $h \geq d$ with the dependence $\mu(\rho)$ given by Eq. (47) at $h \leq d$. The latter dependence allows for both dense and dilute precursor with densities approaching those of liquid and vapor. If the latter possibility is ignored, only the right-hand part of the picture, enlarged in the inset, is relevant. This part has qualitatively the same, albeit steeper, tent-like shape as in Fig. 2, allowing for equilibrium between a dense molecular precursor and bulk liquid at $\mu = 0$. Due to the steep dependence of the chemical potential on liquid density, the density depletion relative to ρ^+ in the precursor at equilibrium with bulk fluid remains small even at moderate values of χ.

4.   Dynamic Contact Line

Generalized Cahn–Hilliard Equation

The dynamics of thin films or droplets bounded by a three-phase contact line is described by Eqs. (7), (8) with the potential μ incorporating both interactions with the substrate, defined by Eq. (40) with an appropriately chosen interaction model, and the effect of weak interfacial curvature according to Eq. (35). In the latter, the vapor density can be neglected, while the curvature expressed in lubrication approximation as $\kappa = -\epsilon\nabla^2 h$; the small parameter ϵ due to a different scaling in the vertical and horizontal direction, cannot be excluded from the final form. One can also add here an external potential $V(\mathbf{x})$, e.g. due to gravity. This leads to a generalized Cahn-Hilliard equation, appropriate for the case when the order parameter is conserved:

$$h_t = \nabla \cdot k(h)\nabla W, \quad W = \rho^+ \mu = -\gamma\epsilon\nabla^2 h + \rho^+ \mu_s(h) + V(\boldsymbol{x}). \tag{48}$$

In the abscence of explicit coordinate dependence, this evolution equation can be written in a variational form, e.g.

$$h_t = \nabla \cdot k(h)\nabla\frac{\delta\mathcal{W}}{\delta h}, \quad \mathcal{W} = \int \left[\frac{\gamma\epsilon}{2}|\nabla h|^2 + \rho^+ \int \mu_s(h)\mathrm{d}h\right] \mathrm{d}^2\boldsymbol{x}. \tag{49}$$

A modified non-conservative form of the evolution equation may include evaporation or condensation driven by the difference between the chemical potentials of the fluid layer (μ) and the environment (μ^0):

$$h_t = \nabla \cdot k(h)\nabla W - \beta(W - \mu^0). \tag{50}$$

Static Contact Angle

Under equilibrium conditions (in the absence of external forces), Eq. (48) reduces to

$$W = \gamma\epsilon\nabla^2 h - \rho^+ \mu_s(h) = 0. \tag{51}$$

A contact line can be defined as a transition between a precursor film and a "droplet", or a film of macroscopic thickness. For a straight contact line normal to the x axis, it is convenient to use a "phase plane" representation of this equation obtained by taking the nominal layer thickness h as the independent, and the squared slope $y = (h'(x))^2$ as the dependent variable. The boundary condition $y = 0$ is set at the precursor layer thickness h_0:

$$\frac{\mathrm{d}y}{\mathrm{d}h} = \frac{2}{\gamma\epsilon}\,\mu_s(h), \quad y(h_0) = 0. \tag{52}$$

The rescaled equilibrium contact angle can be computed by integrating Eq. (52) from h_0 to infinity

$$\sqrt{y(\infty)} = \left[\frac{2}{\gamma \epsilon} \int_{h_0}^{\infty} \mu_s(h) dh \right]^{1/2}. \tag{53}$$

This formula can be used to identify the small parameter ϵ with the squared asymptotic contact angle θ; this is done simply by setting $y(\infty) = 1$.

At $\theta \ll 1$, $\cos\theta \approx 1 - \frac{1}{2}\theta^2$, the above result reproduces the standard Young–Laplace formula $\gamma \cos\theta = \gamma_{gs} - \gamma_{ls}$. This can be seen by using the relation $\mu_s = d\gamma/dh$ to evaluate the integral explicitly, and observing that $\gamma(h_0)$ equals to the gas-solid surface tension γ_{gs}, while $\gamma(\infty) = \gamma + \gamma_{gs}$ is the sum of the gas-liquid and liquid-solid surface tensions.

For a finite droplet, the contact line can be nominally defined as the locus of the inflection point of the profile $h(x)$. The slope at the inflection point coincides with the "apparent" contact angle one would measure in a macroscopic experiment, which does not resolve a strongly curved transitional region separating the bulk fluid and the precursor layer. The profile in the transitional region can be easily matched to a macroscopic solution defining the equilibrium shape of the droplet bulk.

Quasistationary motion

As an example of application of the generalized Cahn–Hilliard equation, we consider the case when a droplet is set into slow motion due to either external forces or long-range interactions. We assume that the deviation from equilibrium shape remains weak and can be treated as a small perturbation everywhere. The droplet mobility can be deduced then from integral conditions based on an equilibrium solution. This allows us to avoid solving dynamic equations explicitly and computing a perturbed shape.

We consider *quasistationary* motion without change of form. Let $h(\boldsymbol{x} - \boldsymbol{X})$ be a stationary droplet profile dependent on the position of its center of symmetry $\boldsymbol{X}$. In order to keep the lubrication scaling consistent, one has to assume that the displacement is measurable on the same scale as h, rather than on a longer "horizontal" scale, and denote accordingly $\boldsymbol{U} = \epsilon^{-1/2}\boldsymbol{X}'(t)$. Then the time derivative in Eq. (48) is evaluated as $-\epsilon^{1/2}\boldsymbol{U} \cdot \nabla h$. Following this substitution, Eq. (48) can be integrated to $\epsilon^{1/2}\boldsymbol{U}h = \boldsymbol{j}$, which is just an expression of local mass conservation in quasistationary motion. Using here Eq. (48) yields the quasistationary equation

$$\frac{C h}{k(h)} = \nabla \left[\nabla^2 h - \frac{\mu_s(h) + V(\mathbf{x})}{\gamma \epsilon} \right], \tag{54}$$

where $C = \epsilon^{-3/2}U\eta/\gamma$. It is clear from this expression that dynamic distortion of the droplet shape, governed by competition between viscosity and surface tension is properly characterized by the *rescaled* capillary number, which can be defined after replacing $\epsilon^{1/2} \to \theta$ as $C = |C| = \mathrm{Ca}\,\theta^{-3}$, rather than by the standard capillary number $\mathrm{Ca} = |U|\eta/\gamma$. This dimensionless number combines in a rather non-trivial way the two physically unrelated parameters – velocity and equilibrium contact angle. It characterizes the interplay between the dynamical (viscous) and static (Laplacian) contributions to the apparent contact angle. The well-known Tanner's law in de Gennes' interpretation [5], $\theta \propto \mathrm{Ca}^{1/3}$ follows from a possibility to rescale C to unity in the case when molecular interactions are neglected and θ is reinterpreted as an apparent dynamic contact angle. In the opposite case $C \ll 1$, the contact angle is close to its equilibrium value.

Quasistationary equations may be applicable not uniformly, but only in some constitutive parts of the system. The three distinctive parts, where different scalings are appropriate, are droplet bulk, the contact line region and the precursor. In the first two regions, the standard lubrication scaling, with the ratio of the vertical and horizontal scales equal to θ, is applicable, but the scaling of both h and the horizontal extent of the region is different: the droplet radius a in the droplet bulk, and h_0 in the narrow region near the contact line. In the precursor region, h_0 remains the vertical scale, while the horizontal scale may be extended beyond the standard lubrication ratio, and is defined by the scale of the applied driving force (e.g. by the distance between the droplets when the only driving force is their interaction).

Quasistationary equations can be solved numerically [23, 26] or with the help of analytical matching techniques [27–29] for standard setups, such as advancing or receding menisci or macroscopic droplets on a precursor layer. An example of analytical solution based on integral mobility relations [28] is given in the next Section.

5.   Mobility Relations

Thermodynamic forces

For a single droplet surrounded by the precursor film of thickness h_0, the integral relations defining the droplet mobility are obtained by multiplying Eq. (54) by $h - h_0$ and integrating over a large region centered on the droplet [28]. We shall show that the terms including effects of curvature and molecular interaction with the substrate reduce to a pure divergence, and can be expressed therefore through contour integrals vanishing in an infinite region. The first term in r.h.s. of Eq. (54) is presented as

$$(h - h_0)\nabla\nabla^2 h = \tfrac{1}{2}\nabla \cdot \boldsymbol{T}, \tag{55}$$

where the components of the tensor $\boldsymbol{T}$ are

$$T_{ij} = (h - h_0)(\partial_i\partial_j h + \delta_{ij}\nabla^2 h) - \partial_i h \partial_j h. \tag{56}$$

Integrating this term leads therefore to a contour integral $\frac{1}{2} \oint \boldsymbol{n} \cdot \boldsymbol{T} \, ds$, where $\boldsymbol{n}$ is the normal to the boundary of the integration region. As this boundary lies on a flat precursor layer, the contour integral vanishes.

The contribution of the disjoining potential is presented as

$$(h - h_0)\nabla \mu_s(h) = (h - h_0)\mu_s'(h)\nabla h = \nabla F(h),$$
$$F(h) = \int_{h_0}^{h} (h - h_0)\mu_s'(h)\mathrm{d}h. \tag{57}$$

This leads to a contour integral expressing the driving force due to a weak gradient of the disjoining potential in the precursor layer:

$$\int (h - h_0)\mu_s(h) \, \mathrm{d}\boldsymbol{x} = \oint \boldsymbol{n}F \, \mathrm{d}s. \tag{58}$$

A non-vanishing contour integral yields a driving force due to a weak gradient of the disjoining potential in the precursor layer, which may be caused either by interaction of individual well separated droplets or by variation of the disjoining potential due to weak temperature or concentration gradients or chemical inhomogeneity of the substrate. Still another cause may be the action of boundaries, such as an apparently unexpected migration of a droplet towards the edge of a horizontal substrate [30].

The external potential, being coordinate-dependent, generally, cannot be reduced to a divergence, but in a special case of gravity potential the total force is computed easily. For a plane inclined along the x axis, the gravity potential is $V(x) = g\rho^+(h - \alpha x)$, where the slope is presented as $\alpha\theta \ll 1$ and g is the acceleration of gravity. Multiplying the gradient of the gravity potential, $V'(x) = g\rho^+(\nabla h - \alpha)$, by $h - h_0$ and integrating over a large segment centered on the droplet, we observe that the integral of the first term vanishes whenever the precursor layer is flat in the outlying region, while the second term gives just the total dimensionless volume or gravity force acting upon the droplet proportional to its volume above the precursor layer.

Dissipative integral

The remaining terms in Eq. (54) are those due to the external potential and viscous friction. Whatever is the cause of motion, computation of the droplet velocity requires evaluation of the *dissipative integral*

$$I = \int \frac{h(h - h_0)}{k(h)} \, \mathrm{d}\boldsymbol{x}. \tag{59}$$

As the integrand vanishes in the precursor, the dissipative integral can be evaluated separately in the droplet bulk and the vicinity of the contact line, and matched at a point where the layer thickness lh_0, is much larger than the precursor thickness but much smaller than the maximum droplet height. The exact position of the matching point should eventually fall out without affecting the final result.

We consider first a 2D case. In the contact line region interpolating between the precursor and the bulk, the integration can be carried out using the phase plane solution $y(h)$ obtained by integrating Eq. (52). Using the standard Stokes mobility function, $k(h) = \frac{1}{3}h^3$, the contribution of the contact line region to the dissipative integral is evaluated as

$$I_1 = 3 \int_{h_0}^{lh_0} \frac{h - h_0}{h^2 \sqrt{y(h)}} \, \mathrm{d}h = 3 \ln \frac{l}{b}. \tag{60}$$

The chosen particular form of the disjoining potential which determines the function $y(h)$ affects only the numerical constant b in the last expression.

In the droplet bulk, one can evaluate the dissipative integral using the circular cap solution valid when gravity and other external forces can be neglected, this will be just a spherical cup approximated by a parabola:

$$h = \frac{1}{2}a[1 - (x/a)^2], \tag{61}$$

where a is the droplet radius and x is the distance from its center of symmetry.The contact angle falls out when a, as well as x, is measured on the scale extended relative to the scale of h. The contribution of the bulk region to the dissipative integral is computed in the leading order in the ratio $h_0/a \ll 1$ as

$$I_2 = 3 \int_0^{a - lh_0} \frac{\mathrm{d}x}{h(x)} = 6 \ln \frac{2a}{lh_0}. \tag{62}$$

Adding up both contributions yields

$$I = 2I_1 + I_2 = 6 \ln \frac{2a}{bh_0}. \tag{63}$$

In 3D, the integral I_1 is evaluated exactly as above, since the curvature of the contact line is negligible when the droplet radius in the x-plane far exceeds the thickness of this transitional zone. The direction normal to the droplet boundary does not coincide now, however, with the local direction of U. Therefore the local contribution of the vicinity of the contact line to the dissipative integral given by Eq. (60) should be multiplied by $|\cos \phi|$, where ϕ is the angle between the local radius and the direction of motion. Integrating over the circular droplet boundary yields

$$I_1^\circ = a I_1 \int_0^{2\pi} |\cos \phi| \, \mathrm{d}\phi = 4a \, I_1. \tag{64}$$

The bulk integral is evaluated using the spherical cap solution (61) as

$$I_2^\circ = \int_0^\pi |\cos\phi|\, d\phi \int_0^{a-lh_0} \frac{x dx}{h} = 12a \ln\frac{a}{2lh_0}. \tag{65}$$

The final result is

$$I = I_1^\circ + I_2^\circ = 12a \ln\frac{a}{2bh_0}. \tag{66}$$

Take note that the bulk and contact line contributions, though distinct, cannot be separated in a unique way. The logarithmic factor combining the inner and outer scales is a telltale sign of the matching procedure.

Using the expressions for the dissipative integral, we can readily evaluate the speed of a drop sliding on an inclined plane when the driving force is proportional to the volume. As h_o can be neglected compared with a, the volume is computed, respectively in 2D and 3D, as

$$v_2 = \tfrac{2}{3}a^2, \quad v_3 = \tfrac{1}{4}\pi a^3. \tag{67}$$

In the absence of other forces, the dimensional and dimensionless droplet velocities following from the integral balance of Eq. (55) are

$$U = \frac{c_g\theta^2 g\rho\alpha a^2}{\eta \ln(c_v a/h_m)}, \quad C = \frac{c_g G}{\ln(c_v a/h_m)}, \tag{68}$$

where $G = g\rho a^2\alpha/(\theta\gamma)$ is the Bond number, and the numerical constants are $c_g = 1/9, c_v = 2/b$ in 2D, $c_g = \pi/48, c_v = 1/(2b)$ in 3D. The final expressions retain *logarithmic* dependence on the ration of the macroscopic and microscopic lengths a/h_0. A more interesting problem of motion induced by droplet interaction is considered below.

Interactions Through the Precursor Film

Interaction of droplets is caused by fluxes induced in the precursor. It may result in both motion and volume change. As long as the droplets are far removed, both processes are additive and can be analyzed separately using the integral relations derived in the preceding Section. We consider an assembly of droplets sitting on the precursor film, and neglect both external forces and evaporation. The droplets are assumed to be sufficiently far removed, so that a contour lying in the precursor could be drawn around each of them separately. This requirement is, in fact, not very stringent, since relaxation to a flat precursor outside the contact line region is exponential with a characteristic length of $O(h_0)$. The droplets might have been formed as a result of spinodal dewetting and subsequent coarsening. Then what we are going to describe is a late stage of evolution following the formation of well separated macroscopic droplets.

Interaction of droplets is totally determined under these conditions by the respective contour integrals. Evaluating the latter requires, however, solving Eq. (48) in the precursor. The precursor evolution, as well as droplet motion, can be considered quasistationary when interaction is weak (an *a posteriori* estimate will be given in the end of this subsection). Since the curvature of the precursor is negligible as well, Eq. (48) reduces to the diffusion equation with variable effective diffusivity $D(h)$:

$$\nabla \cdot (D \, \nabla h) = 0, \quad D(h) = \frac{k(h)\mu'_s(h)}{\eta} > 0. \tag{69}$$

Due to a sharp dependence $\mu_s(h)$, like e.g. in Eq. (46), diffusivity may *increase* in this case in thinner layers, in spite of a decrease of mobility $k(h)$.

A single droplet with the radius $a \gg h_0$, which has (in 3D) the form of a spherical cap defined by Eq. (61), is in equilibrium with the precursor film with the thickness

$$\widetilde{h}(a) = h_0(1 + \Delta), \quad \Delta = \frac{2\gamma\theta^2}{ah_0\mu'_s(h_0)} \propto \frac{h_0}{a}. \tag{70}$$

The precursor is *thinner* near larger droplets, which should therefore grow at the expense of smaller ones. The deviations from h_0 are expected be very small, as dependence $\mu_s(h)$, like one defined by Eqs. (46) or (47) is very steep at short distances from the substrate. The diffusion equation (69) can be replaced therefore by the Laplace equation $D\nabla^2 h = 0$ with the effective diffusivity $D(h_0)$.

In 2D (i.e. for a 1D film), the fluxes can be conveniently computed by considering interactions between the nearest neighbors [27], but in 3D exact theory is very complicated, and it is appropriate to use the effective medium approximation, as it is commonly done in theory of coarsening [31]. The total flux out of the droplet can be defined then as exchange with the precursor film of some effective thickness h_c slowly changing with time. The value of h_c is defined by mass conservation requiring the sum of fluxes out of all droplets to vanish. As in the standard Lifshitz–Slyozov theory [31], it can be conveniently defined through the radius a_c of a critical droplet that neither grows nor decays: $h_c = \widetilde{h}(a_c)$

The film thickness distribution created by a number of well separated droplets with radii a_k can be written then as

$$h(\boldsymbol{x}) = \sum (h_k - h_c) \ln(a_k/r_k) = \ln \prod (a_k/r_k)^{h_k - h_c}. \tag{71}$$

where $h_k = \widetilde{h}(a_k)$ and $r_k = |\boldsymbol{x} - \boldsymbol{X}_k|$ is the distance from the center of kth droplet. If all droplets are located in a finite region on an infinite plane, the divergence at large distances is avoided by setting $h_c = \langle h_k \rangle$, which, in view of Eq. (70), means that $a_c = 1/\langle a_k^{-1} \rangle$ is the harmonic average radius.

The flux out of kth droplet is $j_k = 2\pi D(h_k - h_c)$. These fluxes cause coarsening of the droplet size distribution, as large droplets grow and small ones decay. Unlike the standard coarsening problem, the droplets not only grow or decay, but also move under the influence of the gradient of chemical potential in the precursor. The driving force is given by the contour integral in Eq. (58). Due to the fast decay to the local precursor thickness outside the contact line region, the contour in Eq. (58) can be identified, up to negligible $O(h_0)$ deviations, with the circular droplet boundary. The contribution of the constant part of $F(h)$ vanishes upon integration. The driving force stems therefore exclusively from a small circuferential component of ∇F caused by mass fluxes from far removed droplets. Using Eqs. (57), (70), the driving force acting on a test droplet placed at the origin is computed in the leading order as

$$
\begin{aligned}
\boldsymbol{F} &= \oint \boldsymbol{n}\, F(h(s))\, \mathrm{d}s = a \int_0^{2\pi} \widehat{\boldsymbol{r}}(\phi)\, F(h(\phi))\, \mathrm{d}\phi \\
&= a^2 F'(h) \int_0^{2\pi} \widehat{\boldsymbol{r}}\,(\widehat{\boldsymbol{r}} \cdot \nabla h)\, \mathrm{d}\phi \\
&= \pi a^2 (\widetilde{h}(a) - h_0)\mu_s'(h_0)\, \overline{\nabla h} = 2\pi a \gamma \theta^2\, \overline{\nabla h},
\end{aligned}
\tag{72}
$$

where $\widehat{\boldsymbol{r}}$ is the unit vector in the radial direction, ϕ is the angular coordinate, and $\overline{\nabla h}$ is the thickness gradient induced by all other droplets:

$$
\overline{\nabla h} = \sum (h_k - h_c)\, \frac{\boldsymbol{r}_k}{r_k^2}.
\tag{73}
$$

Using Eq. (66), the velocity of the droplet is computed then as

$$
\boldsymbol{U} = -\frac{\theta}{\eta}\, \frac{\boldsymbol{F}}{I} = -\frac{\pi}{6}\, \frac{\gamma \theta^3}{\eta \ln(a/2bh_0)} \sum (h_k - h_c)\, \frac{\boldsymbol{r}_k}{r_k^2}.
\tag{74}
$$

The test droplet is, respectively, attracted and repelled by the droplets above and below the critical size. This can be attributed to the action of weak fluxes sucked in by larger and blown out by smaller droplets. For a pair of droplets considered in the preceding subsection, this causes both droplets to migrate in the direction of the larger droplet. The smaller droplet moves somewhat faster due to a smaller logarithmic factor in the dissipative integral, and therefore has some chance to catch up and coalesce before disappearing.

Conclusion

The above theory blends hydrodynamic and thermodynamic relations into a common formalism in the framework of lubrication approximation. This provides a common framework for discussion of different physical effects. It still does not alleviate major computational difficulties stemming from large scale separation under conditions when a simple quasistationary theory fails and the form of the moving film or droplet becomes strongly distorted in the vicinity of the contact line. Major unresolved difficulties remain outside the applicability limits of lubrication approximation. For practical applications, taking account of surface inhomogeneities is essential. This would introduce into the problem an whole continuum of scales spanning the gap between nanoscale molecular interactions and and mesoscopic hydrodynamic scale. Decades of studies of the dynamic contact line have lead so far to understanding these major difficulties rather than resolving them, and a new level of detail in both experiment and computations is necessary to attain abilities to effectively control the three-phase boundary under realistic dynamic conditions.

Acknowledgement

This work has been supported by Israel Science Foundation (grant 55/02). The author is grateful to Yves Pomeau and Arik Yochelis for numerous discussions.

References

[1] J.D. van der Waals, Z. f. Phys. Chem. **13** 657 (1894); English translation J.S. Rowlinson, J. Stat. Phys. **20** 197 (1979).

[2] J.W. Cahn and J.E. Hilliard, J. Chem. Phys. **31** 688 (1959).

[3] J.S. Rowlinson and B. Widom, *Molecular Theory of Capillarity*, Oxford University Press, 1982.

[4] S. Dietrich, Wetting phenomena, in *Phase Transitions and Critical Phenomena*, edited by C. Domb and J. Lebowitz, Vol.12, p.1, Academic, London, 1988.

[5] P.G. de Gennes, Rev. Mod. Phys. **57** 827 (1985).

[6] O. V. Voinov, Fluid Dynamics **11**, 714 (1976).

[7] A. Oron, S.H. Davis, and S.G. Bankoff, Rev. Mod. Phys. **69** 931 (1997).

[8] D.M. Anderson, G.B. McFadden, and A.A. Wheeler, Annu. Rev. Fluid. Mech. **30** 139 (1998).

[9] D. Jasnow and J. Viñals, Phys. Fluids **8** 660 (1996).

[10] P. Seppecher, Int. J. Engng Sci. **34** 977 (1996).

[11] D. Jacqmin, J. Fluid Mech. **402** 57 (2000).

[12] H. Muller-Krumbhaar, H. Emmerich, E. Brener, and M. Hartmann, Phys. Rev. E **63** 026304 (2001).

[13] L.M. Pismen and Y. Pomeau, Phys. Rev. E **62**, 2480 (2000).

[14] L.M. Pismen, Phys. Rev. E **64** 021603 (2001).

[15] C. Huh and L.E. Scriven, J. Coll. Int. Sci. **35**, 85 (1971).

[16] E.B. Dussan V and S.H. Davis, J. Fluid Mech. **65**, 71 (1974).

[17] J. Koplik, J.R. Banavar and J.F. Willemsen, Phys. Fluids A **1**, 781 (1989).

[18] T. Qian, X.-P. Wang, and P. Sheng, Phys. Rev. Lett. **93**, 094501 (2004).

[19] P. Joseph and P. Tabeling, Phys. Rev. E **71**, 035303 (2005).

[20] H. Lamb, *Hydrodynamics*, Dover, 1932.

[21] T.D. Blake and J.M. Haynes, J. Colloid Interface Sci. **30**, 421 (1969).

[22] E. Ruckenstein and C.S. Dunn, J. Coll. Interface Sci. **59**, 135 (1977).

[23] L.M. Pismen and B.Y. Rubinstein, Langmuir, **17** 5265 (2001).

[24] L.D. Landau, and E.M. Lifshitz, v. V, *Statistical Physics*, Part I, Pergamon Press, 1980.

[25] B.V. Derjaguin, N.V. Churaev, and V.M. Muller, *Surface Forces*, Consultants Bureau, New York, 1987.

[26] U. Thiele and E. Knobloch, Phys. Fluids **15**, 892 (2003).

[27] K.B. Glasner and T.P. Witelski, Phys. Rev. E **67**, 016302 (2003).

[28] L.M. Pismen and Y. Pomeau, Phys. Fluids **16**, 2604 (2004).

[29] J. Eggers, Phys. Rev. Lett. **93**, 094502 (2004); Phys. Fluids **16**, 3491 (2004).

[30] A. Marmur and M.D. Lelah, J. Coll. Int. Sci. **78**, 262 (1980).

[31] I.M. Lifshitz and V.V. Slyozov, J. Phys. Chem. Solids **19**, 35 (1961).

DYNAMICS OF THERMAL POLYMERIZATION WAVES

V.A. Volpert
Department of Engineering Sciences and Applied Mathematics
Northwestern University, Evanston, IL 60208, USA
v-volpert@northwestern.edu

Abstract   Frontal polymerization (FP) is a process in which a spatially localized reaction zone propagates through monomer, converting it into polymer. The process is of interest from both fundamental and applied viewpoints. Two different types of FP are known, thermal and isothermal. The mechanism of propagation of thermal FP waves is similar to that in combustion, namely, the heat released locally by the exothermic combustion reactions diffuses ahead, where it accelerates the reactions, and the process repeats. In isothermal FP, wave propagation is due to mass diffusion and auto-catalysis, specifically, the gel effect. This chapter reviews mathematical works on thermal frontal polymerization. We first discuss the kinetics of free-radical polymerization and then formulate a mathematical model of free-radical FP. We observe that the famous gasless combustion (GC) model is a limiting case of the FP model. We describe in detail the methods used to study the GC model. Specifically, we discuss asymptotic methods, reaction front approximation and step-function approaches. These methods are applied to the GC model to determine uniformly propagating one-dimensional waves, to study their stability, and to perform bifurcation studies near the stability threshold. We then discuss the same questions for the FP model, comparing the results with those for the GC model. Finally, we briefly describe extensions of the base FP model discussed in the literature.

Keywords: Modeling, gasless combustion, frontal polymerization, free-radical polymerization, traveling wave, stability, bifurcation

1.   Introduction

This chapter discusses propagation of polymerization waves. In a polymerization wave, a spatially localized reaction zone, in which the polymerization reactions occur, propagates into initial reactants (the monomer) leaving the reaction product (the polymer) in its wake. Two types of polymerization waves, thermal and isothermal, have been observed experimentally, and the mechanism of wave propagation for each is markedly different. Thermal polymeriza-

A. A. Golovin and A. A. Nepomnyashchy (eds.),
Self-Assembly, Pattern Formation and Growth Phenomena in Nano-Systems, 195–245.

tion waves propagate due to diffusion of heat released in exothermic polymerization reactions. Isothermal polymerization waves are due to mass diffusion of the species coupled with the gel effect. The chapter focuses on thermal polymerization waves which, in many respects, are similar to combustion waves. It turns out that the nondimensional parameters that determine the structure of a reaction wave are of the same order of magnitude in many polymerization processes as in combustion. As a result, thermal polymerization and combustion waves have similar structures (Figure 1). The wave consists of a narrow reaction zone which separates high-temperature products from initial reactants. This also explains why methods of combustion theory are useful in the study of polymerization waves, an area far less developed than combustion.

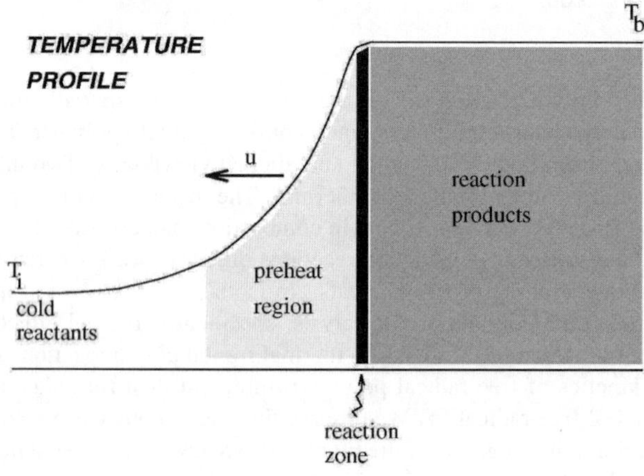

Figure 1.   Schematic of the propagating thermal wave.

The simplest experimental setup for a frontal polymerization (FP) study consists of a test tube filled with monomer and initiator. When a heat source is applied at one end of the tube, a reaction wave forms and propagates through the tube. The heat source locally intensifies the chemical reactions. Initiator decomposition, which produces the active free radicals that start the polymer chains, is typically a weakly endothermic reaction. The polymerization reactions that lengthen the polymer chains are significantly exothermic. The heat that is released due to polymerization diffuses into adjacent layers of the reactant mixture and intensifies the reactions there. The process repeats, and in this way, a self-sustained wave propagates along the tube. Despite the aforementioned similarity in the structure of polymerization and combustion waves, they behave quite differently. The propagation velocity of such polymerization waves is of the order of 1 cm/min, and the temperature increase in the system can be as much as 200K depending on the particular polymerization process, while these characteristics of combustion waves are much higher.

Polymerization waves are interesting from two points of view. First, they are used to produce polymers. This process strongly resembles another technological process occurring in a frontal regime, namely combustion synthesis (CS) of materials (also known as the self-propagating high-temperature synthesis, SHS), in which combustion waves are used to make ceramics and intermetallic compounds [53, 54, 57, 60, 92]. Second, polymerization waves are interesting from a fundamental point of view as another example of reaction wave propagation.

Frontal polymerization studies began in 1972 [13]. This work was performed as a polymerization analog of SHS. Using methyl methacrylate as the monomer, the authors demonstrated the phenomenon of polymerization wave propagation, and they showed how the front velocity and the polymer composition are influenced by the choice of initiator, its concentration and pressure change [10–13]. Subsequent experimental work [42, 68, 75, 76] demonstrated FP for a variety of monomers. Most of these experiments used liquid undiluted monomers. However, FP has also been carried out in systems with solid monomers [3, 73], in dispersions [71], and in solutions [69]. Along with demonstrating the feasibility of FP with particular monomers, experimentalists have also studied how experimental parameters affect the process and how to produce polymers with specific desired properties, e.g. functionally gradient polymeric materials [15], temperature-sensitive hydrogels [99], thermochromic composites [61, 62], conductive composites [89], nanocomposites [16], polymer-dispersed liquid-crystal materials [27] and others.

The chemical mechanism in FP is usually free-radical polymerization which, in the simplest case, includes three kinetic steps – initiation, propagation, and termination [64]. However, the process has also been successfully applied to epoxy curing [14], ring-opening metathesis [49], thiol-ene systems [78] and some others, all of which have different chemistries. Another way to classify FP processes concerns the number of different monomers used in the same experiment. Homopolymerization refers to systems that contain one type of monomer. Binary and copolymerization are systems that contain two (or more in the case of copolymerization) types of monomers. Free-radical binary polymer chains grow by successively adding the same type of monomer radical to the end of the chain and are analogous to forming two homopolymers in the same system. Binary polymerization has been demonstrated using a free-radical monomer and an epoxy monomer, which forms a step polymer [70]. In contrast to free-radical binary polymerization, free-radical copolymerization forms polymer chains that contain units of both monomers in the final product [64]. Frontal copolymerization has been achieved for a variety of monomer pairs [67, 90].

Mathematical models of frontal polymerization have also been developed. They are discussed below. These works study the velocity of the reaction front and determine which factors influence the behavior of the wave [2, 28–31, 90, 65–67]. The above references assume a uniformly propagating wave. However, experiments indicate that instabilities can develop in the wave thus destroying its uniformity [35, 77]. Relevant theoretical studies focus on linear [81, 87] and nonlinear [17, 32, 74] stability analyses for frontal homopolymerization. The paper [66] extends the analysis to two monomer systems, namely frontal copolymerization and binary frontal polymerization.

In this chapter, we review the theoretical work on thermal FP. We begin with a derivation of a base mathematical model which governs propagation of free-radical polymerization waves. We demonstrate that this model reduces in a limiting case to the famous gasless combustion (GC) model, which has been extensively studied in the context of SHS. We discuss some theoretical approaches to the simpler GC model and then apply them to the base model of free-radical FP. Some extensions of the base model are also discussed.

2. Mathematical model

The base mathematical model of free-radical FP was first proposed in [31], then studied in more detail in [86], and employed in a number of subsequent papers. In what follows, we formulate the model, referring the reader to the papers cited above for additional information regarding the derivation of the model. Reference [86] addresses most completely the validity of the assumptions that are made when deriving the model.

Free-radical polymerization involves a number of kinetics steps [64]. There are two species in the initial mixture: the monomer and the initiator. First, the initiator decomposes producing primary radicals, which are needed to start the growth of the polymer chain: a primary radical reacts with a monomer molecule to produce a polymer radical. The chain growth step is carried out by the successive addition of monomer molecules to the active polymer radical. Finally, an inactive polymer is formed by the combination of two radicals. These kinetics steps can be summarized as follows:

$$
\begin{aligned}
&\textbf{(1)} \quad \mathrm{I} \xrightarrow{k_d} f \times 2\mathrm{R} && \text{(initiator decomposition)} \\
&\textbf{(2)} \quad \mathrm{R} + \mathrm{M} \xrightarrow{k_p} \mathrm{P}_1^* && \text{(chain initiation)} \\
&\textbf{(3)} \quad \mathrm{P}_n^* + \mathrm{M} \xrightarrow{k_p} \mathrm{P}_{n+1}^* && \text{(chain growth)} \\
&\textbf{(4)} \quad \mathrm{P}_n^* + \mathrm{R} \xrightarrow{k_t} \mathrm{P} && \text{(primary radical termination)} \\
&\textbf{(5)} \quad \mathrm{P}_n^* + \mathrm{P}_m^* \xrightarrow{k_t} \mathrm{P} && \text{(polymer radical termination)}
\end{aligned}
$$

where I, R, M, P, and P_n^*, $n = 1, 2, \ldots$ represent initiator, primary radical, monomer, dead polymer and a polymer radical containing n monomer molecules, respectively. Here, combination is the assumed mode of termination (i.e., only one polymer molecule is formed as a result of the termination reactions), and f is the initiator efficiency factor (defined as the ratio of the number of primary radicals in polymer molecules to the total number of primary radicals formed by initiator). The reaction rate parameters have the usual form of Arrhenius exponentials

$$k_j = k_j(T) = k_j^0 \exp\left(-\frac{E_j}{R_g T}\right) \quad (j = d, p, t),$$

where R_g is the universal gas constant, T is the temperature of the medium, and k_j^0 and E_j are the pre-exponential factor and activation energy, respectively, of the reaction designated by j. Thus, the kinetic equations that describe the change in the concentrations of the species with time t due to the species production and consumption in the chemical reactions can be written as

$$\frac{\partial I}{\partial t} = -k_d I, \tag{2.1}$$

$$\frac{\partial R}{\partial t} = 2f k_d I - k_p R M - k_t R P^*, \tag{2.2}$$

$$\frac{\partial M}{\partial t} = -k_p R M - k_p M P^*, \tag{2.3}$$

$$\frac{\partial P^*}{\partial t} = k_p R M - k_t R P^* - k_t P^{*2}, \tag{2.4}$$

$$\frac{\partial P}{\partial t} = k_t R P^* + k_t P^{*2}. \tag{2.5}$$

Here I, R, M, P^*, and P denote the concentrations in mol/L of the corresponding species, and P^* is the total polymer radical concentration, i.e., $P^* = \sum_n P_n^*$.

These kinetic equations must be supplemented by the energy balance in the system, which accounts for the thermal diffusion and the net heat release. Reaction (**1**) is typically endothermic, whereas reactions (**2**) through (**5**) are exothermic. Although all of the reactions contribute somewhat to the net enthalpy of the polymerization process, the heat release due to the chain initiation and growth is most significant [46]. Thus, we account for the heat release due to reactions (**2**) and (**3**) and neglect all other heats of reaction. The energy balance takes the form

$$\frac{\partial T}{\partial t} = \kappa \tilde{\nabla}^2 T + q k_p (R M + M P^*),$$

where κ is the thermal diffusivity of the mixture and q is the rise in temperature induced per unit concentration of reacted monomer. We assume that κ is constant. We consider the problem in two dimensions: the $\tilde{x}$-axis coincides with the direction of propagation of the polymerization wave (i.e., the axis of the tube), and the y-axis points in the direction perpendicular to that of propagation. Thus, $\tilde{\nabla}^2$ is the two-dimensional Laplacian with respect to the spatial variables $\tilde{x}$ and y. We could formulate a three-dimensional model as well, i.e., to consider polymerization wave propagation in a 3D cylinder (simply replacing the 2D Laplacian by a 3D), but for the most part of this chapter, the two-dimensional formulation is sufficient.

We remark that we have not included diffusion terms in the mass balance equations, because mass diffusion in condensed phase is negligible compared to heat diffusion. We also have not accounted for convective flow of reactants in the equations, because in many polymerizing systems the medium is so viscous that the flow can be neglected.

The characteristic scale of the polymerization wave (i.e., the spatial region over which the major variation of the temperature and the species concentrations occurs) is typically much smaller than the length of the tube. Thus, on the scale of the polymerization wave the tube can be considered infinite, $-\infty < \tilde{x} < \infty$. It is convenient to introduce a moving coordinate $x = \tilde{x} - \varphi(t, y)$, where φ is the position of a characteristic point of the wave. The specific choice of φ will be described later. Expressed in the moving coordinate system, the mass and energy balance equations become

$$\frac{\partial I}{\partial t} - \varphi_t \frac{\partial I}{\partial x} + k_d I = 0, \tag{2.6}$$

$$\frac{\partial T}{\partial t} = \kappa \nabla^2 T + \varphi_t \frac{\partial T}{\partial x} + q k_p M (R + P^*), \tag{2.7}$$

$$\frac{\partial P}{\partial t} - \varphi_t \frac{\partial P}{\partial x} - k_t R P^* - k_t P^{*2} = 0, \tag{2.8}$$

$$\frac{\partial R}{\partial t} - \varphi_t \frac{\partial R}{\partial x} - 2 f k_d I + k_p R M + k_t R P^* = 0, \tag{2.9}$$

$$\frac{\partial P^*}{\partial t} - \varphi_t \frac{\partial P^*}{\partial x} - k_p R M + k_t R P^* + k_t P^{*2} = 0, \tag{2.10}$$

$$\frac{\partial M}{\partial t} - \varphi_t \frac{\partial M}{\partial x} + k_p M (R + P^*) = 0. \tag{2.11}$$

Here, the Laplacian ∇^2 in the moving coordinate system is given by

$$\nabla^2 = \left(1 + \varphi_y^2\right) \frac{\partial^2}{\partial x^2} + \frac{\partial^2}{\partial y^2} - \varphi_{yy} \frac{\partial}{\partial x} - 2 \varphi_y \frac{\partial^2}{\partial x \partial y}, \tag{2.12}$$

and φ_t, φ_y, φ_{yy} denote the corresponding derivatives.

The boundary conditions for the system (2.6)–(2.11) are the initial state of the mixture far ahead of the polymerization wave ($x \to -\infty$) and the final state far behind the wave ($x \to \infty$), where all of the reactions have come to completion

$$x \to -\infty: \quad T = T_0, \ I = I_0, \ M = M_0, \ R = P = P^* = 0, \qquad (2.13)$$

$$x \to +\infty: \quad \frac{\partial T}{\partial x} = 0. \qquad (2.14)$$

Here, I_0 and M_0 denote the concentrations of the initiator and the monomer initially present in the mixture, and T_0 is the initial temperature. In addition to the boundary conditions (2.13)-(2.14), we need boundary conditions in y. Let us assume that y varies over a finite interval, $0 < y < \ell$, and either periodic or no-flux conditions in y are prescribed. In most of the discussions below, we are interested in one-dimensional traveling waves, i.e., the solutions that depend only on x. The boundary conditions in y stated above allow for such y-independent solutions. Unless the dependence of the solution on y is really an issue, we will not even refer to the boundary conditions in y when discussing these mathematical models.

To further simplify the system of kinetic equations, we use the steady-state assumption (SSA) for the total radical concentration, i.e., the sum of the primary (R) and polymer (P^*) radical concentrations. The SSA asserts that the rate of change of the concentration of the radicals is negligible compared to the rates of their production and consumption, and it involves setting the time derivative of the total radical concentration to zero. It mathematically means that, appropriately nondimensionalized, the equation for the total radical concentration has a small parameter in front of the time derivative. Setting this parameter to zero means that we consider only the outer solution and thus disregard a short transient from the initial state to the steady state. This assumption, which has been studied in the context of polymerization waves in [86], reduces equations (2.9)-(2.11) to the single equation

$$\frac{\partial M}{\partial t} - \varphi_t \frac{\partial M}{\partial x} + k_{eff}(T)\sqrt{I}M = 0, \qquad (2.15)$$

and the energy balance to

$$\frac{\partial T}{\partial t} = \kappa \nabla^2 T + \varphi_t \frac{\partial T}{\partial x} + q k_{eff}(T)\sqrt{I}M, \qquad (2.16)$$

where the effective reaction rate $k_{eff}(T)$ is given by

$$k_{eff}(T) = k_{eff}^0 \exp\left(-\frac{E_{eff}}{R_g T}\right),$$

$$k^0_{eff} = k^0_p \left(\frac{2 f k^0_d}{k^0_t} \right)^{1/2}, \quad E_{eff} = E_p + \frac{E_d - E_t}{2}.$$

Our mathematical model consists of the mass balances in (2.6) and (2.15), the energy balance (2.16), the boundary conditions (2.13) and (2.14) in x, the boundary conditions in y discussed above and appropriate initial conditions. For convenience, we make a change of variables for the initiator concentration, $I = J^2$, and the equations then become

$$\frac{\partial J}{\partial t} - \varphi_t \frac{\partial J}{\partial x} + J k_1(T) = 0, \tag{2.17}$$

$$\frac{\partial M}{\partial t} - \varphi_t \frac{\partial M}{\partial x} + J M H(J_0 - J) k_2(T) = 0, \tag{2.18}$$

$$\frac{\partial T}{\partial t} = \kappa \nabla^2 T + \varphi_t \frac{\partial T}{\partial x} + q J M H(J_0 - J) k_2(T), \tag{2.19}$$

and the boundary conditions ahead of ($x = -\infty$) and behind ($x = +\infty$) the polymerization wave are

$$x = -\infty : \ M = M_0, \ T = T_0, \ J = J_0 \equiv \sqrt{I_0}, \tag{2.20}$$

$$x = +\infty : \ \frac{\partial T}{\partial x} = 0. \tag{2.21}$$

Here

$$k_1(T) = \frac{1}{2} k_d(T) = k_{01} \exp\left(-\frac{E_1}{R_g T} \right), \quad k_{01} = \frac{1}{2} k^0_d, \ E_1 = E_d,$$

$$k_2(T) = k_{eff}(T) = k_{02} \exp\left(-\frac{E_2}{R_g T} \right), \quad k_{02} = k^0_{eff}, \ E_2 = E_{eff},$$

and $H(J_0 - J)$ is the Heaviside function, i.e., it equals zero if $J \geq J_0$, and unity if $J < J_0$. Using the SSA results in a spurious nonzero value of the polymerization reaction rate for the initial state of the wave (i.e., where $J = J_0$). Introducing the Heaviside function fixes this problem.

Finally, our mathematical model that describes the polymerization wave propagation consists of equations (2.17)-(2.21).

3.   Gasless combustion

It is useful to consider certain limiting cases in which the multistep kinetics reduces to a one-step reaction scheme. One such limiting situation is when appreciable consumption of the initiator occurs in the wake of the polymerization wave and, therefore, does not affect the wave propagation. This is the case, e.g., if the decomposition rate constant is sufficiently small. In such a situation, we can approximate J in (2.18) and (2.19) by its initial value J_0, the Heaviside function – by unity, and disregard the equation (2.17) for I. The equations then become

$$\frac{\partial M}{\partial t} - \varphi_t \frac{\partial M}{\partial x} + Mk(T) = 0, \tag{3.22}$$

$$\frac{\partial T}{\partial t} - \varphi_t \frac{\partial T}{\partial x} - \kappa \nabla^2 T - qMk(T) = 0, \tag{3.23}$$

where ∇^2 is as defined in (2.12), the heat release parameter q is the same as in the previous model, and $k(T)$ is an Arrhenius exponential,

$$k(T) = k_0 \exp(-E/(R_g T)), \quad k_0 = k_{02} \sqrt{I_0}, \quad E = E_2.$$

The boundary conditions are

$$x \to -\infty: \quad T = T_0, \quad M = M_0, \qquad x \to \infty: \quad \frac{\partial T}{\partial x} = 0. \tag{3.24}$$

This is the famous model of gasless combustion (GC) [63]. GC is a combustion process in which both the initial reactants and the final products are condensed, and intermediate gaseous components, even if they are present, do not play a significant role in the process. The story of the discovery of such combustion processes is interesting. In the 60s, a group of researchers in the Institute of Chemical Physics in Chernogolovka, USSR, were studying the mechanism of combustion of gasifying condensed systems (such as explosives and propellants). The process was so complicated by the production of gases that no convincing theoretical model was proposed. In order to better understand the mechanism of combustion, the researchers looked for a "toy" system, which would have many features of the real process except for gas production. Such a system was found (the iron-aluminum thermite) [43], and its behavior was consistent with the predictions of the simplest combustion theory based on (3.22) and (3.23). This success attracted attention of the researchers, and a search for new gasless combustion systems eventually led to a remarkable discovery of the combustion synthesis process.

3.1 Uniformly propagating GC wave

We begin with a discussion of a one-dimensional traveling wave solution which is a mathematical representation of the uniformly propagating reaction wave that we are discussing. This discussion is a bit technical but gives a good insight into the combustion theory.

Since the system (3.22), (3.23) is already written in a moving coordinate system, the traveling wave solution is a stationary solution of this problem, i.e., a solution of the problem

$$\kappa\frac{d^2\widehat{T}}{dx^2} - u\frac{d\widehat{T}}{dx} + q\widehat{M}k(\widehat{T}) = 0, \tag{3.25}$$

$$u\frac{d\widehat{M}}{dx} + \widehat{M}k(\widehat{T}) = 0, \tag{3.26}$$

$$x \to -\infty: \quad \widehat{T} = T_0, \quad \widehat{M} = M_0, \qquad x \to \infty: \quad \frac{d\widehat{T}}{dx} = 0, \tag{3.27}$$

where $u = -\varphi_t$ is the propagation velocity of the wave that has to be determined in the course of solution along with the temperature and concentration profiles. The hat denotes the stationary solution.

We observe that eliminating the reaction term from (3.25), (3.26) yields

$$\kappa\frac{d^2\widehat{T}}{dx^2} - u\frac{d\widehat{T}}{dx} - qu\frac{d\widehat{M}}{dx} = 0, \tag{3.28}$$

Integrating (3.28) and using the boundary conditions (3.27) as $x \to -\infty$, we obtain the first integral of the system

$$\kappa\frac{d\widehat{T}}{dx} = u(\widehat{T} - T_0) + qu(\widehat{M} - M_0). \tag{3.29}$$

Evaluating (3.29) as $x \to \infty$ allows us to determine the limiting value T_a of the temperature as $T_a = T_0 + qM_0$. This is the adiabatic temperature in the system which ensues due to the temperature rise by q degrees as a unit concentration of M reacts.

It is convenient to nondimensionalize the problem (3.25) – (3.27) using the following quantities

$$\theta = \frac{T - T_a}{T_a - T_0}, \quad a = \frac{M}{M_0}, \quad \xi = x\sqrt{\frac{k(T_a)}{\kappa Z}},$$

$$v = u\sqrt{\frac{Z}{\kappa k(T_a)}}, \quad Z = \frac{E(T_a - T_0)}{R_g T_a^2}, \quad \delta = \frac{T_a - T_0}{T_a}.$$

Thus, the nondimensional adiabatic and initial temperatures are equal to zero and negative one, respectively. The nondimensional concentration varies from zero to one. Next, Z is the so-called Zeldovich number, which can be thought of as a nondimensional activation energy of the reaction and is ~ 10 in both combustion and FP problems. It is the large parameter in our asymptotic studies below. Finally, the parameter δ is very close to unity in combustion problems because $T_a \gg T_0$, while in FP problem, it is about 0.5.

The nondimensional problem has the form

$$\frac{d^2\theta}{d\xi^2} - v\frac{d\theta}{d\xi} + Za\exp\left(\frac{Z\theta}{1+\delta\theta}\right) = 0,\tag{3.30}$$

$$-v\frac{da}{d\xi} - Za\exp\left(\frac{Z\theta}{1+\delta\theta}\right) = 0,\tag{3.31}$$

$$\theta(-\infty) = -1,\quad \theta(+\infty) = 0,\quad a(-\infty) = 1.\tag{3.32}$$

The first integral (3.29) becomes

$$\frac{d\theta}{d\xi} = v(\theta + a).\tag{3.33}$$

3.1.1 Phase plane approach. Equations (3.31), (3.33) can be reduced to a single phase-plane equation

$$v^2\frac{da}{d\theta} = -\frac{Za\exp\left(\frac{Z\theta}{1+\delta\theta}\right)}{\theta + a},\quad -1 < \theta < 0,\quad a(-1) = 1,\ a(0) = 0\tag{3.34}$$

for the concentration a as a function of θ. Note that there are two boundary conditions for a first order equation. The additional condition allows us to determine the nondimensional propagation velocity v.

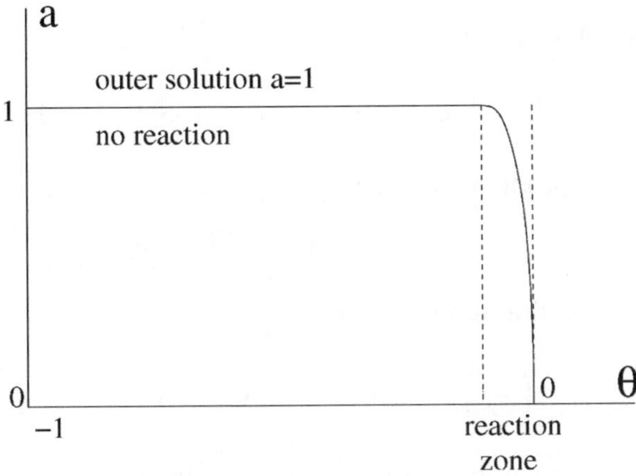

Figure 2. Schematic of the phase-plane solution.

The structure of the solution $a(\theta)$ can be understood by analyzing the reaction term in the equation. For all θ, $-1 < \theta < 0$, that are not too close to zero, the reaction term is exponentially small for $Z \gg 1$. It becomes appreciable in an $O(1/Z)$ vicinity of zero. Thus, the entire range of θ can be divided into two regions (see Figure 2). One region is where θ is not too close to zero. In this case the reaction term can be neglected, and we obtain the *outer* solution $a(\theta) = 1$ taking into account the boundary condition at $\theta = -1$. The other region is a boundary layer where θ is close to zero. This is the reaction zone. Here, a drops from unity to zero at $\theta = 0$ due to the reaction. To determine the solution in this region, the *inner* solution, we stretch the reaction zone by introducing the inner variables

$$a(\theta) = A(\tau), \quad \tau = Z\theta.$$

Then (3.34) reduces to

$$v^2 \frac{dA}{d\tau} = -\frac{A \exp\left(\frac{\tau}{1+\delta\tau/Z}\right)}{\tau/Z + A}, \quad -\infty < \tau < 0, \quad A(-\infty) = 1, \quad A(0) = 0.$$

We remark that the inner variable τ varies over the negative semiaxis, and the above condition at $-\infty$ represents the matching condition for the inner and outer solutions.

Expanding both unknowns, namely, the concentration and the propagation velocity, as

$$A(\tau) = A_0(\tau) + \frac{1}{Z} A_1(\tau) + \dots, \quad v = v_0 + \frac{1}{Z} v_1 + \dots$$

we obtain at the leading order

$$v_0^2 \frac{dA_0}{d\tau} = -e^\tau, \quad -\infty < \tau < 0, \quad A_0(-\infty) = 1, \quad A_0(0) = 0.$$

The solution of this problem is given by

$$v_0 = 1, \quad A_0(\tau) = 1 - \exp\tau,$$

and the complete solution has the form

$$v = 1 + O\left(\frac{1}{Z}\right), \quad a(\theta) = 1 - \exp(Z\theta) + O\left(\frac{1}{Z}\right).$$

Equation (3.33), which at the leading order takes the form,

$$\frac{d\theta}{d\xi} = \theta + 1 - \exp(Z\theta), \tag{3.35}$$

allows us to determine the spatial distribution of the nondimensional temperature $\theta(\xi)$. It is easy to find an implicit solution of (3.35). However, the integral that appears in the course of solution cannot be evaluated exactly. It could be evaluated asymptotically for large Z. Instead, we will dwell upon the asymptotic solution of the differential equation. We first observe that the equation is invariant with respect to translations in ξ, i.e., if $\theta(\xi)$ is a solution of the equation, then $\theta(\xi + c)$ is also a solution for any constant c. Thus, we can assume without loss of generality that θ takes on a prescribed value at $\xi = 0$ (the value must be between -1 and 0 because θ varies over this interval, see (3.32)). For convenience, we take

$$\theta(0) = \theta_0 \equiv -\frac{1}{Z}\ln 2. \tag{3.36}$$

Consider first $\xi > 0$. In this region, θ increases from θ_0 to zero as ξ goes from zero to infinity. Thus, θ is of order $1/Z$. It is not difficult to check that the ξ-region where the major increase in θ occurs is also of order $1/Z$. Thus, we introduce

$$\eta = Z\xi, \quad \tau(\eta) = Z\theta(\xi). \tag{3.37}$$

The initial value problem (3.35), (3.36) takes the form

$$\frac{d\tau}{d\eta} = \frac{1}{Z}\tau + 1 - \exp(\tau), \quad \tau(0) = -\ln 2.$$

The solution of this problem is

$$\tau(\eta) = -\ln(1 + e^{-\eta}) + O\left(\frac{1}{Z}\right). \tag{3.38}$$

Let us now consider the region $\xi < 0$. Here, θ takes on both order one and order $1/Z$ values, so we consider two regions. One region is adjacent to the origin, where both ξ and θ are of order $1/Z$ (inner region), and the other region is where both ξ and θ are of order one (outer region). In the inner region, we use the same scaling (3.37) and as a result get the same solution (3.38). In the outer region, the solution of (3.35) that satisfies $\theta(-\infty) = -1$ is

$$\theta(\xi) = -1 + c\exp(\xi),$$

where c is a constant that has to be found upon matching the outer and the inner solutions. To match the solutions, we expand the outer solution for small ξ as

$$\theta(\xi) \sim -1 + c + c\xi,$$

the inner solution for large negative η as

$$\tau(\eta) \sim \eta \quad \Longrightarrow \quad \theta(\xi) \sim \xi,$$

and equate the expansions to obtain $c = 1$. The composite solution [34] is

$$\theta(\xi) = -1 + e^{\xi} - \frac{1}{Z}\ln(1 + e^{-Z\xi}) - \xi.$$

Finally, upon simple manipulations, we obtain

$$\theta = \begin{cases} -1 + \exp(\xi) - \frac{1}{Z}\ln(1 + \exp(Z\xi)), & \xi < 0 \\ -\frac{1}{Z}\ln(1 + \exp(-Z\xi)), & \xi > 0 \end{cases}. \tag{3.39}$$

For future reference, we present here the corresponding dimensional propagation velocity for $Z \gg 1$

$$u^2 = \frac{\kappa}{Z}k(T_a) = \kappa \frac{R_g T_a^2}{E(T_a - T_0)} k_0 \exp\left(-\frac{E}{R_g T_a}\right). \tag{3.40}$$

3.1.2 Estimates of velocity. Let us discuss the accuracy of the formula for the propagation velocity that we derived above for large Z. Equation (3.34) can be written in the form

$$v^2 = -\frac{Za \exp\left(\frac{Z\theta}{1+\delta\theta}\right)}{(\theta + a)da/d\theta} \equiv B(a(\theta), \theta).$$

If $a(\theta)$ is the solution of the problem, then $B(a(\theta), \theta)$, which is a function of θ, is in fact a constant equal to the square of the propagation velocity. Suppose we consider $B(\rho(\theta), \theta)$, where $\rho(\theta)$ is an arbitrary monotonically decreasing function that satisfies the same boundary conditions as $a(\theta)$. Then it has been proved in [97] that

$$\min_{-1 < \theta < 0} B(\rho(\theta), \theta) \leq v^2 \leq \max_{-1 < \theta < 0} B(\rho(\theta), \theta). \tag{3.41}$$

This is an interesting result because choosing *any* function ρ that has the above properties will give an estimate of the propagation velocity. Of course, in order to derive accurate estimates, we have to choose a $\rho(\theta)$ that is in some sense close to the actual solution $a(\theta)$. A convenient choice is

$$\rho(\theta) = 1 - \frac{G(\theta)}{G(0)}, \quad G(\theta) = \int_{-1}^{\theta} \frac{1}{1+s} \exp\left(\frac{Zs}{1+\delta s}\right) ds$$

(this choice as well as related issues are discussed in [97]). Then the inequality (3.41) yields

$$ZG(0) \leq v^2 \leq \frac{ZG(0)}{1 - G(0)}. \tag{3.42}$$

We evaluate the integral $G(0)$ for large Z using the Laplace method to obtain

$$G(0) = \int_{-1}^{0} \frac{1}{1+s} \exp\left(\frac{Zs}{1+\delta s}\right) ds = \frac{1}{Z} + O\left(\frac{1}{Z^2}\right).$$

Thus, as expected

$$v^2 = 1 + O\left(\frac{1}{Z}\right).$$

While this proves our asymptotic result, the estimate (3.42) in fact gives more than just an asymptotic formula, because we can evaluate the integral to get numerical estimates of v as well as asymptotic estimates. It turns out that for physically relevant values of the parameters, $v = 1 \pm 0.02$.

3.1.3 Further analysis and reaction front approach. We now return to the problem (3.30)-(3.32). We have already solved the problem using the phase-plane approach and gained some understanding of the behavior of the solution. We want to recover the solution by solving the problem directly, i.e., without reducing the problem to a phase-plane equation. The goal is to learn how to handle such problems in preparation for solving complete, time- and space-dependent problems like (3.22)-(3.24) that cannot be reduced to the phase- plane equation.

Let us begin with analyzing the reaction term

$$W = Za \exp\left(\frac{Z\theta}{1 + \delta\theta}\right).$$

Using (3.33) (with $v = 1$) and (3.35), we obtain

$$W = Z\left(\frac{d\theta}{d\xi} - \theta\right) \exp\left(\frac{Z\theta}{1 + \delta\theta}\right) = Z(1 - e^{Z\theta}) \exp\left(\frac{Z\theta}{1 + \delta\theta}\right).$$

Let us assume that $\delta = 0$ in the exponential. This assumption, which does not change the leading order behavior of the reaction term, makes the subsequent calculations more transparent. Thus, we consider

$$W = Ze^{Z\theta}(1 - e^{Z\theta}).$$

Next, we use (3.39) to determine the dependence of W on ξ. For $\xi > 0$, we obtain

$$W = Z\frac{e^{-Z\xi}}{(1 + e^{-Z\xi})^2}.$$

For $\xi < 0$, the reaction term is exponentially small for $|\xi| = O(1)$ and reduces to

$$W = Z\frac{e^{Z\xi}}{(1 + e^{Z\xi})^2}$$

for small $|\xi|$ (note that the above expressions for W for $\xi > 0$ and for $\xi < 0$ are equal). Thus, the reaction term is significant only in a narrow region of order $1/Z$ in ξ, and the structure of the solution is as follows. Ahead of the narrow reaction zone, there is a preheat region where the reaction term is negligibly small, and the temperature increases from its initial value to almost the adiabatic value. Behind the reaction zone, there is a product region where the reaction term is also negligible (see Figure 1). In both the product region ($\xi > 0$) and the preheat region ($\xi < 0$) the reaction term can be neglected in (3.30), (3.31), and the reactionless equations

$$\frac{d^2\theta}{d\xi^2} - v\frac{d\theta}{d\xi} = 0, \quad v\frac{da}{d\xi} = 0$$

have to be solved. The solution that satisfies the boundary conditions (3.32) (as well as the condition $a(\infty) = 0$) is

$$\theta(\xi) = \begin{cases} -1 + c\exp(v\xi), & \xi < 0 \\ 0, & \xi > 0 \end{cases}, \tag{3.43}$$

$$a(\xi) = \begin{cases} 1, & \xi < 0 \\ 0, & \xi > 0 \end{cases}. \tag{3.44}$$

To determine the solution in the reaction zone, we introduce the inner variables

$$A(\eta) = a(\xi), \quad \tau(\eta) = Z\theta(\xi), \quad \eta = Z\xi.$$

Then (3.30) and (3.31) take the form

$$\frac{d^2\tau}{d\eta^2} - \frac{v}{Z}\frac{d\tau}{d\eta} + A\exp\left(\frac{\tau}{1 + \delta\tau/Z}\right) = 0,$$

$$v\frac{dA}{d\eta} + A\exp\left(\frac{\tau}{1 + \delta\tau/Z}\right) = 0,$$

and reduce at the leading order in $Z \gg 1$ to

$$\frac{d^2\tau}{d\eta^2} + A\exp(\tau) = 0, \quad v\frac{dA}{d\eta} + A\exp(\tau) = 0. \tag{3.45}$$

Using the matching conditions to the solution in the product region

$$\tau(\infty) = 0, \quad A(\infty) = 0, \tag{3.46}$$

we obtain the first integral of the system (3.45)

$$\frac{d\tau}{d\eta} = vA,$$

which, combined with the temperature equation in (3.45), yields

$$\frac{d^2\tau}{d\eta^2} + \frac{1}{v}\frac{d\tau}{d\eta}e^\tau = 0.$$

The solution of this equation that satisfies (3.46) and (3.36) is given by

$$\tau(\eta) = -\ln(1 + e^{-\eta/v})$$

(the latter condition is used for consistency with the phase-plane approach). To match the inner solution to the outer solution in the preheat region, we expand the outer solution for small ξ

$$\theta(\xi) = -1 + c\exp(v\xi) \sim -1 + c + cv\xi$$

and the inner solution for large negative η

$$\tau(\eta) = -\ln(1 + e^{-\eta/v}) \sim \eta/v \quad \theta(\xi) \sim \xi/v.$$

Comparing the expansions, we obtain $c = 1$, $v = 1$. Thus, the leading order solution, which is obtained by adding the inner and outer solutions and subtracting the common parts, is

$$\theta = \begin{cases} -1 + \exp(v\xi) - \frac{1}{Z}\ln(1 + \exp(Z\xi/v)), & \xi < 0 \\ -\frac{1}{Z}\ln(1 + \exp(-Z\xi/v)), & \xi > 0 \end{cases}, \qquad (3.47)$$

$$v = 1.$$

This solution coincides with the solution (3.39) obtained by the phase-plane approach. It can be easily checked that the inner and outer solutions for the concentration $a(\xi)$ also match.

An important observation is that, in the limit $Z \to \infty$, the solution (3.47) reduces to

$$\theta = \begin{cases} -1 + \exp(v\xi), & \xi < 0 \\ 0, & \xi > 0 \end{cases}, \qquad (3.48)$$

This is a leading order outer solution, i.e., it is the solution in the preheat and product regions where the reaction term is neglected (cf. (3.43)). This solution is continuous at $\xi = 0$, i.e.,

$$\theta|_{\xi \to 0+} = \theta|_{\xi \to 0-}, \qquad (3.49)$$

and it satisfies

$$\left.\frac{d\theta}{d\xi}\right|_{\xi \to 0+} - \left.\frac{d\theta}{d\xi}\right|_{\xi \to 0-} = -v. \qquad (3.50)$$

The last condition means that the heat flux undergoes a jump across the reaction zone due to heat production by the reaction.

This observation is important, because it suggests a simplified approach to the problem, i.e., an approximate solution can be constructed to a large extent by just looking at the preheat and product regions where the reaction term is insignificant and then matching the two solutions, so that the resulting solution is continuous (3.49), and an appropriate heat balance (3.50) is satisfied. Of course, this solution is not (and cannot be) complete, because the propagation speed v cannot be found unless we have accounted for the processes in the reaction zone. Thus, solution (3.48) has to be supplemented by some analysis of the processes in the reaction zone that allows to determine the propagation velocity. We perform this analysis for the original problem (3.22), (3.23) rather than for the nondimensional uniformly propagating wave problem (3.30), (3.31). This analysis is guided by the above analysis of the equations in the reaction zone that demonstrated which terms are most significant in the reaction zone. The significant terms in (3.22), (3.23) are the reaction rates, the heat conduction in the direction of wave propagation x, and the convective term in the concentration equation. Thus, equations (3.22), (3.23) in the reaction zone reduce to

$$-\varphi_t \frac{\partial M}{\partial x} + Mk(T) = 0, \tag{3.51}$$

$$\kappa(1 + \varphi_y^2)\frac{\partial^2 T}{\partial x^2} + qMk(T) = 0. \tag{3.52}$$

One could arrive at this result by means of a more systematic analysis, namely, by nondimensionalizing the equations and introducing stretched inner variables as we did before. We will restrict ourselves to using this less formal approach.

Eliminating the reaction term from (3.51) and (3.52), we obtain

$$\kappa(1 + \varphi_y^2)\frac{\partial^2 T}{\partial x^2} = -q\varphi_t \frac{\partial M}{\partial x}. \tag{3.53}$$

We integrate this equation across the reaction zone, taking into account that as the reaction zone shrinks to an interface, (i) the temperature gradient at the left end of the reaction zone must be equal to the gradient of the preheat region temperature at $x = 0$, and (ii) the temperature gradient at the right end of the reaction zone must be equal to the gradient of the product region temperature at $x = 0$. In addition, we note that the concentration M of the reactant varies from its initial value M_0 to zero across the reaction zone. As a result, we obtain a jump condition for the gradient of the outer temperature solution

$$\kappa\left[\frac{\partial T}{\partial x}\right] = -qM_0 \frac{\varphi_t}{1 + \varphi_y^2}. \tag{3.54}$$

Here

$$[f] = f\big|_{x \to 0^+} - f\big|_{x \to 0^-} \tag{3.55}$$

denotes the jump in the quantity f across the reaction zone located at $x = 0$. Condition (3.54) is a dimensional heat balance at the reaction zone that has the same physical meaning as (3.50).

Next, integrating (3.53) from a current point in the reaction zone to the product end of it, we obtain

$$\kappa(1 + \varphi_y^2)\frac{\partial T}{\partial x} = -q\varphi_t M.$$

Combining this equation with (3.51), we obtain

$$\frac{\varphi_t^2}{1 + \varphi_y^2}\frac{\partial M}{\partial x} = -\frac{\kappa}{q}k(T)\frac{\partial T}{\partial x}. \tag{3.56}$$

Integrating (3.56) across the reaction zone yields

$$\frac{\varphi_t^2}{1 + \varphi_y^2} = \frac{\kappa}{qM_0}\int_{T_0}^{T_b} k(T)\,dT. \tag{3.57}$$

Here, T_b is the temperature at the right end of the reaction zone. The lower limit of integration in (3.57) is taken to be T_0 instead of the actual temperature at the left end of the reaction zone, because the main contribution to the integral comes from the temperature range close to T_b so that changing the lower limit does not significantly change the integral. Finally, we observe that if the Zeldovich number is large, the integral can be approximately evaluated (using the Laplace method). Employing the same nondimensional quantities as above (but using T_b as the reference temperature instead of T_a), i.e., introducing

$$\theta = \frac{T - T_b}{T_b - T_0}, \quad Z = \frac{E(T_b - T_0)}{R_g T_b^2}, \quad \delta = \frac{T_b - T_0}{T_b},$$

we obtain

$$\int_{T_0}^{T_b} k(T)\,dT = \int_{-1}^{0} k(T_b)\exp\left(\frac{Z\theta}{1 + \delta\theta}\right)(T_b - T_0)\,d\theta \sim$$

$$\int_{-1}^{0} k(T_b)\exp(Z\theta)\,(T_b - T_0)\,d\theta \sim \frac{T_b - T_0}{Z}k(T_b) = \frac{R_g T_b^2}{E}k(T_b).$$

As a result, (3.57) reduces to

$$\frac{\varphi_t^2}{1 + \varphi_y^2} = F_c(T_b) \equiv \kappa\frac{R_g T_b^2}{EqM_0}k(T_b). \tag{3.58}$$

We have arrived at the reaction front formulation of the GC model. It consists of the reactionless equations

$$\frac{\partial M}{\partial t} - \varphi_t\frac{\partial M}{\partial x} = 0, \quad \frac{\partial T}{\partial t} - \varphi_t\frac{\partial T}{\partial x} - \kappa\nabla^2 T = 0, \tag{3.59}$$

that have to be solved in the preheat ($x < 0$) and product ($x > 0$) regions subject to the boundary conditions (3.24), appropriate initial conditions, the condition that the temperature is continuous at $x = 0$, the heat balance condition (3.54), and, finally, the formula (3.57) (or (3.58)) for the velocity.

3.1.4 Step function approach. There is another approach to solving the GC and FP problems, which we find quite useful. It involves the use of step functions in the reaction rate terms. Such an approach has a long and glorioushistory, beginning with works by Le Chatelier and many others who introduced the ignition temperature in the 'pre-Arrhenius' times and were severely criticized by subsequent researchers (see [100] for a more detailed discussion). By no means are we trying to revive the old theories. Our use of step functions is due to the understanding that the Arrhenius exponential and an *appropriately chosen* step function are close to one another in the sense of distributionsand, therefore, yield nearly the same results. Let us discuss it in greater detail. Consider the temperature dependence of the reaction rate in (3.30), i.e., the function

$$\tilde{k}(\theta) \equiv Z \exp\left(\frac{Z\theta}{1 + \delta\theta}\right),\tag{3.60}$$

where, as before, $-1 < \theta < 0$. This function is depicted in Figure 3 for different values of Z (left figure).

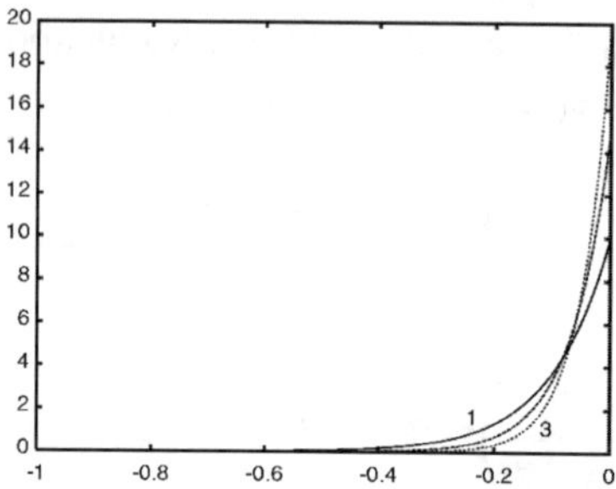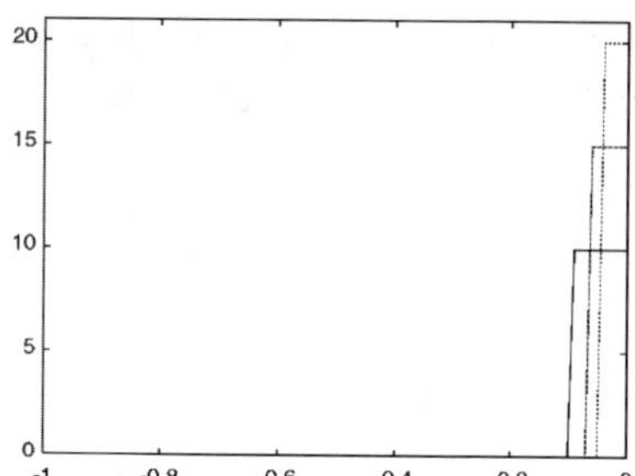

Figure 3.   Comparison of the Arrhenius (left) and step-function (right) kinetics.

We make two observations regarding $\tilde{k}(\theta)$. First, as Z increases, the region in θ where $\tilde{k}(\theta)$ is appreciable decreases (in fact, it behaves as $1/Z$). Second, the integral value of this function over the entire interval of variation of θ is approximately 1. Thus, $\tilde{k}(\theta)$ can be thought of as an approximation to the Dirac delta-function $\delta(\theta)$ (in fact, it would be more correct to refer to it as a 'one-sided' delta-function because $\theta < 0$). One might expect that the reason why combustion front approximations provide results similar to those with Arrhenius kinetics is exactly that the Arrhenius exponential is an approximation to the δ-function. If this is the case, then a different approximation to the δ-function should yield results similar to those with Arrhenius kinetics. As a simple approximation, one can use step-functions (see Figure 3)

$$\tilde{K}(\theta) = ZH(\theta - 1/Z),$$

where H is the Heaviside function. We note that the integral value of $\hat{k}(\theta)$ is the same as that of $\tilde{k}(\theta)$, and their maximum values are equal. Let us apply this approach to the dimensional problem (3.22), (3.23).

We replace the Arrhenius temperature dependence $k(T)$ of the reaction rate in (3.22), (3.23) by the step function

$$K(T) = k(T_b)H(T - T_*), \tag{3.61}$$

where H is the Heaviside function, T_b is a characteristic temperature in the reaction zone, and T_* is chosen in such a way that

$$\int_{T_0}^{T_b} K(T)\, dT = \int_{T_0}^{T_b} k(T)\, dT. \tag{3.62}$$

Computing the integrals in (3.62) as

$$\int_{T_0}^{T_b} K(T)\, dT = \int_{T_*}^{T_b} k(T_b)\, dT = (T_b - T_*)k(T_b),$$

$$\int_{T_0}^{T_b} k(T)\, dT \approx \frac{R_g T_b^2}{E} k(T_b),$$

we obtain

$$T_* = T_b(1 - \epsilon), \qquad \epsilon = \frac{RT_b}{E} \ll 1. \tag{3.63}$$

To illustrate this approach, we consider the problem (3.25)-(3.27). We replace the Arrhenius exponential k in this problem by the step function (3.61), (3.63). In addition, to simplify the notation, we introduce $a = M/M_0$. Thus, we want to solve the equations

$$\kappa \frac{d^2\widehat{T}}{dx^2} - u\frac{d\widehat{T}}{dx} + qaK(\widehat{T}) = 0, \tag{3.64a}$$

$$u\frac{d\widehat{M}}{dx} + aK(\widehat{T}) = 0, \tag{3.64b}$$

subject to the boundary conditions

$$x \to -\infty: \quad \widehat{T} = T_0, \; a = 1, \qquad x \to \infty: \quad \frac{d\widehat{T}}{dx} = 0. \tag{3.65}$$

As the characteristic temperature T_b, we take

$$T_b = \widehat{T}(+\infty). \tag{3.66}$$

We also assume that $\widehat{T}(0) = T_*$ which can be done due to translational invariance of the problem. We obtain the following equations

$$x < 0: \quad \kappa\frac{d^2\widehat{T}}{dx^2} - u\frac{d\widehat{T}}{dx} = 0, \quad -u\frac{da}{dx} = 0, \tag{3.67a}$$

$$x > 0: \quad \kappa\frac{d^2\widehat{T}}{dx^2} - u\frac{d\widehat{T}}{dx} + qak(T_b) = 0, \quad -u\frac{da}{dx} - k(T_b)a = 0. \tag{3.67b}$$

The matching conditions at $x = 0$ are

$$\widehat{T}(0^-) = \widehat{T}(0^+) = T_*, \quad a(0^-) = a(0^+), \tag{3.68a}$$

$$\frac{d\widehat{T}(0^-)}{dx} = \frac{d\widehat{T}(0^+)}{dx}. \tag{3.68b}$$

Solving equations (3.67) subject to the matching conditions (3.68a) and the boundary conditions (3.65) yields

$$x < 0: \quad \widehat{T}(x) = T_0 + (T_* - T_0)e^{ux/\kappa}, \quad a(x) = 1, \tag{3.69a}$$

$$x > 0: \quad \widehat{T}(x) = T_* + \frac{qu^2}{\kappa k(T_b) + u^2}(1 - e^{-k(T_b)x/u}), \quad a(x) = e^{-k(T_b)x/u}. \tag{3.69b}$$

Applying (3.66) and the matching conditions (3.68b) to the solution (3.69) results in

$$T_b = T_* + \frac{qu^2}{\kappa k(T_b) + u^2}, \quad (T_* - T_0)\frac{u}{\kappa} = \frac{qu^2}{\kappa k(T_b) + u^2}\frac{k(T_b)}{u}.$$

Solving these equations yields $T_b = T_0 + q$ (i.e., the temperature far in the product region is the adiabatic temperature as expected) and

$$u^2 = \frac{\kappa k(T_b)}{Z - 1}. \tag{3.70}$$

Since $Z \gg 1$, (3.70) simplifies to

$$u^2 = \frac{\kappa k(T_b)}{Z}, \tag{3.71}$$

which is exactly the formula we derived earlier (see (3.40)).

The step function approach typically gives accurate results (in some cases even more accurate than the reaction front approximation). This method, however, involves more algebraic manipulations than the reaction front approximation, and it is quite often advantageous to use the reaction front approximation. The step function approach is particularly useful when the problem at hand involves several different Arrhenius exponentials. In this case, asymptotic analysis may yield nonuniform and even irrelevant results (see the discussion in [55]).

3.2 Linear stability analysis of gasless combustion wave

The uniformly propagating wave that we have found may be either stable or unstable depending on the parameter values. If the wave is unstable, it cannot be observed experimentally, and other regimes of propagation occur. Before we can talk about these other regimes of propagation, we have to perform a linear stability analysis of the uniformly propagating wave. We consider the gasless combustion wave under the reaction front approximation.

We perturb the stationary solution as

$$T = \widehat{T}(x) + \delta \exp(\omega t + iky)\widetilde{T}(x), \tag{3.72}$$

$$M = \widehat{M}(x) + \delta \exp(\omega t + iky)\widetilde{M}(x), \tag{3.73}$$

$$\varphi = -ut + \delta \exp(\omega t + iky). \tag{3.74}$$

Here, k is the wavenumber, ω is the temporal frequency of perturbations, and δ is a small magnitude of perturbations.

Substituting the expansions (3.72)-(3.74) into equations (3.59) (separately, for $x > 0$ and $x < 0$), into the boundary conditions (3.24), and linearizing (i.e., keeping only linear terms in $\delta \ll 1$) we obtain

$$(\omega + \kappa k^2)\widetilde{T} + u\widetilde{T}' - \kappa\widetilde{T}'' = (\omega + \kappa k^2)\widehat{T}', \tag{3.75}$$

$$\omega\widetilde{M} + u\widetilde{M}' = 0, \tag{3.76}$$

$$\widetilde{T}(-\infty) = 0, \quad \widetilde{T}'(+\infty) = 0, \quad \widetilde{M}(-\infty) = 0. \tag{3.77}$$

Substituting the expansions (3.72)-(3.74) into the matching conditions and linearizing, we obtain

$$\left[\widetilde{T}\right] = 0, \tag{3.78}$$

$$\frac{\kappa}{u}\left[\widetilde{T}'\right] = \omega\frac{\widehat{T}_b - T_0}{u}, \tag{3.79}$$

$$-\frac{\omega}{u} = \frac{z\widetilde{T}(0)}{2(\widehat{T}_b - T_0)}, \tag{3.80}$$

where

$$z = \frac{F'(\widehat{T}_b)(\widehat{T}_b - T_0)}{F(\widehat{T}_b)} \equiv (\widehat{T}_b - T_0)\frac{\partial \ln u}{\partial \widehat{T}_b}. \tag{3.81}$$

The solution of (3.75), (3.77) is

$$\widetilde{T}(x) = \begin{cases} c_1 \exp(r_+ x) + \frac{u}{\kappa}(\widehat{T}_b - T_0)\exp\left(\frac{u}{\kappa}x\right), & x < 0 \\ c_2 \exp(r_- x), & x > 0 \end{cases}, \tag{3.82}$$

where

$$r_\pm = u(1 \pm d)/(2\kappa), \quad d = \sqrt{1 + 4\Omega + 4s^2},$$

with $s = \kappa k/u$ as the non-dimensional wavenumber and $\Omega = \kappa\omega/u^2$ as the non-dimensional frequency of the perturbation. Substituting this into the linearized matching conditions gives

$$c_1 + \frac{u}{\kappa}(\widehat{T}_b - T_0) = c_2,$$

$$\frac{\kappa}{u}\left(c_2 r_- - c_1 r_+ - \left(\frac{u}{\kappa}\right)^2 (\widehat{T}_b - T_0)\right) = \frac{(\widehat{T}_b - T_0)\omega}{u},$$

$$-\frac{\omega}{u} = \frac{z c_2}{2(\widehat{T}_b - T_0)}.$$

Eliminating c_1 and c_2 from the above equations yields the dispersion relation

$$4\Omega^3 + \left(1 + 4s^2 + 2z - \frac{1}{4}z^2\right)\Omega^2 + \frac{1}{2}z(1 + 4s^2)\Omega + \frac{1}{4}s^2 z^2 = 0. \quad (3.83)$$

Instability occurs when a pair of complex conjugate eigenvalues crosses the imaginary axis as the parameters vary. At the stability boundary

$$z_c = 4 - 4s^2(1 + 4s^2)^{-1} + 2\sqrt{(2 - 2s^2(1 + 4s^2)^{-1})^2 + 1 + 4s^2}, \quad (3.84)$$

$$\Omega = \pm i\omega_0, \quad \omega_0^2 = \frac{1}{8}z_c(1 + 4s^2). \quad (3.85)$$

The neutral stability curve in the (s, z)-plane has a minimum at $s = 0.5$. The neutral stability curve is shown in Figure 8 (the lowest curve). The uniformly propagating wave is unstable in the region above the curve, i.e., for $z > z_c$, and it is stable below the curve. It can be easily checked that the parameter z is nothing else but the Zeldovich number Z and can be also written in the form

$$z = \frac{t_h}{t_r}, \quad (3.86)$$

where t_r and t_h are the reaction time and the characteristic time of the removal of heat from the reaction zone, respectively. Relation (3.86) provides a simple physical interpretation of this result [98]. If the reaction time in the uniformly propagating wave is much less than the time required to remove the heat from the reaction zone (i.e., z is large), such a wave cannot exist, because the temperature in the reaction zone will grow. One can expect in this case, that the reaction zone will speed up (because higher reaction zone temperatures result in faster propagation) until it hits cold layers of the reactants. Then it will rapidly lose heat due to high temperature gradients. After a period of depression, which is characterized by an almost stationary reaction zone during which the temperature in the reaction zone builds up, the process repeats periodically. This is exactly what is seen in experiments and numerical computations.

3.3 Weakly nonlinear analysis

Previous results, both theoretical and experimental, have indicated that various modes of propagation are possible. In addition to uniformly propagating planar fronts, fronts whose propagation velocities are oscillatory in time (pulsating combustion) were observed. The experiments in [84] exhibited a period doubling in the pulsating mode of propagation. Various types of multi–dimensional combustion modes depending on the geometry of the sample were also shown. In a cylindrical geometry, for example, there have been numerous experimental observations of spinning, radial, and multiple-point combustion waves [22, 23, 88, 44, 45, 58]. In the spinning mode of propagation, which was first found in filtration combustion [59], in which a gas participates in the reaction, one or more luminous points are observed to move in a helical fashion along the surface of the cylindrical sample. In the radial mode, a front whose location depends on the radial variable but not on the angular variable propagates in a pulsating manner. Finally, in multiple–point combustion, one or more luminous spots are observed to appear, disappear and reappear on the surface of the sample.

Numerical simulations were performed on the planar problem in [83], which showed that the transition to the self–oscillatory combustion occurs when a parameter related to the Zeldovich number is increased. Numerical studies [5, 6, 79] of the one-dimensional problem found transitions to relaxation oscillations and period doublings, and these studies demonstrated two routes to chaotic dynamics as the bifurcation parameter related to the Zeldovich number was increased. Numerical studies [36–41, 1, 7, 4] of the two– and three–dimensional model found spinning modes of propagation as well as standing modes, which describe multiple point propagation, and quasi–periodic modes of propagation.

A nonlinear analysis of the one-dimensional problem in [51] described the transition via bifurcation from a uniformly propagating wave to a pulsating combustion wave. Several other theoretical studies have addressed the non-linear stability problem in two and three dimensions [9, 25, 26, 47, 48, 52]. In a cylindrical geometry, bifurcation to stable pulsating, spinning or standing waves were shown as a result of loss of stability of the steady planar mode of combustion [25, 47]. Interactions among several types of pulsating, spinning and standing waves leading to quasi-periodic modes of propagation, were also considered [9, 26, 48]. A continuous band of the wave numbers in the instability region was considered in [52]. Some of these solution branches correspond to the modes of burning which have been experimentally observed, whereas other branches represent new modes of combustion which have not as yet been observed in the laboratory. A related analysis of various types of solutions bifurcating from uniformly propagating traveling wave solutions of a general reaction-diffusion system was performed in [8, 95, 96].

In order to present here some basic results of the weakly stability analysis, we consider below a general reaction-diffusion system of equations. We assume that the problem has a one-dimensional traveling wave solution that loses stability in the same way as the gasless combustion wave as discussed in greater detail below. A study of a general reaction-diffusion system rather than a specific model is useful, because it allows us to focus on general properties of the solution, independent of a particular model.

We consider the general reaction-diffusion system

$$\frac{\partial u}{\partial t} = A\left(\frac{\partial^2 u}{\partial \tilde{x}^2} + \frac{\partial^2 u}{\partial y^2}\right) + f(u; \mu), \quad t > 0, \quad -\infty < \tilde{x} < +\infty, \quad 0 < y < \ell,$$

$$(3.87)$$

where $u(\tilde{x}, y, t)$ and $f(u; \mu)$ are vector-functions,

$$u = (u_1, u_2, \ldots, u_m), \quad f(u; \mu) = (f_1(u; \mu), f_2(u; \mu), \ldots, f_m(u; \mu)),$$

and u is assumed to be bounded as $\tilde{x} \to \pm\infty$ and periodic in y. The matrix A is diagonal with nonnegative elements a_j $(j = 1, \ldots, m)$ on the diagonal, t, $\tilde{x}$ and y are the temporal and spatial variables, and μ is a real parameter. As related to the gasless combustion model, this problem can be thought of as the combustion wave propagation in a cylindrical shell. The length ℓ is not arbitrary and will be chosen later (see the remark following equation (3.92)). We assume that (3.87) exhibits a traveling wave solution $\hat{u}(x)$, $x = \tilde{x} + ct$, which propagates along the $\tilde{x}$-axis with constant velocity c. Note that in general, $\hat{u}(x)$ and c depend on μ, so we will write $\hat{u}_\mu$ and c_μ when it is necessary to point out this dependence. The traveling wave solution, which we refer to as the basic solution, satisfies the system of equations

$$A\hat{u}'' - c\hat{u}' + f(\hat{u}; \mu) = 0, \tag{3.88}$$

where the prime denotes differentiation with respect to x.

We assume that the basic solution loses stability when μ exceeds a critical value, and we seek solutions appearing as a result of this instability. Let us first describe the conditions for the loss of stability. Consider the problem (3.87) linearized about the basic solution and written in the moving coordinate system attached to the traveling wave

$$\frac{\partial \tilde{u}}{\partial t} = A\left(\frac{\partial^2 \tilde{u}}{\partial x^2} + \frac{\partial^2 \tilde{u}}{\partial y^2}\right) - c\frac{\partial \tilde{u}}{\partial x} + B_\mu(\hat{u})\tilde{u}, \tag{3.89}$$

where $\tilde{u}$ is the perturbation of the basic solution and $B_\mu(\hat{u})$ is the Jacobi matrix,

$$B_\mu(\hat{u}) = \left(\frac{\partial f_i(\hat{u}; \mu)}{\partial \hat{u}_j}\right), \quad i, j = 1, \ldots, m.$$

We write $\tilde{u}$ as

$$\tilde{u} = e^{\omega t + iky} g(x), \tag{3.90}$$

where k is the the wave number of the perturbations and $\omega = \omega(k^2, \mu)$ is the growth rate. Substituting (3.90) into (3.89), we obtain

$$Ag'' - cg' - k^2 Ag + B_\mu g = \omega g. \tag{3.91}$$

We seek solutions of (3.91) which are bounded at infinity. Thus, ω is an eigenvalue of the operator defined by the left-hand side of (3.91). Suppose that there exists a critical value of the parameter μ (say $\mu = 0$) such that the following conditions are satisfied:

(i) for $\mu < 0$ and all $k \neq 0$, all eigenvalues ω of (3.91) have negative real parts;

(ii) for $\mu = 0$ and $k = k_0 \neq 0$, there exist a pair of purely imaginary eigenvalues $\omega_\pm$, i.e. $\omega_\pm = \pm i\omega_0 \neq 0$, and all other eigenvalues corresponding to $k \neq 0$ have negative real parts;

(iii) for each $\mu > 0$ (and sufficiently small), there is a pair of complex conjugate eigenvalues $\omega^\pm(k_0^2, \mu)$ such that $\Re(\omega^\pm(k_0^2, \mu)) > 0$ and $\omega^\pm(k_0^2, 0) = \omega_\pm$. The real parts of all other eigenvalues with $k \neq 0$ are negative;

(iv) for $k = 0$ and all values of μ, there exists an eigenvalue $\omega = 0$ which, as can be easily seen by differentiating (3.88) with respect to x, corresponds to the eigenfunction $\hat{u}'$; all other eigenvalues for $k = 0$ are assumed to have negative real parts.

The conditions stated above mean that the basic solution is stable when $\mu < 0$, and it loses stability when μ passes through the critical value $\mu = 0$ via a Hopf bifurcation. (This is exactly the case with the gasless combustion wave as discussed earlier.)

Therefore, for $\mu = 0$, the system of equations (3.89) has the following solutions

$$e^{\pm i(\omega_0 t + k_0 y)} g_0(x), \quad e^{\pm i(\omega_0 t - k_0 y)} g_0(x), \quad \hat{u}_0', \tag{3.92}$$

where g_0 is a solution of (3.91) for $k = k_0$, $\mu = 0$, $\omega = i\omega_0$. In order for these modes to satisfy the periodic boundary conditions in y, we will assume that the length ℓ of the interval is

$$\ell = \frac{2\pi N}{k_0}, \tag{3.93}$$

where N is a positive integer. It is worth noting that the exponentials

$$\exp(\pm i k_0 y)$$

arise in the course of solution of (3.89) by separation of variables, and they are nothing else but the eigenfunctions of the operator $\partial^2/\partial y^2$ on the interval $(0, \ell)$ with periodic boundary conditions. In a more general problem, in which the wave propagates along the axis of a cylinder with an arbitrary cross section Ω

and satisfies the no-flux boundary condition at the lateral surface of the cylinder, the same separation of variables process will result in the eigenfunctions of the problem

$$\nabla^2\psi + k_0^2\psi = 0 \quad \text{in} \quad \Omega, \quad \left.\frac{\partial\psi}{\partial n}\right|_{\partial\Omega} = 0, \tag{3.94}$$

in place of the exponentials.

We observe that the solutions we seek are not traveling wave solutions in the strict sense of the word. For a traveling wave solution u, we must have $u = u(x)$, where $x = \tilde{x} + ct$ for some constant speed c. It can be also written as $x = \tilde{x} - \varphi(t)$ with $\varphi(t) = -ct$. Here, $\varphi(t)$ is the coordinate of the front (or any other characteristic point of the solution). We seek solutions with $\varphi = \varphi(y, t)$. That is, we look for solutions u of the problem (3.87) having the form

$$u(\tilde{x}, y, t) = v(x, y, t), \quad x = \tilde{x} - \varphi(y, t), \tag{3.95}$$

which bifurcate from the basic solution when it loses stability. The function φ is to be determined as part of the solution process. We observe that the specific choice of coordinate to which φ corresponds will not affect any of the results obtained below. Substituting (3.95) into (3.87), we obtain the set of equations for v and φ

$$\frac{\partial v}{\partial t} - \frac{\partial\varphi}{\partial t}\frac{\partial v}{\partial x} = A\nabla^2 v + f(\mu; v), \tag{3.96}$$

where v is a bounded vector-function, and ∇^2 is the Laplacian in the moving coordinate system,

$$\nabla^2 \equiv \frac{\partial^2}{\partial x^2} + \left(\frac{\partial}{\partial y} - \frac{\partial\varphi}{\partial y}\frac{\partial}{\partial x}\right)^2. \tag{3.97}$$

We observe that $v = \hat{u}(x), \quad \varphi = -ct$ is a steady state solution of the problem (3.96)-(3.97).

We seek solutions near the stability threshold, i.e., when the parameter μ is near the stability boundary, and the solution itself is close to the uniformly propagating wave. Therefore, we introduce the small parameter ϵ by defining

$$\mu = \nu\epsilon^2, \tag{3.98}$$

and represent the solution as

$$v = \hat{u} + w, \quad \varphi = -c_\mu t + \Psi, \tag{3.99}$$

where w and Ψ are small. Substituting (3.99) into (3.96), we obtain

$$\frac{\partial w}{\partial t} + c_\mu \frac{\partial w}{\partial x} - \frac{\partial \Psi}{\partial t}\left(\frac{d\hat{u}}{dx} + \frac{\partial w}{\partial x}\right) = A\nabla^2 w + f(\hat{u}+w;\mu) - f(\hat{u};\mu). \quad (3.100)$$

The following expansion of the nonlinear term f about the basic solution is used

$$f(\hat{u}+w;\mu) \sim f(\hat{u};\mu) + B_\mu(\hat{u})w + \frac{1}{2}a_\mu(\hat{u};w,w) + \frac{1}{6}b_\mu(\hat{u};w,w,w) + \dots, \quad (3.101)$$

where a_μ and b_μ are bilinear and trilinear forms in w, respectively.

Since we seek solutions that are close to periodic solutions but not necessarily periodic, we introduce slow time scales

$$t_1 = \epsilon t, \quad t_2 = \epsilon^2 t, \quad (3.102)$$

in addition to the fast time scale t and expand as

$$w \sim \sum_{j=1}^{\infty} \epsilon^j w_j(t,t_1,t_2,y,x), \quad \Psi \sim \sum_{j=1}^{\infty} \epsilon^j \Psi_j(t,t_1,t_2,y). \quad (3.103)$$

Upon formal substitution of (3.98), (3.101)-(3.103) into (3.100), we equate coefficients of like powers of ϵ and obtain a sequence of problems for the recursive determination of w_j, $j = 1, 2, 3, \dots$

$$Lw_j + M\Psi_j \frac{d\hat{u}}{dx} = f_j, \quad (3.104a)$$

$$Lw \equiv \frac{\partial w}{\partial t} + c_0\frac{\partial w}{\partial x} - A\frac{\partial^2 w}{\partial x^2} - A\frac{\partial^2 w}{\partial y^2} - B_0 w, \quad (3.104b)$$

$$M\Psi \equiv A\frac{\partial^2 \Psi}{\partial y^2} - \frac{\partial \Psi}{\partial t}, \quad (3.104c)$$

where c_0 and B_0 are the values of c_μ and B_μ at $\mu = 0$. The functions f_j for $j = 1, 2, 3$ are given by

$$f_1 = 0,$$

$$f_2 = -\frac{\partial w_1}{\partial t_1} - M\Psi_1\frac{\partial w_1}{\partial x} + \frac{\partial \Psi_1}{\partial t_1}\frac{d\hat{u}}{dx} - 2A\frac{\partial \Psi_1}{\partial y}\frac{\partial^2 w_1}{\partial x\partial y}$$

$$+ A\left(\frac{\partial \Psi_1}{\partial y}\right)^2 \frac{d^2\hat{u}}{dx^2} + \frac{1}{2}a_0(w_1,w_1),$$

$$f_3 = -\frac{\partial w_2}{\partial t_1} - \frac{\partial w_1}{\partial t_2} + a_0(w_1,w_2) + \frac{1}{6}b_0(w_1,w_1,w_1) + \nu T w_1 -$$

$$M\Phi_1\frac{\partial w_2}{\partial x} - M\Phi_2\frac{\partial w_1}{\partial x} - \frac{\partial \Psi_1}{\partial t_1}\frac{\partial w_1}{\partial x} + \frac{\partial \Psi_2}{\partial t_1}\frac{d\hat{u}}{dx} + \frac{\partial \Psi_1}{\partial t_2}\frac{d\hat{u}}{dx} + \nu\frac{\partial \Psi_1}{\partial t_0}\frac{d\dot{\hat{u}}}{dx}$$

$$-2A\left(\frac{\partial\Psi_2}{\partial y}\frac{\partial^2 w_1}{\partial x\partial y}+\frac{\partial\Psi_1}{\partial y}\frac{\partial^2 w_2}{\partial x\partial y}-\frac{\partial\Psi_1}{\partial y}\frac{\partial\Psi_2}{\partial y}\frac{d^2\hat{u}}{dx^2}-\frac{1}{2}\left(\frac{\partial\Psi_1}{\partial y}\right)^2\frac{\partial^2 w_1}{\partial x^2}\right),$$

where a_0 and b_0 are the values of a and b at $\mu=0$, and

$$Tw=\dot{B}_0 w-\dot{c}_0\frac{\partial w}{\partial x},\quad \dot{B}_0=\left.\frac{\partial B_\mu}{\partial\mu}\right|_{\mu=0},\quad \dot{c}_0=\left.\frac{\partial c_\mu}{\partial\mu}\right|_{\mu=0},\quad \dot{\hat{u}}=\left.\frac{\partial\hat{u}}{\partial\mu}\right|_{\mu=0}.$$

Since the homogeneous equation

$$Lw=0,\quad -\infty<x<\infty,\quad 0<y<\ell,\quad t>0 \tag{3.105}$$

has nontrivial solutions that are ℓ-periodic in y, $2\pi/\omega_0$-periodic in time and decay as $x\to\pm\infty$, it is necessary that solvability conditions be satisfied in order for solutions of the nonhomogeneous problem (3.104) to exist. To derive the solvability conditions, we introduce the inner product

$$\langle f,g\rangle=\int_0^{2\pi/\omega_0}dt\int_0^\ell dy\int_{-\infty}^\infty f\bar{g}dx,$$

for functions f and g that are bounded continuous functions of x on $(-\infty,\infty)$. We assume as well that the product $f\bar{g}\to 0$ as x tends to $\pm\infty$ in such a way that the above integrals exist. Furthermore, we assume that both functions are $2\pi/\omega_0$-periodic in time and ℓ-periodic in y.

The formally adjoint operator L^* is defined in the usual way, i.e.,

$$\langle Lf,g\rangle=\langle f,L^*g\rangle,$$

and has the form

$$L^*\phi\equiv\frac{\partial\phi}{\partial t}+c\frac{\partial\phi}{\partial x}+A\frac{\partial^2\phi}{\partial x^2}+A\frac{\partial^2\phi}{\partial y^2}+B_0^*\phi=0, \tag{3.106}$$

$$\phi\to 0\quad\text{as}\quad x\to\pm\infty,$$

and the adjoint homogeneous solutions, which are $2\pi/\omega_0$-periodic in time and ℓ-periodic in y, are

$$\phi_0=\theta_0(x),\quad \phi_1=\theta_1(x)e_1,\quad \phi_2=\theta_1(x)e_2, \tag{3.107}$$

where

$$e_1\equiv e^{i(\omega_0 t+k_0 y)},\quad e_2\equiv e^{i(\omega_0 t-k_0 y)},$$

and θ_0 and θ_1 are solutions of

$$A\theta_0''+c_0\theta_0'+B_0^*\theta_0=0,$$

$$A\theta_1'' - Ak^2\theta_1 + c_0\theta_1' + B_0^*\theta_1 + i\omega_0\theta_1 = 0.$$

Here, the matrix B_0^* is adjoint to B_0. The solvability conditions can then be written in the form

$$\langle f_j, \phi_k \rangle = 0 \quad j = 1, 2, 3, \ldots; \quad k = 0, 1, 2. \tag{3.108}$$

Now, we determine the solutions of (3.104) for $j = 1$ and $j = 2$. For $j = 1$ the solution can be written in the form

$$w_1 = w_1^h + w_1^p, \tag{3.109a}$$

where w_1^p is a particular solution that corresponds to $M\Psi_1 \hat{u}'$, and w_1^h is the homogeneous solution. It can be verified directly that

$$w_1^p = \Psi_1 \hat{u}'. \tag{3.109b}$$

The homogeneous solution has the form

$$w_1^h = (R_1 e_1 + S_1 e_2)g_0(x) + c.c. - H_1 \hat{u}', \tag{3.109c}$$

where $c.c.$ denotes complex conjugate. This is due to the fact that the operator L is the same operator as the one we get in the linear stability analysis for $\mu = 0$, and therefore, w_1^h is a linear combination of the linearly unstable modes (3.92) with the amplitudes that depend on the slow times t_1 and t_2. Next, the form of the front Φ_1 is also represented as a superposition of the same modes

$$\Psi_1 = R_1 e_1 + S_1 e_2 + c.c. + H_1. \tag{3.109d}$$

We remark that we wrote the same H_1 in (3.109c) and in (3.109d) so that w_1 does not contain the mode $\hat{u}'$. This ought to be the case, because in the moving coordinate system (3.95), the system is no longer invariant with respect to translations in x, and therefore the mode $\hat{u}'$ which appears due to translation invariance should be absent.

For $j = 2$, the three solvability conditions (3.108) imply that

$$\frac{\partial R_1}{\partial t_1} = \frac{\partial S_1}{\partial t_1} = 0, \quad \frac{\partial H_1}{\partial t_1} = m(\mid R_1 \mid^2 + \mid S_1 \mid^2), \tag{3.110}$$

where m is a constant, which can be easily computed. If the solvability conditions are satisfied, the solution w_2 can be written in the form

$$w_2 = w_2^h + w_2^p + \Psi_2 \hat{u}', \tag{3.111}$$

where w_2^h is a solution of the homogeneous problem, so that

$$w_2^h = (R_2 e_1 + S_2 e_2)g_0(x) + c.c. - H_2 \hat{u}', \tag{3.112}$$

and

$$\Psi_2 = R_2 e_1 + S_2 e_2 + c.c. + H_2.$$

The amplitudes of the modes depend on the slow times. Next, w_2^p is a particular solution that corresponds to the right–hand side f_2. To determine the particular solution, we expand f_2 as a polynomial in e_1, e_2, and their complex conjugates and find that w_2^p is given by

$$w_2^p = (R_1^2 e_1^2 + S_1^2 e_2^2) g_1(x) + R_1 S_1 e_1 e_2 g_2 + R_1 \overline{S}_1 e_1 \overline{e}_2 g_3 + \tag{3.113}$$

$$+ e_1 g_4 + e_2 g_5 + c.c. + g_6.$$

The functions $g_1, \ldots, g_6$ satisfy the equations that can be easily derived by substituting (3.112) into equation (3.104). These equations are not presented here (cf. [93]).

Next, applying the solvability conditions (3.108) to (3.104) with $j = 3$, we obtain a coupled set of Landau equations

$$\frac{\partial R_1}{\partial t_2} = \nu R_1 \chi + \beta_1 R_1 |R_1|^2 + \beta_2 R_1 |S_1|^2, \tag{3.114a}$$

$$\frac{\partial S_1}{\partial t_2} = \nu S_1 \chi + \beta_1 S_1 |S_1|^2 + \beta_2 S_1 |R_1|^2. \tag{3.114b}$$

We do not present here explicit expressions for the coefficients β_1, β_2 and χ. Though computing these expressions is a straightforward calculation, it is not very instructive, and we restrict ourselves by mentioning that these coefficients are some complex-valued quantities. Amplitude equations of this form have been derived in many problems, e.g., in the gasless combustion problem [47]. This is not surprising, because the form of the amplitude equations is typically a consequence of symmetries of the original problem.

Let us discuss some solutions of the Landau amplitude equations. It is convenient to write the amplitudes R_1 and S_1 of Ψ_1, which determines the shape of the front, in the form

$$R_1(t_2) = a(t_2) \exp(i\theta_a(t_2)), \quad S_1(t_2) = b(t_2) \exp(i\theta_b(t_2)). \tag{3.115}$$

Substituting (3.115) into (3.114) and separating real and imaginary parts results in

$$\frac{da}{dt_2} = \nu \chi_r a + \beta_{1r} a^3 + \beta_{2r} a b^2 \tag{3.116a}$$

$$\frac{db}{dt_2} = \nu \chi_r b + \beta_{1r} b^3 + \beta_{2r} b a^2 \tag{3.116b}$$

$$\frac{d\theta_a}{dt_2} = \nu \chi_i + \beta_{1i} a^2 + \beta_{2i} b^2, \tag{3.116c}$$

$$\frac{d\theta_b}{dt_2} = \nu\chi_i + \beta_{1i}b^2 + \beta_{2i}a^2. \tag{3.116d}$$

Here, the subscripts r and i denote the real and imaginary parts of the respective coefficients. In order to determine the steady state solutions of (3.116), which in the original problem correspond to a superposition of waves traveling along the front, we set $da/dt_2 = db/dt_2 = 0$. This leads to

$$a(\nu\chi_r + \beta_{1r}a^2 + \beta_{2r}b^2) = 0, \quad b(\nu\chi_r + \beta_{1r}b^2 + \beta_{2r}a^2) = 0. \tag{3.117}$$

There are four critical points

$$a_1 = b_1 = 0, \quad a_2 = 0,\, b_2 = w_t, \quad a_3 = w_t,\, b_3 = 0, \quad a_4 = b_4 = w_s,$$

where

$$w_t = (-\nu\chi_r/\beta_{1r})^{1/2}, \quad w_s = (-\nu\chi_r/(\beta_{1r} + \beta_{2r}))^{1/2}.$$

In the case of the first critical point, the amplitudes R_1 and S_1 are identically equal to zero, which corresponds to the uniformly propagating wave in the original problem. The second and third critical points correspond to waves traveling along the front, which are right- and left-traveling waves (spinning modes), respectively. The last critical point corresponds to a standing wave.

It can be shown that $\chi_r > 0$ for all parameter values. Thus, from the expression for w_t, we conclude that left- and right-traveling waves exist for $\nu > 0$ (the so-called supercritical bifurcation) if $\beta_{1r} < 0$ and for $\nu < 0$ (the subcritical bifurcation) if $\beta_{1r} > 0$. In a similar way, the supercritical bifurcation of standing waves occurs if $\beta_{1r} + \beta_{2r} < 0$, and the subcritical bifurcation occurs if $\beta_{1r} + \beta_{2r} > 0$. All the subcritical bifurcations are known to produce locally unstable regimes. The supercritical bifurcation can lead to either stable or unstable solutions depending on the parameter values. Specifically, the supercritical bifurcation of traveling waves (which occurs if $\beta_{1r} < 0$) is stable if $\beta_{2r} < \beta_{1r}$ and unstable otherwise. The supercritical bifurcation of standing waves (which occurs if $\beta_{1r} + \beta_{2r} < 0$) is stable if $\beta_{2r} > \beta_{1r}$ and unstable otherwise.

Let us provide some interpretations of the bifurcating regimes based on the magnitude of ℓ (equivalently, the value of N in (3.93)). Consider a spinning mode. If $N = 1$, then Ψ_1 has one maximum in y that travels along the front with time. This mode has been observed in GC experiments and is referred to as a one-headed spinning mode. The maximum manifests itself as a bright spot. Similarly, $N = 2$ and $N = 3$ correspond to two- and three-headed spinning modes, respectively, that have also been experimentally observed. Spinning modes with more than three bright spots have not been seen in the GC experiment. Standing modes can be given a similar interpretation (they are discussed in greater detail in another chapter of this book).

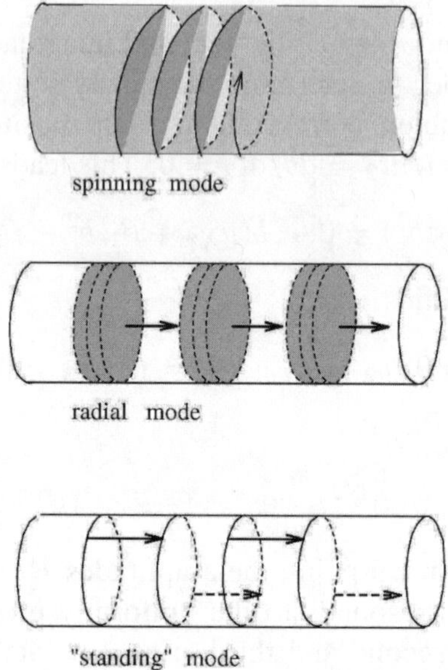

Figure 4. Propagation modes in a cylindrical sample.

A similar bifurcation analysis has been done for other geometries. Consider, for example, a circular cylinder of radius R with no-flux boundary conditions at the lateral surface. The wave $\hat{u}$ propagates along the axis of the cylinder. The form of the front in this case is

$$\Psi_1 = R_1\psi_1 + S_1\psi_2 + c.c. + H_1,$$

where ψ_1 and ψ_2 are the eigenfunctions of the problem (3.94) given by

$$\psi_{1,2} = e^{\pm in\theta} J_n(\sigma_{nm}r/R), \quad n, m = 1, 2, 3, \ldots.$$

Here, (r, θ) are polar coordinates in the cross section, σ_{nm} is the mth positive zero of the derivative of the Bessel function J_n, and the radius R must be such that

$$\sigma_{nm}/R = k_0.$$

The amplitudes R_1 and S_1 satisfy the same Landau equations as in the previous case (with different values of the coefficients). The value of n gives the number of maxima in the angular direction, while the value of m gives the number of maxima in the radial direction. Let $R_{nm} = \sigma_{nm}/k_0$. Since

$$\sigma_{11} < \sigma_{21} < \sigma_{01} < \sigma_{31} < \sigma_{12} < \ldots,$$

we have

$$R_{11} < R_{21} < R_{01} < R_{31} < R_{12} < \ldots,$$

and we can expect the following bifurcating spinning modes as R increases. For $R = R_{11}$, it is a one-headed spinning mode with the bright spot rotating along the front at the lateral surface; for $R = R_{21}$ – a similar two-headed spinning mode; for $R = R_{01}$, the mode is quite different. It does not have any angular dependence, and a bright circle periodically appears and disappears at the surface (contracting to the axis of the cylinder when it disappears). In GC experiments, this mode was named the limiting mode; we refer to it as the radial mode (see Figure 4). We remark that σ_{01} is a simple eigenvalue of (3.94) rather than a double, so a single Landau equation governs the amplitude of the unstable mode. For $R = R_{31}$, there is a three-headed spinning mode at the surface. For $R = R_{12}$, there are two bright spots rotating in the angular direction – one along the lateral surface of the cylinder and the other inside.

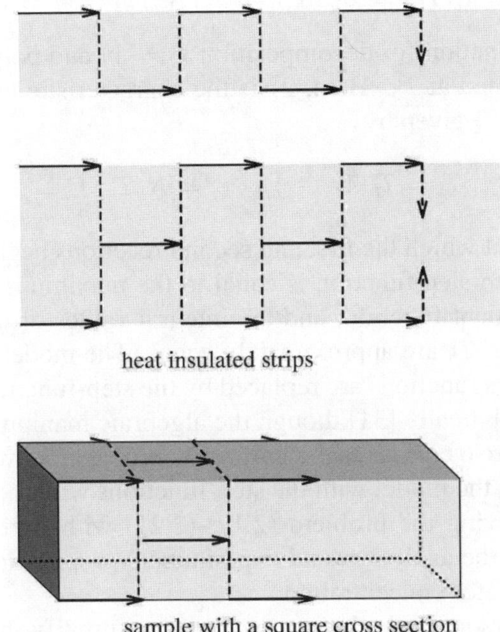

Figure 5. Propagation modes in rectangular geometries.

Consider next rectangular geometries, such as a strip with no-flux boundary conditions in y or a cylinder with a square cross section. Here, the unstable modes are different from the ones we discussed above, and the amplitude equations are not Landau equations [93, 94], which is not surprising given a different symmetry of the problem. We illustrate some of the propagation modes in Figure 5.

4.　　Analysis of base FP model

We now return to the base model (2.17)-(2.21) of frontal polymerization. We want to find uniformly propagating FP waves and perform linear and non-linear stability analyses, as we did in the case of the gasless combustion model. Before we study the model, we would like to reformulate it using the reaction front approximation.

First, we use the step-function approach and replace the Arrhenius dependence of the reaction rate on temperature $k_i(T)$ with the step function (cf. Section 3.1.4)

$$K_i(T) = A_i H(T - T_i), \tag{4.118}$$

where

$$A_i = k_i(T_b) = k_{0i} \exp[-E_i/(R_g T_b)], \quad i = 1, 2$$

is the Arrhenius function for decomposition ($i = 1$) and polymerization ($i = 2$) reactions evaluated at T_b. Here, T_b is the characteristic temperature in the reaction zone, and T_i given by

$$T_i = T_b(1 - \epsilon_i), \quad \epsilon_i = R_g T_b / E_i$$

is the temperature at which the first and second reactions begin. As before, the height of the chosen step function is equal to the maximum of the Arrhenius temperature-dependent function, and the integral values of the two functions over the range T_0 to T_b are approximately equal. The model (2.17)-(2.21), in which the Arrhenius functions are replaced by the step-functions (4.118), is already tractable analytically [31] though the algebraic manipulations are rather cumbersome. Figure 6 compares the uniformly propagating wave derived analytically by solving the model with the step-functions with the wave computed numerically by solving the problem (2.17)-(2.21) with Arrhenius functions. Figure 7 compares the analytical and experimental propagation velocities as a function of the initiator concentration.

To avoid cumbersome calculations, we further simplify the model making use of the fact that the activation energies of the decomposition and polymerization reactions are large, which results in narrow reaction zones that can be replaced by a reaction front (cf. Section 3.1.3). In this case, equations (2.17)-(2.19) must be solved without the reaction term both ahead of and behind the reaction front and matched at the reaction front by satisfying matching conditions. In order to derive the matching conditions, we study the equations in the reaction zone similar to what we did in the case of the gasless combustion

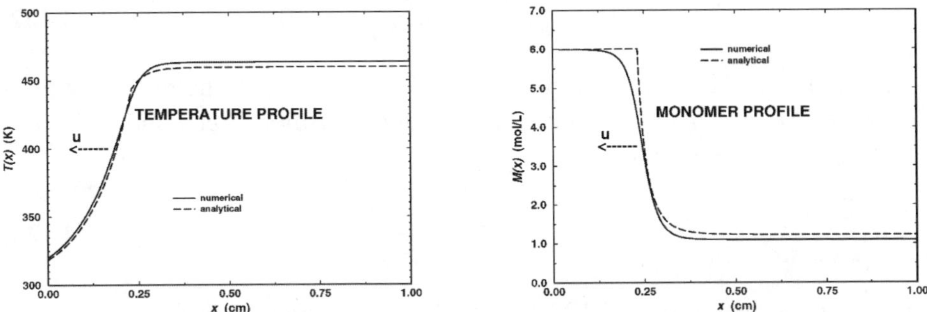

Figure 6. Temperature and monomer profiles in the uniformly propagating polymerization wave obtained analytically in the step-function model and numerically in the full model.
Adapted from [31].

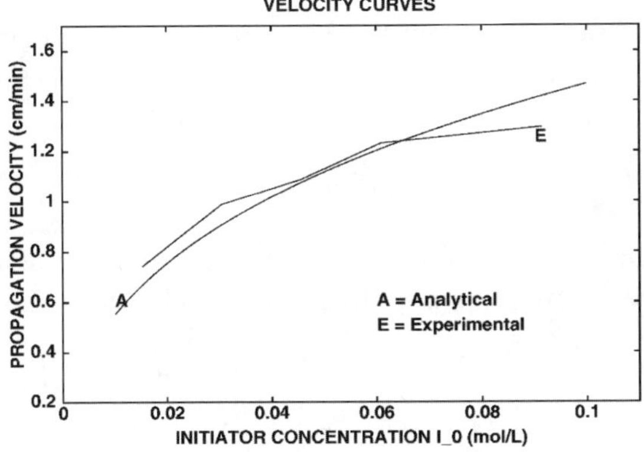

Figure 7. Propagation velocity as a function of the initiator concentration. The analytical
result is a solution of the step-function model. Adapted from [31].

model. The dominant balance in the reaction zone equations is between the reaction term and the highest spatial derivative (which could be formally derived
by stretching the spatial scale in the direction of propagation x)

$$-\varphi_t \frac{\partial J}{\partial x} + J A_1 = 0, \tag{4.119}$$

$$-\varphi_t \frac{\partial M}{\partial x} + J M A_2 = 0, \tag{4.120}$$

$$\kappa(1 + \varphi_y^2)\frac{\partial^2 T}{\partial x^2} + q J M A_2 = 0. \tag{4.121}$$

Combining equations (4.120) and (4.121) results in

$$\kappa(1 + \varphi_y^2)\frac{\partial^2 T}{\partial x^2} + q\varphi_t\frac{\partial M}{\partial x} = 0. \tag{4.122}$$

Integrating (4.122) from a point x in the reaction zone to the product end of it, where $M = M_b$ (the final concentration of the monomer), and $\partial T/\partial x$ is essentially zero, we obtain

$$\kappa\frac{\partial T}{\partial x} = -\frac{q(M - M_b)\varphi_t}{1 + \varphi_y^2}. \tag{4.123}$$

Furthermore, integrating (4.122) across the reaction zone gives

$$\kappa\left[\frac{\partial T}{\partial x}\right] = \frac{q(M_0 - M_b)\varphi_t}{1 + \varphi_y^2}, \tag{4.124}$$

where the square brackets denote the jump in a quantity across the reaction zone (see (3.55)). As the reaction zone is very thin, we can assume continuity of temperature across the front

$$[T] = 0. \tag{4.125}$$

Eliminating x from (4.119) and (4.120) and integrating the resulting equation from a point in the reaction zone to the product end of it (where $M = M_b$ and J is essentially zero) gives

$$M = M_b\exp\left(\frac{JA_2}{A_1}\right). \tag{4.126}$$

Evaluating (4.126) at the left end of the reaction zone (where $M = M_0$ and $J = J_0$) yields

$$M_b = f(T_b) \equiv M_0\exp(-j_0), \quad j_0 = \frac{J_0 A_2}{A_1}. \tag{4.127}$$

Finally, we derive an expression for the propagation velocity of the polymerization front analogous to formula (3.58) derived for the gasless combustion wave. We substitute (4.126) into (4.123) to obtain

$$\kappa\frac{\partial T}{\partial x} = -\frac{qM_b\left(\exp\left(\frac{JA_2}{A_1}\right) - 1\right)\varphi_t}{1 + \varphi_y^2}. \tag{4.128}$$

Dividing (4.128) by (4.119) and integrating over the reaction zone, we obtain

$$\kappa A_1(T_b - T_i) = \frac{\varphi_t^2}{1 + \varphi_y^2}qM_b\int_0^{j_0}\frac{\exp\eta - 1}{\eta}d\eta.$$

Using the definitions

$$T_i = T_b \left(1 - \frac{R_g T_b}{E_1} \right), \quad A_1 = k_{01} \exp \left(-\frac{E_1}{R_g T_b} \right),$$

and inserting (4.127) in place of M_b, we reduce the above equation to

$$\frac{\varphi_t^2}{1 + \varphi_y^2} = F_p(T_b), \tag{4.129a}$$

$$F_p(T_b) \equiv \frac{\kappa k_{01} R_g T_b^2}{q M_0 E_1} \exp(j_0 - \frac{E_1}{R_g T_b}) \left(\int_0^{j_0} \frac{e^\eta - 1}{\eta} d\eta \right)^{-1}. \tag{4.129b}$$

Note that using the definition of j_0, we can rewrite (4.129b) as

$$F_p(T_b) = \frac{\kappa J_0 k_{02} R_g T_b^2}{q M_0 E_1} \exp \left(-\frac{E_2}{R_g T_b} \right) \Phi(j_0), \tag{4.129c}$$

$$\Phi(j_0) = \left(j_0 e^{-j_0} \int_0^{j_0} \frac{e^\eta - 1}{\eta} d\eta \right)^{-1}. \tag{4.129d}$$

Under the reaction front approximation, we solve the reactionless equations on either side of the reaction front (i.e. all the reaction rates are neglected in (2.17)-(2.19))

$$\frac{\partial T}{\partial t} - \varphi_t \frac{\partial T}{\partial x} = \kappa \nabla^2 T, \tag{4.130}$$

$$\frac{\partial M}{\partial t} - \varphi_t \frac{\partial M}{\partial x} = 0, \tag{4.131}$$

$$\frac{\partial J}{\partial t} - \varphi_t \frac{\partial J}{\partial x} = 0 \tag{4.132}$$

subject to the boundary conditions (2.20), (2.21) in x, appropriate boundary conditions in y, appropriate initial conditions and the conditions (4.124), (4.125), (4.127), and (4.129) derived from the reaction zone analysis.

It is interesting to compare the reaction front formulations of the gasless combustion and frontal polymerization models. First of all, there is an additional differential equation in the FP model, namely, equation (4.132). The equation, however, can be easily solved yielding

$$J = \left\{ \begin{array}{ll} J_0, & x < 0 \\ 0, & x > 0 \end{array} \right.,$$

which agrees with our physical understanding that, due to the absence of any transport of the species, J takes on a constant initial concentration value ahead of the reaction zone, and it is completely consumed in the reaction zone. Therefore, the J-equation can be disregarded. In both reaction front formulations, the temperature is continuous across the reaction zone, the heat flux undergoes a jump (due to heat production in the reaction zone), and there is an expression for the propagation velocity in terms of the temperature T_b in the reaction zone. The main difference between the models is the presence of the quantity M_b in the FP model. This is the final value of the concentration of M, i.e., the concentration of the unreacted monomer. It is not surprising that $M_b > 0$ in the FP model, while in the gasless combustion model, the final concentration of the reactant is zero. Indeed, in the gasless combustion model, M must be zero behind the reaction zone – otherwise, the reaction would not stop. In the FP model, the reactions stop, because the initiator is completely consumed, so M_b does not have to be zero, and, in fact, it is not. In the limit $M_b \to 0$ (equivalently, $j_0 \to \infty$, from which $\Phi \to 1$), the FP model simplifies to the gasless combustion model (with a small modification in F_c).

4.1 Uniformly propagating wave

In this section, we determine stationary solutions of the above problem, which correspond to uniformly propagating one-dimensional traveling waves in the laboratory coordinate system. We solve the following reactionless system ahead of $(x < 0)$ and behind $(x > 0)$ the front

$$\frac{d\widehat{M}}{dx} = 0, \quad \kappa \frac{d^2\widehat{T}}{dx^2} - u\frac{d\widehat{T}}{dx} = 0, \tag{4.133}$$

subject to the boundary conditions

$$x = -\infty: \quad \widehat{T} = T_0, \quad \widehat{M} = M_0, \tag{4.134}$$

$$x = +\infty: \quad \frac{d\widehat{T}}{dx} = 0, \tag{4.135}$$

and the matching conditions

$$\left[\widehat{T}\right] = 0, \quad \kappa \left[\frac{d\widehat{T}}{dx}\right] = -qu(M_0 - \widehat{M_b}), \tag{4.136}$$

$$u^2 = F_p(\widehat{T_b}), \quad \widehat{M_b} = M_0 \exp(-\widehat{j_0}(\widehat{T_b})), \tag{4.137}$$

$$\widehat{j_0}(\widehat{T_b}) \equiv 2\sqrt{I_0}\frac{k_{02}}{k_{01}} \exp\frac{E_1 - E_2}{R\widehat{T_b}}.$$

Here, the quantities with the hat denote the stationary solution, and $u = -\varphi_t$ is the propagation velocity of the uniformly propagating wave. The stationary solution that satisfies (4.133)-(4.136) is given by

$$\widehat{T}(x) = \begin{cases} T_0 + (\widehat{T_b} - T_0)\exp\left(\frac{u}{\kappa}x\right), & x < 0 \\ \widehat{T_b}, & x > 0 \end{cases}, \tag{4.138}$$

$$\widehat{M}(x) = \left\{ \begin{array}{ll} M_0, & x < 0 \\ \widehat{M_b}, & x > 0 \end{array} \right. , \qquad (4.139)$$

where u, $\widehat{M_b}$ and $\widehat{T_b}$ are determined from

$$\widehat{T_b} = T_0 + q(M_0 - \widehat{M_b}), \quad \widehat{M_b} = f(\widehat{T_b}), \quad \widehat{u}^2 = F(\widehat{T_b}). \qquad (4.140)$$

In order to complete the solution of the stationary problem, we have to determine from (4.140) the propagation velocity u, the temperature $\widehat{T_b}$ in the reaction zone, and the concentration $\widehat{M_b}$ of the unreacted monomer. The first two equations determine $\widehat{T_b}$ and $\widehat{M_b}$, and then the last equation gives u. It is easy to find an approximate solution for $\widehat{T_b}$ and $\widehat{M_b}$ in the case when $\widehat{M_b}$ is small. Then $\widehat{T_b}$ is close to the adiabatic temperature T_a and approximately

$$\widehat{M_b} = f(T_a), \quad \widehat{T_b} = T_0 + q(M_0 - \widehat{M_b}).$$

We remark that if a more accurate solution is needed, the above equations can be easily converted into a convergent iteration procedure.

4.2 Linear stability analysis of frontal polymerization

As was the case with gasless combustion, the uniformly propagating polymerization wave may become unstable as parameters are varied. We perform a linear stability analysis of the uniformly propagating polymerization wave by employing the reaction front approximation.

We perturb the stationary solution as we did in the case of gasless combustion as

$$T = \widehat{T}(x) + \delta \exp(\omega t + iky)\widetilde{T}(x), \qquad (4.141)$$

$$M = \widehat{M}(x) + \delta \exp(\omega t + iky)\widetilde{M}(x), \qquad (4.142)$$

$$J = \widehat{J}(x) + \delta \exp(\omega t + iky)\widetilde{J}(x), \qquad (4.143)$$

$$\varphi = -ut + \delta \exp(\omega t + iky). \qquad (4.144)$$

As before, k is the wavenumber, ω is the temporal frequency, and δ is the small magnitude of the perturbations. All of the steps of the analysis (and even many intermediate results) here are the same as for gasless combustion, and, therefore, we omit the details. We substitute these expansions into the reactionless equations, boundary conditions and matching conditions. We then linearize the problem and solve the linear problem to obtain the dispersion relation

$$4\Omega^3 + \left(1 + 4s^2 + 2z - \left(\frac{z}{2} - P\right)^2\right)\Omega^2 + \frac{z}{2}(1 + 4s^2 + P)\Omega + s^2\frac{z^2}{4} = 0. \qquad (4.145)$$

Here, we used the same nondimensional quantities as in (3.83), and

$$P = qf'(\widehat{T}_b) \equiv q\frac{\partial \widehat{M}_b}{\partial \widehat{T}_b}.$$

An instability occurs when a pair of complex conjugate eigenvalues crosses the imaginary axis as the parameters vary. At the stability boundary,

$$z_c = 4 + 2P - 4s^2 \left(1 + 4s^2 + P\right)^{-1}$$

$$+2\sqrt{\left(2 + P - 2s^2 \left(1 + 4s^2 + P\right)^{-1}\right)^2 - P^2 + 1 + 4s^2},$$

$$\Omega = \pm i\omega_0, \quad \omega_0^2 = \frac{1}{8}z_c(1 + 4s^2 + P). \tag{4.146}$$

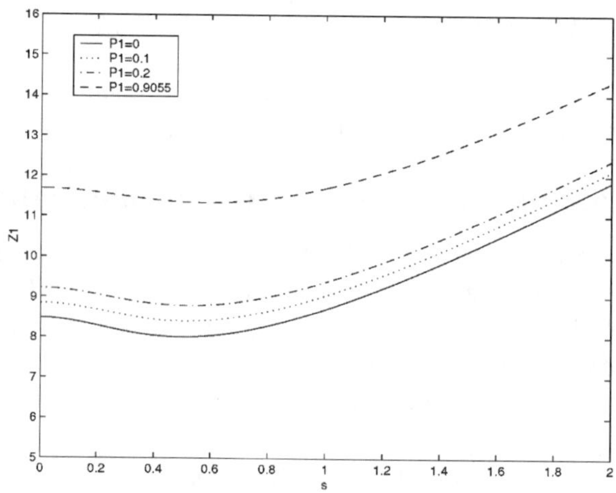

Figure 8.   Neutral stability curves in the (s, z) plane for various values of P. Above respective curves, uniform propagation is unstable.

The neutral stability curve in the (s, z)-plane has a minimum at $s = s_m > 0$ for all $P \geq 0$. Neutral stability curves are shown in Figure 8 for physically meaningful values of P. We remark that the parameter P is related to the concentration of the unreacted monomer. Thus, it is not surprising that the dispersion relation for $P = 0$ coincides with the dispersion relation in the GC model (3.83).

4.3 Weakly nonlinear analysis of frontal polymerization

There are interesting experimental results concerning the nonlinear dynamics of polymerization waves [24, 50, 72, 74, 77]. In [77] for example, non-standard geometries such as conical test tubes are used. Theoretical work on

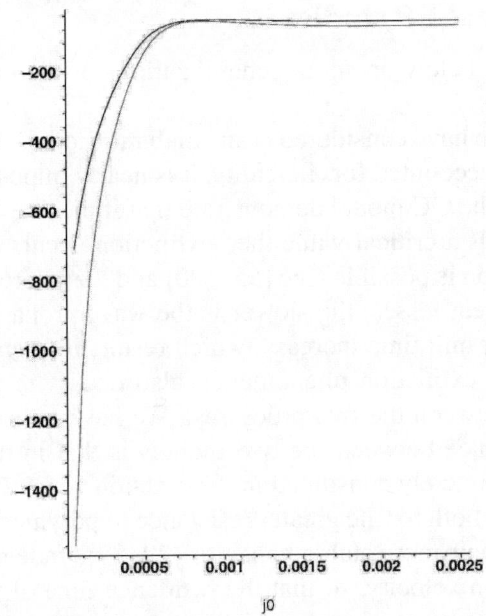

Figure 9.　　Graphs of β_{1r} (upper curve) and β_{2r} (lower curve) versus the nondimensional initial concentration of the initiator j_0. Adapted from [17].

this subject is limited. The only nonlinear analysis studies are those of one-dimensional pulsating modes of propagation [32] and spinning and standing modes in a cylindrical shell [17, 18].

In the latter case, the authors derived Landau equations (3.114) for the amplitudes of the unstable modes and determined the dependence of the coefficients of these equations on the parameters of the FP problem. Figure 9 depicts the real parts of the coefficients β_1 and β_2 as functions of j_0 (which is proportional to the initial concentration of the initiator) for the typical parameter values [81] $(E_1 - E_2)/(R_g q M_0) = 19.79$ and $E_1/(R_g q M_0) = 58.4$. The wavenumber $s = 0.55$, which is close to the value s_m at which the neutral stability curve has a minimum.

We see that both quantities are negative, which implies that both traveling and standing waves appear as a result of a supercritical bifurcation. For the parameter values chosen, $\beta_{1r} > \beta_{2r}$, so the traveling waves are stable while the standing waves are unstable. This observation agrees with the experimental data in [72], where spinning waves have been observed while standing waves have not been seen.

5.   Other thermal FP studies

We briefly comment below on some generalizations of the base FP model discussed in this chapter.

The FP model that we have considered is an adiabatic model – no heat losses to the environment are accounted for. In reality, it is nearly impossible to avoid heat losses. Studies of the GC model demonstrate that if the rate of heat loss to the environment exceeds a critical value then extinction occurs, i.e., no combustion wave propagation is possible (see [55, 100] and the references therein). Indeed, the larger are heat losses, the slower is the wave propagation, so that effective heat losses per unit time increase, which results in even slower wave propagation, etc. This extinction phenomenon also occurs in FP. However, there are differences between the two processes. We have seen that the principal qualitative difference between the two models is that in the FP model, the monomer is not completely consumed in the reaction zone. The remaining monomer is responsible both for the greater resistance of polymerization waves to heat losses and for their more stable behavior [29, 87]. Indeed, heat losses decrease the propagation velocity, so that the residence time of the monomer in the reaction zone is greater. Therefore, in the nonadiabatic FP model, the monomer reacts to a higher conversion than in the adiabatic problem, thus producing more heat and helping the wave to survive. In the GC model, this survival mechanism is absent because conversion is complete.

Mathematical models of the frontal copolymerization process were developed, studied and compared with experimental data in [67, 90]. An interesting observation was that the propagation speed of the copolymerization wave was not necessarily related to the propagation speeds in the two homopolymerization processes, in which the same two monomers were polymerized separately. For example, the propagation speeds in the homopolymerization processes could be 1 cm/min in each, but in the copolymerization process, the speed could be 0.5 cm/min. Mathematical models of free-radical binary frontal polymerization were presented and studied in [66, 91]. Another model in which two different monomers were present in the system (thiol-ene polymerization) was discussed in [21]. A mathematical model that describes both free-radical binary frontal polymerization and frontal copolymerization was presented in [65]. The paper was devoted to the linear stability analysis of polymerization waves in two monomer systems. It turned out that the dispersion relation for two monomer systems was the same as the dispersion relation for homopolymerization. In fact, this dispersion relation held true for N-monomer systems provided that there is only one reaction front, and the final concentrations of the monomers could be written as a function of the reaction front temperature.

Sandwich-type FP models were considered in [19, 20]. The geometry of the problem involves two adjacent layers that can exchange heat. One layer contains the reactive mixture that includes a monomer and an initiator, while the other layer is comprised either of a heat conducting inert material [20] or a chemically active mixture [19]. The mathematical model is similar to that in GC [82]. A motivation to study this geometry came from experiments on fabrication of polymer-dispersed liquid-crystal materials where this geometry has been utilized. An interesting feature of the problem is the existence of several propagating waves for the same parameter values.

Some other works that we would like to mention studied the kinetics effects in FP [85], the influence of the gel effect on the propagation of thermal frontal polymerization waves [28], the use of complex initiators as a means to increase the degree of conversion of the monomer [30], and FP of metal-containing monomers [3].

In the frontal polymerization studies discussed in this chapter, the focus was on the propagation of the polymerization wave, its velocity, stability, the spatial profiles of the species involved, etc. Formation of a polymerization front from relevant initial conditions was presumed. From experimental work, however, it is known that formation of the front does not always occur (this is similar to ignition considerations in GC problems [56, 100]). The problem of formation of a polymerization front by a high temperature heat source was studies in [33, 80]. An asymptotic analysis of the problem resulted in an integral equation. Two parameters govern the qualitative behavior of the solution to the integral equation. Depending on the magnitude of these parameters, the solution exhibits either bounded or unbounded behavior indicating the formation or non-formation of a polymerization wave. Paper [33] contains a numerical analysis of the model.

6.   Conclusion

Frontal polymerization, the process of propagation of a polymerization wave, is important from both fundamental and applied viewpoints. In this chapter, we reviewed theoretical results on the base model of free-radical frontal polymerization. Based on the analogy of the gasless combustion model and using the methods developed in combustion theory, we determined uniformly propagating polymerization waves and discussed their linear and nonlinear stability.

Frontal polymerization can be used to synthesize valuable products, e.g. nanocomposites and liquid crystals, and theoretical studies represent an important tool for understanding the process.

References

[1] A. P. ALDUSHIN, A. BAYLISS, AND B. J. MATKOWSKY, *Dynamics in layer models of solid flame propagation*, Physica D, 143 (2000), pp. 109–137.

[2] M. APOSTOLO, A. TREDICI, M. MORBIDELLI, AND A. VARMA, *Propagation velocity of the reaction front in addition polymerization systems*, Journal of Polymer Science, Part A: Polymer Chemistry, 35 (1997), pp. 1047–1059.

[3] V. BARELKO, A. POMOGAILO, G. DZHARDIMALIEVA, S. EVSTRATOVA, A. ROZENBERG, AND I. UFLYAND, *The autowave modes of solid phase polymerization of metal-containing monomers in two- and three-dimensional fiberglass-filled matrices*, Chaos, 9 (1999), pp. 342–347.

[4] A. BAYLISS, R. KUSKE, AND B. MATKOWSKY, *A two dimensional adaptive pseudo–spectral method*, J. Comp. Phys., 91 (1990), pp. 174–196.

[5] A. BAYLISS AND B. MATKOWSKY, *Fronts, relaxation oscillations, and period doubling in solid fuel combustion*, J. Comp. Phys., 71 (1987), pp. 147–168.

[6] ———, *Two routes to chaos in condensed phase combustion*, SIAM J. Appl. Math., 50 (1990), pp. 437–459.

[7] A. BAYLISS, B. J. MATKOWSKY, AND A. P. ALDUSHIN, *Dynamics of hot spots in solid fuel combustion*, Physica D, 166 (2002), pp. 104–130.

[8] A. J. BERNOFF, R. KUSKE, B. J. MATKOWSKY, AND V. VOLPERT, *Mean-field effects for counterpropagating traveling-wave solutions of reaction-diffusion systems*, SIAM J. Appl. Math., 55 (1995), pp. 485–519.

[9] M. BOOTY, S. MARGOLIS, AND B. MATKOWSKY, *Interaction of pulsating and spinning waves in condensed phase combustion*, SIAM J. Appl. Math., 46 (1986), pp. 801–843.

[10] N. CHECHILO AND N. ENIKOLOPYAN, *Structure of the polymerization wave front and propagation mechanism of the polymerization reaction*, Dokl. Phys. Chem., 214 (1974), pp. 174–176.

[11] ———, *Effect of the concentration and nature of initiators on the propagation process in polymerization*, Dokl. Phys. Chem., 221 (1975), pp. 392–394.

[12] ———, *Effect of pressure and initial temperature of the reaction mixture during propagation of a polymerization reaction*, Dokl. Phys. Chem., 230 (1976), pp. 840–843.

[13] N. CHECHILO, R. KHVILIVITSKII, AND N. ENIKOLOPYAN, *On the phenomenon of polymerization reaction spreading*, Dokl. Akad. Nauk SSSR, 204 (1972), pp. 1180–1181.

[14] Y. CHEKANOV, D. ARRINGTON, G. BRUST, AND J. POJMAN, *Frontal curing of epoxy resin: Comparison of mechanical and thermal properties to batch cured materials*, J. Appl. Polym. Sci., 66 (1997), pp. 1209–1216.

[15] Y. CHEKANOV AND J. POJMAN, *Preparation of functionally gradient materials via frontal polymerization*, J. Appl. Polym. Sci., 78 (2000), pp. 2398–2404.

[16] S. CHEN, J. SUI, L. CHEN, AND J. A. POJMAN, *Polyurethane-nanosilica hybrid nanocomposites synthesized by frontal polymerization*, J. Polym. Sci. Part A - Polym. Chem., 43 (2005), pp. 1670–1680.

[17] D. COMISSIONG, L. GROSS, AND V. VOLPERT, *Bifurcation analysis of polymerization fronts*, in Nonlinear Dynamics in Polymeric Systems, J. Pojman and Q. Tran-Cong-Miyata, eds., no. 869 in ACS Symposium Series, American Chemical Society, Washington, DC, 2003, pp. 147–159.

[18] D. M. G. COMISSIONG, L. K. GROSS, AND V. A. VOLPERT, *Nonlinear dynamics of frontal polymerization with autoacceleration*, Journal of Engineering Mathematics, (2004).

[19] ———, *Enhancement of the frontal polymerization process*, Journal of Engineering Mathematics, to be submitted, (2005).

[20] ———, *Frontal polymerization in the presence of an inert material*, Journal of Engineering Mathematics, submitted, (2005).

[21] D. E. DEVADOSS, J. A. POJMAN, AND V. A. VOLPERT, *Mathematical modeling of thiol-ene frontal polymerization*, Chem. Eng. Sci., submitted, (2005).

[22] A. V. DVORYANKIN, A. G. STRUNINA, AND A. G. MERZHANOV, *Trends in the spin combustion of thermites*, Combust. Explos. Shock Waves, 18 (1982), pp. 134–139.

[23] ———, *Stability of combustion in thermite systems*, Combust. Explos. Shock Waves, 21 (1985), pp. 421–425.

[24] I. R. EPSTEIN, J. A. POJMAN, AND Q. TRAN-CONG-MIYATA, *Nonlinear dynamics and polymeric systems: an overview*, in Nonlinear Dynamics in Polymeric Systems, J. Pojman and Q. Tran-Cong-Miyata, eds., no. 869 in ACS Symposium Series, American Chemical Society, Washington, DC, 2004, pp. 2–15.

[25] M. GARBEY, H. G. KAPER, G. K. LEAF, AND B. J. MATKOWSKY, *Nonlinear analysis of condensed phase surface combustion*, Euro. J. Appl. Math., 1 (1990), pp. 73–89.

[26] ———, *Quasi-periodic waves and the transfer of stability in condensed phase surface combustion*, SIAM J. Appl. Math., 52 (1992), pp. 384–395.

[27] N. GILL, J. A. POJMAN, J. WILLIS, AND J. B. WHITEHEAD, *Polymer-dispersed liquid-crystal materials fabricated with frontal polymerization*, J. Polym. Sci. Part A - Polym. Chem., 41 (2003), pp. 204–212.

[28] P. GOLDFEDER AND V. VOLPERT, *A model of frontal polymerization including the gel effect*, Math. Problems in Engin., 4 (1998), pp. 377–391.

[29] ———, *Nonadiabatic frontal polymerization*, J. Eng. Math., 34 (1998), pp. 301–318.

[30] ———, *A model of frontal polymerization using complex initiation*, Math. Problems in Engineering, 5 (1999), pp. 139–160.

[31] P. GOLDFEDER, V. VOLPERT, V. ILYASHENKO, A. KHAN, J. POJMAN, AND S. SOLOVYOV, *Mathematical modeling of free-radical polymerization fronts*, J. Phys. Chem. B, 101 (1997), pp. 3474–3482.

[32] L. K. GROSS AND V. A. VOLPERT, *Weakly nonlinear stability analysis of frontal polymerization*, Studies in Applied Mathematics, 110 (2003), pp. 351–375.

[33] A. HEIFETZ, L. RITTER, W. OLMSTEAD, AND V. A. VOLPERT, *A numerical analysis of initiation of polymerization waves*, Mathematical and Computer Modelling, in press, (2004).

[34] M. H. HOLMES, *Introduction to Perturbation Methods*, Springer, New York, 1998.

[35] V. ILYASHENKO AND J. POJMAN, *Single head spin modes in frontal polymerization*, Chaos, 8 (1998), pp. 285–289.

[36] T. P. IVLEVA AND A. G. MERZHANOV, *Mathematical modeling of three-dimensional spinning regimes of gas-free combustion waves*, Dokl. Akad. Nauk, 369 (1999), pp. 186–191.

[37] ———, *Structure and variability of spinning reaction waves in three-dimensional excitable media*, Phys. Rev. E, 6403 (2001), p. art.

[38] ———, *Three-dimensional modeling of solid-flame chaos*, Dokl. Phys. Chem., 381 (2001), pp. 259–262.

[39] ———, *Mathematical simulation of three-dimensional spin regimes of gasless combustion*, Combust. Explos. Shock Waves, 38 (2002), pp. 41–48.

[40] T. P. IVLEVA, A. G. MERZHANOV, AND K. G. SHKADINSKII, *Mathematical model of spin combustion*, Soviet Phys. Dokl., 239 (1978), pp. 255–256.

[41] ———, *Principles of the spin mode of combustion front propagation*, Combust. Explos. Shock Waves, 16 (1980), pp. 133–139.

[42] A. KHAN AND J. POJMAN, *The use of frontal polymerization in polymer synthesis*, Trends Polym. Sci. (Cambridge, U.K.), 4 (1996), pp. 253–257.

[43] E. I. MAKSIMOV, A. G. MERZHANOV, AND V. M. SHKIRO, *Gasless compositions as a simple model for the combustion of nonvolatile condensed systems*, Combust. Explos. Shock Waves, 1 (1965), pp. 15–19.

[44] Y. M. MAKSIMOV, A. G. MERZHANOV, A. T. PAK, AND M. N. KUCHKIN, *Unstable combustion modes of gasless systems*, Combust. Explos. Shock Waves, 17 (1981), pp. 393–400.

[45] Y. M. MAKSIMOV, A. T. PAK, G. B. LAVRENCHUK, Y. S. NAIBORODENKO, AND A. G. MERZHANOV, *Spin combustion of gasless systems*, Combust. Explos. Shock Waves, 15 (1979), pp. 415–418.

[46] G. B. MANELIS, L. P. SMIRNOV, AND N. I. PEREGUDOV, *Nonisothermal kinetics of polymerization processes. finite cylindrical reactor*, Combustion Explosion Shock Wave, 13 (1977), pp. 389–393.

[47] S. B. MARGOLIS, H. KAPER, G. LEAF, AND B. J. MATKOWSKY, *Bifurcation of pulsating and spinning reaction fronts in condensed two-phase combustion*, Comb. Sci. Technol., 43 (1985), pp. 127–165.

[48] S. B. MARGOLIS AND B. J. MATKOWSKY, *New modes of quasi-periodic combustion near a degenerate Hopf bifurcation point*, SIAM J. Appl. Math., 48 (1988), pp. 828–853.

[49] A. MARIANI, S. FIORI, Y. CHEKANOV, AND J. POJMAN, *Frontal ring-opening metathesis polymerization of dicyclopentadiene*, Macromolecules, 34 (2001), pp. 6539–6541.

[50] J. MASERE, F. STEWART, T. MEEHAN, AND J. POJMAN, *Period-doubling behavior in frontal polymerization of multifunctional acrylates*, Chaos, 9 (1999), pp. 315–322.

[51] B. J. MATKOWSKY AND G. I. SIVASHINSKY, *Propagation of a pulsating reaction front in solid fuel combustion*, SIAM J. Appl. Math., 35 (1978), pp. 230–255.

[52] B. J. MATKOWSKY AND V. VOLPERT, *Coupled nonlocal complex Ginzburg-landau equations in gasless combustion*, Physica D, 54 (1992), pp. 203–219.

[53] A. MERZHANOV, *Theory and practice of SHS: worldwide state of the art and the newest results*, Int. J. of SHS, 2 (1993), pp. 113–158.

[54] ———, *Worldwide evolution and present status of SHS as a branch of modern r&d (to the 30th anniversary of SHS)*, Int. J. of SHS, 6 (1997), pp. 119–163.

[55] A. MERZHANOV AND B. KHAIKIN, *Theory of combustion waves in homogeneous media*, Prog. Energy Combust. Sci., 14 (1988), p. 1.

[56] A. G. MERZHANOV AND A. E. AVERSON, *Present state of thermal ignition theory - invited review*, Combust. Flame, 16 (1971), p. 89.

[57] A. G. MERZHANOV AND I. P. BOROVINSKAYA, *New class of combustion processes*, Combust. Sci. Technol., 10 (1975), pp. 195–201.

[58] A. G. MERZHANOV, A. V. DVORYANKIN, AND A. G. STRUNINA, *A new variety of spin burning*, Dokl. Akad. Nauk SSSR, 267 (1982), pp. 869–872.

[59] A. G. MERZHANOV, A. K. FILONENKO, AND I. P. BOROVINSKAYA, *New phenomena in combustion of condensed systems*, Dokl. Phys. Chem., 208 (1973), pp. 122–125.

[60] Z. MUNIR AND U. ANSELMI-TAMBURINI, *Self-propagating exothermic reactions: the synthesis of high-temperature materials by combustion*, Material Science Reports, A Review Journal, 3 (1989), pp. 277–365.

[61] I. NAGY, L. SIKE, AND J. POJMAN, *Thermochromic composite prepared via a propagating polymerization front*, J. Am. Chem. Soc., 117 (1995), pp. 3611–3612.

[62] I. P. NAGY, L. SIKE, AND J. POJMAN, *Thermochromic composites and propagating polymerization fronts*, Adv. Mat., 7 (1995), pp. 1038–1040.

[63] B. NOVOZHILOV, *The rate of propagation of the front of an exothermic reaction in a condensed phase*, Dokl. Akad. Nauk SSSR, 141 (1961), pp. 151–153.

[64] G. ODIAN, *Principles of Polymerization, 2nd edition*, Wiley-Interscience, New York, NY, 1981.

[65] M. F. PERRY AND V. A. VOLPERT, *Linear stability analysis of two monomer systems of frontal polymerization*, Chemical Engineering Science, in press, 59 (2004), pp. 3451–3460.

[66] ——, *Self-propagating free-radical binary frontal polymerization*, Journal of Engineering Mathematics, 49 (2004), pp. 359–372.

[67] M. F. PERRY, V. A. VOLPERT, L. L. LEWIS, H. A. NICHOLS, AND J. A. POJMAN, *Free-radical frontal copolymerization: The dependence of the front velocity on the monomer feed composition and reactivity ratios*, Macromolecular Theory and Simulations, 12 (2003), pp. 276–286.

[68] J. POJMAN, *Traveling fronts of methacrylic acid polymerization*, J. Am. Chem. Soc., 113 (1991), pp. 6284–6286.

[69] J. POJMAN, G. CURTIS, AND V. ILYASHENKO, *Frontal polymerization in solution*, J. Am. Chem. Soc., 115 (1996), pp. 3783–3784.

[70] J. POJMAN, W. ELCAN, A. KHAN, AND L. MATHIAS, *Binary polymerization fronts: A new method to produce simultaneous interpenetrating polymer networks (sins)*, J. Polym. Sci. Part A: Polym Chem., 35 (1997), pp. 227–230.

[71] J. POJMAN, G. GUNN, C. PATTERSON, J. OWENS, AND C. SIMMONS, *Frontal dispersion polymerization*, J. Phys. Chem. B, 102 (1998), pp. 3927–3929.

[72] J. POJMAN, V. ILYASHENKO, AND A. KHAN, *Spin mode instabilities in propagating fronts of polymerization*, Physica D, 84 (1995), pp. 260–268.

[73] J. POJMAN, I. NAGY, AND C. SALTER, *Traveling fronts of addition polymerization with a solid monomer*, J. Am. Chem. Soc., 115 (1993), pp. 11044–11045.

[74] J. POJMAN, S. POPWELL, V. VOLPERT, AND V. VOLPERT, *Nonlinear dynamics in frontal polymerization*, in Nonlinear Dynamics in Polymeric Systems, J. Pojman and Q. Tran-Cong-Miyata, eds., no. 869 in ACS Symposium Series, American Chemical Society, Washington, DC, 2003, pp. 106–120.

[75] J. POJMAN, J. WILLIS, D. FORTENBERRY, V. ILYASHENKO, AND A. KHAN, *Factors affecting propagating fronts of addition polymerization: velocity, front curvature, temperature profile, conversion and molecular weight distribution*, J. Polym. Sci. Part A: Polym Chem., 33 (1995), pp. 643–652.

[76] J. POJMAN, J. WILLIS, A. KHAN, AND W. WEST, *The true molecular weight distributions of acrylate polymers formed in propagating fronts*, J. Polym. Sci. Part A: Polym Chem., 34 (1996), pp. 991–995.

[77] J. A. POJMAN, J. MASERE, E. PETRETTO, M. RUSTICI, D.-S. HUH, M. S. KIM, AND V. VOLPERT, *The effect of reactor geometry on frontal polymerization spin modes*, Chaos, 12 (2002), pp. 56–65.

[78] J. A. POJMAN, B. VARISLI, A. PERRYMAN, C. EDWARDS, AND C. HOYLE, *Frontal polymerization with thiol-ene systems*, Macromolecules, 37 (2004), pp. 691–693.

[79] C. S. RAYMOND, A. BAYLISS, B. J. MATKOWSKY, AND V. A. VOLPERT, *Transitions to chaos in condensed phase combustion with reactant melting*, Int'l J. Self-Propagating High-Temperature Synthesis, 10 (2001), pp. 133–149.

[80] L. RITTER, W. OLMSTEAD, AND V. A. VOLPERT, *Initiation of free-radical polymerization waves*, SIAM J. Appl. Math., 53 (2003), pp. 1831–1848.

[81] D. A. SCHULT AND V. A. VOLPERT, *Linear stability analysis of thermal free radical polymerization waves*, International Journal of Self-Propagating High-Temperature Synthesis, 8 (1999), pp. 417–440.

[82] K. G. SHKADINSKII AND P. M. KRISHENIK, *Steady combustion front in a mixture of fuel and inert material*, Combust. Explos. Shock Waves, 21 (1985), pp. 176–180.

[83] K. G. SHKADINSKY, B. I. KHAIKIN, AND A. G. MERZHANOV, *Propagation of a pulsating exothermic reaction front in condensed phase*, Combust. Explos. Shock Waves, 7 (1971), pp. 15–22.

[84] V. SHKIRO AND G. NERSISYAN, *Structure of fluctuations occurring in the burning of tantalum-carbon mixtures*, Combust. Explos. Shock Waves, 14 (1978), pp. 121–122.

[85] S. SOLOVYOV, V. ILYASHENKO, AND J. POJMAN, *Numerical modeling of self-propagating fronts of addition polymerization: The role of kinetics on front stability*, Chaos, 7 (1997), pp. 331–340.

[86] C. SPADE AND V. VOLPERT, *On the steady state approximation in thermal free radical frontal polymerization*, Chem. Eng. Sci., 55 (2000), pp. 641–654.

[87] C. A. SPADE AND V. A. VOLPERT, *Linear stability analysis of non-adiabatic free-radical polymerization waves*, Combust. Theory Modelling, 5 (2001), pp. 21–39.

[88] A. G. STRUNINA, A. V. DVORYANKIN, AND A. G. MERZHANOV, *Unstable regimes of thermite system combustion*, Combust. Explos. Shock Waves, 19 (1983), pp. 158–163.

[89] J. SZALAY, I. NAGY, I. BARKAI, AND M. ZSUGA, *Conductive composites prepared via a propagating polymerization front*, Die Ang. Makr. Chem., 236 (1996), pp. 97–109.

[90] A. TREDICI, R. PECCHIINI, AND M. MORBIDELLI, *Self-propagating frontal copolymerization*, Journal of Polymer Science, Part A: Polymer Chemistry, 36 (1998), pp. 1117–1126.

[91] A. TREDICI, R. PECCHINI, A. SLIEPCEVICH, AND M. MORBIDELLI, *Polymer blends by self-propagating frontal polymerization*, J. Appl. Polym. Sci., 70 (1998), pp. 2695–2702.

[92] A. VARMA, A. ROGACHEV, A. MUKASYAN, AND S. HWANG, *Combustion synthesis of advanced materials: principles and applications*, Advances in Chemical Engineering, 24 (1998), pp. 79–226.

[93] A. I. VOLPERT, V. A. VOLPERT, AND V. A. VOLPERT, *Traveling Wave Solutions of Parabolic Systems*, American Mathematical Society, Providence, RI, 1994.

[94] V. A. VOLPERT, A. V. DVORYANKIN, AND A. G. STRUNINA, *Non-one-dimensional combustion of specimens with rectangular cross-section*, Combust. Explos. Shock Waves, 19 (1983), pp. 377–380.

[95] V. A. VOLPERT, A. I. VOLPERT, AND A. G. MERZHANOV, *Analysis of nonunidimensional combustion regimes by bifurcation theory methods*, Dokl. Phys. Chem., 263 (1982), pp. 239–242.

[96] ———, *Application of bifurcation theory to the investigation of spin waves of combustion*, Dokl. Phys. Chem., 262 (1982), pp. 55–58.

[97] V. A. VOLPERT AND V. A. VOLPERT, *Propagation velocity estimation for condensed phase combustion*, SIAM J. Appl. Math., 51 (1991), pp. 1074–1089.

[98] ———, *Condensed phase reaction waves with variable thermo-physical characteristics*, Combust. Sci. Technol., 127 (1997), pp. 29–53.

[99] R. P. WASHINGTON AND O. STEINBOCK, *Frontal polymerization synthesis of temperature-sensitive hydrogels*, J. Am. Chem. Soc., 123 (2001), pp. 7933–7934.

[100] Y. B. ZELDOVICH, G. I. BARENBLATT, V. B. LIBROVICH, AND G. M. MAKHVILADZE, *The Mathematical Theory of Combustion and Explosion*, Consultants Bureau, New York, 1985.

SPATIOTEMPORAL PATTERN FORMATION IN SOLID FUEL COMBUSTION

Alvin Bayliss
Bernard J. Matkowsky
Vladimir A. Volpert
Department of Engineering Sciences and Applied Mathematics
Northwestern University, Evanston, IL 60208, USA

a-bayliss@northwestern.edu
b-matkowsky@northwestern.edu
v-volpert@northwestern.edu

Abstract

We consider the gasless combustion model of the SHS (Self-Propagating High Temperature Synthesis) process in which combustion waves, referred to as solid flames, are employed to synthesize desired materials. Specifically, we consider the combustion of a solid sample in which combustion occurs on the surface of a cylinder of radius R. In addition to uniformly propagating planar waves there are many other types of waves. The study of different waves is important since the nature of the wave determines the structure of the product material. The model depends, in particular, on the Zeldovich number Z, a nondimensional measure of the activation energy of the reaction, and the nondimensional radius R of the sample. In our analytical study we consider $R \gg 1$ and derive coupled nonlocal complex Ginzburg–Landau type equations for the amplitudes of counterpropagating waves along the front as functions of slow temporal and spatial variables. The equations are written in characteristic variables and involve averaged terms which reflect the fact that on the slowest time scale, the effect on one wave, of a second wave traveling with the group velocity in the opposite direction on the intermediate time scale, enters only through its average. Solutions of the amplitude equations in the form of traveling, standing, and quasiperiodic waves are found. In our numerical study we describe various types of propagating solid flames as parameters of the problem are varied, including (i) uniformly propagating planar flames, (ii) pulsating propagating planar flames, and flames exhibiting more complex spatiotemporal dynamics. These include (iii) spin modes in which one or several symmetrically spaced hot spots rotate around the cylinder as the flame propagates along the cylindrical axis, thus following a helical path, (iv) counterpropagating (CP) modes, in which spots propagate in opposite angular directions around the cylinder, executing various types of dynamics, (v) alternating spin CP modes (ASCP), where

A. A. Golovin and A. A. Nepomnyashchy (eds.),
Self-Assembly, Pattern Formation and Growth Phenomena in Nano-Systems, 247–282.
© 2006 *Springer. Printed in the Netherlands.*

rotation of a spot around the cylinder is interrupted by periodic events in which a
new spot is spontaneously created ahead of the rotating spot, (vi) modulated spin
waves consisting of rotating spots which exhibit a periodic modulation in speed
and temperature as they rotate (vii) asymmetric spin waves in which two spots of
unequal strength and not separated by angle π, rotate together as a bound state,
(viii) modulated asymmetric spin waves in which the two asymmetric spots oscillate in a periodic manner as they rotate, alternately approaching each other
and then moving apart periodically in time, and others.

Keywords: Solid fuel combustion, SHS, Spatiotemporal pattern formation

1.   Introduction

In the SHS (Self Propagating High Temperature Synthesis) process of materials synthesis reactants are ground into a powder, cold pressed and ignited
at one end. A high temperature combustion wave then propagates through the
sample converting reactants into products. When gas plays no significant role
in the process, the resulting gasless combustion wave is referred to as a "solid
flame". The process was pioneered in the former Soviet Union, and has subsequently been the focus of a great deal of research e.g., [24, 25, 28]. The
SHS process enjoys a number of advantages over conventional technology, in
which the sample is placed into a furnace and "baked" until it is "well done".
The advantages include (i) simpler equipment, (ii) significantly shorter synthesis times, (iii) greater economy, since the internal energy of the chemical
reactions is employed rather than the costly external energy of the furnace,
(iv) greater product purity, due to volatile impurities being burned off by the
very high combustion temperatures of the propagating combustion wave, and
(v) no intrinsic limit on the size of the sample to be synthesized, as exists in
conventional technology.

The objective of this paper is to examine the roles of the Zeldovich number
Z and the nondimensional sample size R on the different modes of propagation
possible in SHS. In particular, we expand on results in [6] for surface modes
on a cylinder of radius R as well as chaotic modes occurring in planar SHS
combustion. The characteristics of the resulting combustion wave depend significantly on the mode of propagation and impact on the nature of the product
synthesized and indeed the ability of the combustion wave to propagate in large
samples.

We note that burning sometimes occurs throughout the sample, while under
other circumstances it occurs only on the surface of the sample, though this
is generally due to the effect of limited oxidizer filtration through the sides of
the sample, which is not accounted for here. In this chapter we nevertheless
restrict consideration to surface burning so that the process can be modeled in
two dimensions. Our model thus describes solid flame propagation in a thin
cylindrical annulus between two coaxial cylinders, as in the synthesis of hollow

tubes. It also provides insight into the behavior of combustion throughout the sample since visual observation is limited to views of the surface of the sample. Since the activation energy of the combustion reaction is typically large, reaction is restricted to a thin zone. No appreciable reaction occurs ahead of the zone since the temperature is not sufficiently high and no reaction occurs behind the zone since the reactants have been consumed. In analytical studies the thin reaction zone is often approximated by an interface separating the fresh unburned mixture from the burned product. The resulting model, referred to as the reaction sheet model, employs surface delta function kinetics with reaction confined to the interface. In numerical computations there is no interface, but rather a thin reaction zone which we refer to as the combustion front. The model employs Arrhenius reaction kinetics.

It is known that in many instances the combustion wave does not propagate in a uniform spatial and temporal manner, but rather nonuniformities can develop in the front speed and in the temperature along the front as well. Since the mode of propagation determines the microstructure of the product, i.e., the nature of the final product, a study of different modes of propagation is important for technological applications of the SHS process. In particular, it is known that for sufficiently large activation energies (more precisely Zeldovich numbers, defined below) the uniformly propagating combustion wave is unstable. In this case the only stable planar mode is the planar pulsating mode in which there is no spatial structure along the front, i.e., the combustion wave remains planar while the front speed and temperature oscillate in time in a periodic, quasiperiodic or chaotic fashion, e.g., [3, 8, 9, 18, 21, 26, 31, 32].

Other modes of propagation involve spatial as well as temporal structure. The modes described in this chapter involve one or more localized hot spots (temperature maxima) along the front which exhibit a variety of dynamics. The spots become highly localized as the sample size increases. We will see that, in a sense, the spots behave like particles, which execute interesting dynamics. The modes described here involve dynamical behavior of one or more hot spots. The modes of propagation include spin combustion, in which one or more hot spots move on a helical path along the surface of the cylinder, and multiple point combustion, in which hot spot(s) repeatedly appear, disappear and then reappear. Spinning waves were observed in [26]. Subsequent observations of spinning waves and other nonplanar modes are described in, e.g., [13, 19, 24, 25, 34]. Nonplanar modes, including spinning modes, have also been described analytically, e.g., [19, 33], and numerically, e.g., [1, 5, 11, 14, 15].

Another family of nonplanar modes of propagation involves pairs of counterpropagating (CP) hot spots along the front. Such modes of solid flame propagation have been observed experimentally in [27] and have been described via numerical computations in [11]. It was found in [27] that under low pressure conditions the desired product was formed by the high temperature spots. Thus, the spots play a crucial role in the synthesis process. The observations in [27] indicated that the spots appeared to be annihilated at a collision, only to be regenerated further along the sample. The spots can be regenerated at the same angle where the collision occurred or on the opposite side of the cylinder, e.g., $180°$ away from the collision site, e.g., [11, 27]. Furthermore, for some parameters the angles corresponding to creation and annihilation sites can spontaneously change as the mode propagates along the cylinder ([11]). This behavior was described in [11] employing the model for surface combustion described below. However, it was shown that the annihilation and creation of spots was only apparent. After collision the spots would stay at a relatively low undetectable level until they would be amplified to a detectable level further along the sample. The results in [11] suggest that CP modes develop as transitions from standing waves. In this chapter we show that CP behavior can also be associated with spinning modes.

An important issue for both the theory and applications of SHS is the behavior of unsteady solid flames in large scale systems, e.g., cylinders whose diameter significantly exceeds the scale of the combustion wave, i.e., the size of the preheat zone, which is of the order of millimeters. Technological applications of SHS are generally associated with systems whose scale is much larger since often the desired goal is to synthesize large samples of material. In addition, the problem is of theoretical interest since it involves a much wider range of scales than is typically encountered in analysis of SHS problems. Virtually all analytical studies of solid flame propagation employ the Zeldovich number $Z = N(1 - \sigma)/2$, where N is the suitably nondimensionalized activation energy and $\sigma = T_u/T_b$ where T_u and T_b are the unburned and burned temperatures far ahead of and far behind the front, respectively, as a control parameter. We also employ the radius R of the sample as a control parameter, as in [11]. Here, we consider that the cylindrical radius increases up to $O(10)$ times the size of the preheat zone. Furthermore, the scale of the patterns, e.g., the extent of the localized hot spots associated with spin combustion, becomes progressively narrower as R increases and can be smaller than the size of the preheat zone. Thus, combustion waves for large diameter samples involve the effect of small scale behavior on the large scale dynamics, i.e., on the scale of the sample size. One characteristic that we use to categorize these branches is the mean axial flame speed, V, which is both a readily measurable quantity and which is related to the ability of the flames to survive when heat losses are accounted for.

We first describe some analytical results and then describe results obtained from computations. We analytically derive coupled nonlocal complex Ginzburg–Landau type equations for the amplitudes of counterpropagating waves along the front as functions of slow temporal and spatial variables. The equations are written in characteristic variables and involve averaged terms which reflect the fact that on the slowest time scale, the effect on one wave, of a second wave traveling with the group velocity in the opposite direction on the intermediate time scale, enters only through its average. Solutions of the amplitude equations in the form of traveling, standing, and quasiperiodic waves are found.

Our numerical studies concern the behavior of solutions for a fixed R, with Z employed as the control parameter. Planar front solutions always exist. The planar solutions can be either uniformly propagating, where the front temperature and front speed are constant, or planar pulsations, where the temperature on the front and the front speed oscillate in time, often exhibiting complex dynamics, e.g., [9, 31]. It is known that as Z increases past a critical value the uniformly propagating solutions lose stability to pulsating flames if R is sufficiently small and to spinning flames if R is sufficiently large, e.g., [21]. In the case of planar pulsations, as Z increases further the pulsations become increasingly relaxational, with temporally localized large temperature spikes alternating with long periods of relatively low temperatures. A sequence of period doubling transitions is known to occur as Z increases further, e.g, [9]. The pulsating planar solutions have a lower mean front speed than the uniformly propagating flame and the mean front speed generally decreases as the pulsating solutions become more relaxational, i.e., as Z increases.

In our computations we fix Z at a value beyond the stability boundary so that the uniformly propagating planar solution is unstable and the pulsating planar solution has become relaxational and undergone a transition from a singly periodic solution to a period doubled $(2T)$ solution. Furthermore, the mean speed of the pulsating planar solution is reduced considerably from the adiabatic, uniformly propagating flame speed. While we do not consider heat losses in this chapter, it is well known that slowly propagating flames are more prone to extinction via heat losses than more rapidly propagating flames. However, heat losses are also a source of instability. Thus, on the route to extinction interesting dynamics occur (see e.g., [16, 30].

We find that the pulsating planar solution is stable for small R, but becomes unstable to nonplanar perturbations for larger values of R. Our results show that the transition from pulsating planar, i.e., one dimensional behavior, to spot behavior is a jump transition, i.e., the spots enter with finite amplitude as R increases, consistent with the phenomenological description in [2]. This is not surprising since instability of the uniformly propagating solution is also expected to lead to instability of spinning solutions which differ infinitesimally from it.

We next summarize the most important modes that we find. All modes are characterized by hot spots on the front exhibiting various dynamical behavior. These modes are described in more detail in section 4. As R increases the first nonplanar mode that we find corresponds to a pair of counterpropagating

(CP) hot spots, with the transition occurring at $R = R_{cp}$. The front speed V associated with these modes near the transition point is larger than that of the unstable, uniformly propagating planar mode. CP modes are characterized by two counterpropagating hot spots which either collide and pass through each other unchanged except for a phase shift, as in the behavior of solitons, or are weakened in the collision and subsequently strengthened. The latter can occur in various ways as described in [11]. A pair of CP spots will have two collision sites. Near the transition the behavior at each collision site, i.e., of each collision, is symmetrical. As R increases, the collisions of the counterpropagating spots become asymmetrical, so that the amplitudes of the spots at one collision site differ from those at the other. Thus, there is apparent creation or annihilation at the sites, consistent with the observations in [27]. This increasingly complex dynamical behavior is accompanied by a reduction in the mean flame speed V. We provide only a limited description of the CP modes as they were discussed in detail in [11], albeit for different parameter values.

Spin modes enter at $R = R_{tw1} > R_{cp}$. Our results indicate that, as with the CP modes, the spin modes enter with finite amplitude, consistent with the phenomenological description in [2]. This is not surprising since instability of the uniformly propagating solution is also expected to lead to instability of spinning solutions which differ infinitesimally from it. The first spin mode is a traveling wave with one hot spot (TW1). As R increases, the spot becomes more localized and the temperature of the spot increases. This increasing localization and relaxational behavior of the spin mode is associated with a reduction in the mean flame speed V. We find a transition to modulated traveling waves (MTWs) at $R_{mtw} > R_{tw1}$, in which the intensity and speed of the hot spot oscillate in time as the wave propagates.

We next find a transition to a family of modes which does not appear to have been previously observed. For $R_{ascp} > R_{mtw}$ we find Alternating Spin CP modes (ASCP). For most of the time there is a single hot spot spinning around the cylinder. Call this spot S_1 and say that it is spinning clockwise. At certain times (periodic in an appropriately moving coordinate system), a new spot is spontaneously created ahead of S_1. Call this spot S_2. The new spot S_2 splits into two counterpropagating spots, $S_{2,c}$ and $S_{2,cc}$ which propagate in the clockwise and counterclockwise directions, respectively. The spots S_1 and $S_{2,cc}$ subsequently collide and are eventually mutually annihilated after the collision while the spot $S_{2,c}$ continues to propagate. In contrast to the CP modes described above, where the annihilation is only apparent, i.e., a weak spot emanates from the collision, we find that actual annihilation occurs here. The spot $S_{2,c}$ continues rotating clockwise until a new spot is created ahead of it and the process repeats periodically in time. The rotation rate is nonuniform, although it is roughly uniform away from the CP events. We believe that the behavior of the ASCP mode is a consequence of the increasing localization of the spots as R increases. As the spot is localized, a considerable region of the combustion front is nearly planar. Since planar fronts are unstable for the parameters that we consider, additional spots are expected to form, leading to the pattern described above. The ASCP branch evolves from a branch of 1-headed spin solutions, which is stable up to $R = O(10)$ in units of the preheat zone length.

We have also found other spin modes. For any value of R for which a TW1 exists, a TW2 will exist for $2R$, simply by replicating the TW1 solution. Thus, since stable TW1 modes are found in the interval (R_{tw1}, R_{mtw}), TW2 modes exist and may be stable in the interval $(2R_{tw1}, 2R_{mtw})$. Furthermore, the mean front speed V should be the same as for the corresponding TW1 solution. Thus, rapid axially propagating TW2 modes should exist for R near $2R_{tw1}$. We find that the replicated TW2 modes are stable but only for a subset of the interval $(2R_{tw1}, 2R_{mtw})$. The replicated TW2 mode clearly involves two hot spots of equal intensity which are symmetric, by which we mean that the two identical spots are symmetrically located on the circle, i.e., are separated by angle $\psi = \pi$. In addition to the TW2 modes we have found a family of asymmetric traveling waves consisting of 2 rotating hot spots which are of unequal temperature and are asymmetric, i.e., separated by angle $\psi < \pi$. Nevertheless, the two spots rotate together as a traveling wave in which the two spots are bound together. We call these modes ATW2 for Asymmetric TW2 modes. We find such modes for relatively large values of R, and their axial propagation speeds are smaller than for the replicated TWs. We also find modulated ATW2 modes (MATW2) in which the two spots alternately approach each other and move apart as they propagate.

We have also investigated 3–headed spins. We find that replicated TW3 modes are not stable. In a limited range of R we have found stable 3-headed spin modes that are not pure TWs. Rather, each spot alternately speeds up and slows down so that it alternately approaches and separates from its neighbors. We find that for R near $3R_{tw}$ the behavior is that of a modulated traveling wave (MTW3) exhibiting quasiperiodic dynamics. For larger values of R, the behavior is apparently chaotic. At seemingly random times two of the spots nearly touch, after which one spot is rapidly propelled away from the other as they rotate. Upon increasing R further we find that two spots collide during the transient, leading to annihilation of one spot and a collapse of the solution from three heads to two.

In section 2 we describe the model and briefly describe the numerical method employed. In section 3 we present our analytical results. In section 4 we discuss the numerical results in detail.

2.   Mathematical Model

We consider a model which accounts for diffusion of heat and one-step, irreversible Arrhenius kinetics. Since the reactants are solid we neglect diffusion of mass. We consider the case where combustion occurs on the surface of a cylindrical sample of radius $\tilde{R}$. We further assume that there is a deficient component of the solid mixture and that all other components are present in sufficiently large quantities that the evolution of only the deficient component needs to be followed in the model.

We let $\tilde{\ }$ denote dimensional quantities and let $\tilde{x}$ and ψ denote the axial and angular cylindrical coordinates, respectively, and assume that the combustion wave propagates in the $-\tilde{x}$ direction. Let $\tilde{T}$ and $\tilde{Y}$ denote the temperature and mass fraction of the deficient component respectively. The model is given by

$$\tilde{T}_{\tilde{t}} = \tilde{\lambda}\tilde{T}_{\tilde{x}\tilde{x}} + \tilde{\lambda}\frac{1}{\tilde{R}^2}\tilde{T}_{\psi\psi} + \tilde{\beta}\tilde{A}\tilde{Y}\,\exp\left(-\frac{\tilde{E}}{\tilde{R}_g\tilde{T}}\right), \qquad (2.1)$$

$$\tilde{Y}_{\tilde{t}} = -\tilde{A}\tilde{Y}\,\exp\left(-\frac{\tilde{E}}{\tilde{R}_g\tilde{T}}\right),$$

where $\tilde{\lambda}$ is the thermal diffusivity, $\tilde{A}$ is the frequency factor, $\tilde{\beta}$ is the scaled heat of reaction, $\tilde{E}$ is the activation energy (all assumed constant) and $\tilde{R}_g$ is the gas constant. The boundary conditions are

$$\tilde{Y} \to \tilde{Y}_u, \quad \tilde{T} \to \tilde{T}_u, \quad \text{as} \quad \tilde{x} \to -\infty,$$
$$\tilde{T}x \to 0, \quad \text{as} \quad \tilde{x} \to \infty.$$

We note that $\tilde{T} \to \tilde{T}_b$ as $\tilde{x} \to \infty$ where the subscripts u and b refer to unburned and burned respectively, however, we employ the Neumann condition in our model. The solution is periodic in ψ with period 2π. We observe that T_b is derivable from the time-independent solution of the problem as $\tilde{T}_b = \tilde{T}_u + \tilde{\beta}\tilde{Y}_u$.

We nondimensionalize as in [21] by introducing

$$Y = \frac{\tilde{Y}}{\tilde{Y}_u}, \quad \Theta = \frac{\tilde{T} - \tilde{T}_u}{\tilde{T}_b - \tilde{T}_u}, \quad t = \frac{\tilde{t}\tilde{U}^2}{\tilde{\lambda}}, \quad x = \frac{\tilde{x}\tilde{U}}{\tilde{\lambda}},$$

$$\sigma = \frac{\tilde{T}_u}{\tilde{T}_b}, \quad N = \frac{\tilde{E}}{\tilde{R}_g\tilde{T}_b}.$$

Here,

$$\tilde{U}^2 = \frac{\tilde{\lambda}\tilde{A}}{2Z}\exp(-N),$$

where $Z = N(1 - \sigma)/2$ is the Zeldovich number, $\tilde{U}$ is the velocity of the uniformly propagating front for asymptotically large Z ([18]), and lengths are scaled by the size of the preheat zone. We next introduce the moving coordinate

$$z = x - \phi(t),$$

where $\phi(t)$ is defined by $Y(\phi(t), \psi = 0, t) = 0.5$. Here, the choice of the angle ψ at which $Y = 0.5$ is arbitrary; $\psi = 0$ is chosen merely for convenience. The velocity ϕ_t and the location $x = \phi(t)$ do not model the front velocity and location when there are nonplanar disturbances, since they, unlike these functions, also depend on ψ. However, for all angles the reaction zone will be localized in a neighborhood of $z = 0$. Thus, ϕ_t is the approximate velocity of the wave, so that the transformation to the moving coordinate system enables us to localize the reaction zone to a neighborhood of $z = 0$.

In terms of the nondimensionalized quantities, the system (2.1) becomes

$$
\begin{aligned}
\Theta_t &= \phi_t \Theta_z + \Theta_{zz} + \frac{1}{R^2} \Theta_{\psi\psi} + 2ZY \, \exp\left(\frac{N(1-\sigma)(\Theta-1)}{\sigma + (1-\sigma)\Theta} \right), \\
Y_t &= \phi_t Y_z - 2ZY \, \exp\left(\frac{N(1-\sigma)(\Theta-1)}{\sigma + (1-\sigma)\Theta} \right),
\end{aligned}
\tag{2.2}
$$

where

$$R = \frac{\tilde{R}\tilde{U}}{\tilde{\lambda}}.$$

The coefficient of the reaction term arises from the use of the asymptotic planar adiabatic burning velocity $\tilde{U}$ in the nondimensionalization. The use of the asymptotic value of $\tilde{U}$ with finite activation energy affects the length and time scales of the computation (for example, uniformly propagating, planar solutions at finite activation energy have a front velocity slightly different from unity), but does not change any of the resulting spatiotemporal patterns.

In addition, we introduce a cutoff function $g(\Theta)$ which multiplies the Arrhenius term so that the reaction term vanishes for sufficiently small temperatures. The function $g(\Theta)$ is defined by

$$g(\Theta) = 0, \ \ \Theta < \Theta_{cut}, \ \ g(\Theta) = 1, \ \ \Theta > \Theta_{cut}.$$

We have tested both discontinuous and smooth cutoff functions and find that the different possibilities have virtually no effect on the computed solution. The cutoff function is employed to avoid the "cold boundary difficulty" which arises because the Arrhenius model for the reaction term does not vanish far ahead of the front, which is incompatible with the boundary condition $T = T_u$ as $x \to \infty$. In addition, in practice no significant reaction occurs ahead of the reaction zone. For the computations presented here $\Theta_{cut} = .03$. We have found that the results are insensitive to variations in Θ_{cut} as long as its value is of this order.

The boundary conditions are specified at finite points far from the reaction zone which is in the vicinity of $z = 0$. The computations presented here were obtained with the boundary conditions imposed at $z = \pm 12$. There is virtually no effect of further increasing the size of the computational domain. Note that no boundary condition is imposed on Y in the burned region.

Since we consider distributed Arrhenius kinetics, there is strictly speaking no combustion front but rather a narrow reaction zone in which the chemical reaction terms are significant. However, a front can be defined in a number of ways, e.g., as the curve of maximum reaction rate or as the curve where the reactant mass fraction decreases to half its initial value, etc. These procedures treat the front as a curve for each value of time, $z_f(\psi, t)$, along which temperature, mass fraction and reaction rate vary. In order to present our results, we have developed a procedure to approximate the combustion front and the temperature on the front as a function of time t and the cylindrical angle ψ. We define the front for each value of ψ as the z location where the reaction term is maximal. Thus, for each value of ψ and t we compute a value $z = z_f(\psi, t)$, where the reaction term takes a maximum and define the resulting function as the combustion front. That is, with

$$W(z, \psi, t) = Y \, \exp\left(\frac{N(1 - \sigma)(\Theta - 1)}{\sigma + (1 - \sigma)\Theta} \right),$$

we let z_i denote the axial collocation points, and compute W at all collocation points. For each value of ψ and t we find i^m, the value of i where $W(z_i, \psi, t)$ attains its maximum. We then locally fit W to a quadratic in z using the points $W(z_{i^m}, \psi, t)$, $W(z_{i^{m+1}}, \psi, t)$ and $W(z_{i^{m-1}}, \psi, t)$. We next compute the value of z which maximizes this quadratic and use this value as $z_f(\psi, t)$, our approximation to the front location. Finally, we compute $\Theta_f(\psi, t)$, the temperature at $z_f(\psi, t)$, by Chebyshev interpolation. We refer to $\Theta_f(\psi, t)$ obtained in this manner as the temperature at the front.

In our model for the front it does not follow that Θ is maximal at the front. In many instances the location of maximum temperature does in fact correlate with the front location as defined above. However, under certain circumstances the temperature can increase behind the front. Axial profiles of temperature for a one dimensional problem are shown in [9].

For the numerical computations we employ an adaptive Chebyshev pseudospectral method in z that we previously developed, e.g., [8,4], together with a Fourier pseudo-spectral method in ψ. In order to better resolve the reaction zone in which the solution changes rapidly, we adaptively transform the coordinate z. The transformation has the effect of expanding the reaction zone so that in the new coordinate system the solution varies more gradually and is therefore easier to compute. The method is described in detail in other references, e.g., [11, 6].

3.   Analytical Results

For the analytical studies we employ the reaction sheet approximation, in which the distributed Arrhenius kinetics is replaced by a surface delta function corresponding to the reaction being confined to the interface $\phi(\psi, t)$. For simplicity we employ Cartesian coordinates. We introduce the moving coordinate system (ξ, y) where $\xi = x - \phi(y, t)$. The mass fraction Y of the deficient reaction component is given by $Y = 1$ for $\xi < 0$, and $Y = 0$ for $\xi > 0$, which corresponds to complete consumption of the reactant at the interface. The temperature $\Theta \equiv \frac{T - T_u}{T_b - T_u}$ then satisfies

$$\frac{\partial \Theta}{\partial t} - \frac{\partial \phi}{\partial t} \frac{\partial \Theta}{\partial \xi} = \Delta \Theta, \tag{3.3}$$

subject to the boundary conditions $\Theta \to 0$ as $\xi \to -\infty$ and $\Theta \to 1$ as $\xi \to \infty$. The interface ϕ satisfies

$$\frac{\partial \phi}{\partial t} = -\left(1 + \left(\frac{\partial \phi}{\partial y}\right)^2\right)^{1/2} e^{Z[\Theta(t,0,y)-1]}. \tag{3.4}$$

Here, the Laplacian Δ is expressed in the moving coordinate system. At the interface the temperature is continuous, i.e., $[\Theta] = 0$, while its derivative jumps, i.e., $[\frac{\partial \Theta}{\partial \xi}] = \frac{\partial \phi}{\partial t} / \left(1 + \left(\frac{\partial \phi}{\partial y}\right)^2\right)$. Here, $[f] = f(\xi = 0^+) - f(\xi(0^-)$ denotes the jump in the quantity f across the interface.

The problem possesses a solution which describes a uniformly propagating planar interface. This solution, which we term the basic solution, is given by

$$\hat{\Theta}(\xi) = \begin{cases} 1, & \xi > 0 \\ e^{\xi}, & \xi < 0 \end{cases}, \tag{3.5}$$

$$\hat{\Phi}(t) = -t. \tag{3.6}$$

The stability of this solution depends on the parameters of the problem. We first note that the Zeldovich number Z can be interpreted as the ratio of the diffusion time scale t_D and the reaction time scale t_r. Thus, for Z sufficiently small all the heat released in the reaction can be diffused away from the interface. However, for Z sufficiently large this is not the case. The heat released in the reaction which exceeds the amount of heat that can be carried away by diffusion necessarily raises the temperature at the interface. This, in turn, leads to an increase in the thermal gradient into the fresh fuel (leading to the interface speeding up), which lowers the interface temperature (leading to the interface slowing down), and the process repeats. This is the mechanism for the onset of oscillatory propagation as Z exceeds a critical value Z_c.

A linear stability analysis of the basic solution yields the dispersion relation

$$4(i\omega)^3 + (i\omega)^2(1 + 4Z - Z^2 + 4k^2) + i\omega Z(1 + 4k^2) + Z^2 k^2 = 0, \quad (3.7)$$

where $i\omega$ is the growth rate and k is the wave number of the perturbation. The basic solution is linearly stable (unstable) if $\Re(i\omega) < 0$ ($\Re(i\omega) > 0$). The neutral stability curve in the (k, Z)-plane is determined by

$$(1 + 4k^2)Z^2 - 4(1 + 3k^2)Z - (1 + 4k^2)^2 = 0.$$

The basic solution (3.5)-(3.6) is linearly stable (unstable) in the region below (above) the curve. The curve has a minimum at $k = k_0 = 0.5$, $Z = Z_c = 4$. The frequency of oscillation for a solution on the neutral stability curve at the wavenumber k is given by $\omega^2 = (1 + 4k^2)Z/4$. We observe that $\omega_0 = \omega(k_0) = \sqrt{2}$. We note that due to translation invariance the dispersion relation (3.7) also admits, for all values of Z, the trivial solution $\omega = k = 0$, which corresponds to the eigenfunction $\hat{\Theta}'$. Since $0 < \psi < 2\pi$, only discrete points on the neutral stability curve are permitted, with their location determined by the sample size R. For R sufficiently small all the points except for the point corresponding to $(k = 0, Z = 2 + \sqrt{5})$ which corresponds to planar pulsations, occur very high up on the curve. Thus, for R sufficiently small only the planar pulsating solution is expected to be observed, in agreement with experimental observations. As R is increased the points high up on the curve begin to march downward, and additional, more complex spatiotemporal modes of propagation are expected to be observed, e.g., spin modes which have been experimentally observed for larger R.

An analysis of the bifurcation from uniformly propagating planar modes to pulsating propagating planar modes was carried out in [21], which showed that the bifurcation is a supercritical Hopf bifurcation. Weakly nonlinear analyses of the problem in 3D, in which the effect of melting ahead of the reaction interface occurred, exhibited the bifurcation of spinning as well as pulsating planar propagation [19]. Similar analyses near codimension-two points [12, 20] described the interaction of the pulsating and spinning modes. The effect of heat losses leading to extinction, was considered in [16]. Finally, a description of the transition to a spiral mode of propagation was considered in [22]. In this analysis, gasless solid fuel combustion in a thin disk ignited at a point was considered. Consider the behavior on the upper face of the disk. It was shown that a circular interface with a uniform temperature distribution propagated outward from the ignition point at a uniform rate. When a critical radius was reached, a hot spot (local temperature maximum) appeared on the circular interface. As the interface continued to propagate outward, the hot spot rotated around the expanding circle at a uniform rate, leading to the hot spot following a trajectory which is an Archimedean spiral. Furthermore, additional hot spots appeared, sometimes leading to the appearance of a luminous circle.

In [23] a weakly nonlinear analysis for the case of counterpropagating waves was considered. In this case coupled nonlocal complex Ginzburg-Landau equations for the amplitudes of the counterpropagating waves were derived. We now summarize the derivation. Again, for simplicity, we consider the problem in Cartesian coordinates. Moreover, though the derivation in [23] was for the case when the effect of melting was accounted for, here we do not include melting effects.

We introduce the perturbation Ψ as

$$\phi = -t + \Psi. \tag{3.8}$$

We then formally introduce an operator $\aleph$ which acts on any smooth function $v(\xi)$ as

$$\aleph(v) = (\exp(\Psi \frac{\partial}{\partial \xi}))v(\xi) \equiv \sum_{j=0}^{\infty} \frac{1}{j!} (\Psi \frac{\partial}{\partial \xi})^j v(\xi), \tag{3.9}$$

and define the perturbation w of the basic solution $\hat{\Theta}$ as

$$\Theta = \aleph(\hat{\Theta} + w). \tag{3.10}$$

We observe that the operator (3.9) represents a shift in ξ. We choose the perturbation to be of this form since it will simplify the ensuing calculations. Substituting (3.8), (3.10) into equations (3.3)–(3.4) for Θ and ϕ subject to the jump conditions, we find that w and Ψ satisfy

$$\aleph(Lw) \equiv \aleph \left(\frac{\partial w}{\partial t} + \frac{\partial w}{\partial \xi} - \frac{\partial^2 w}{\partial \xi^2} - \frac{\partial^2 w}{\partial y^2} \right) = 0, \tag{3.11}$$

$$\frac{\partial \Psi}{\partial t} = 1 - \left(1 + \left(\frac{\partial \phi}{\partial y} \right)^2 \right)^{1/2} \exp(Z\aleph(\hat{\Theta} + w)(t, 0, y) - 1), \tag{3.12}$$

$$[\aleph(\hat{\Theta}+w)] = 0, \quad \left[\frac{\partial \aleph(\hat{\Theta} + w)}{\partial \xi} \right] = \left(-1 + \frac{\partial \Psi}{\partial t} \right) / \left(1 + \left(\frac{\partial \Psi}{\partial y} \right)^2 \right), \tag{3.13}$$

$$w \to 0 \text{ as } \xi \to -\infty, \tag{3.14}$$

$$| w | < \infty \text{ as } \xi \to +\infty, \tag{3.15}$$

$$| w | < \infty \text{ as } y \to \pm\infty. \tag{3.16}$$

We seek solutions with slowly–varying amplitudes when the parameter Δ is near the stability boundary. Therefore we introduce the small parameter ϵ by defining

$$Z = Z_c(1 + \sigma\epsilon^2). \tag{3.17}$$

We then introduce the slow time and space variables

$$t_1 = \epsilon t, \quad t_2 = \epsilon^2 t, \tag{3.18}$$

$$y_1 = \epsilon y, \tag{3.19}$$

and expand as

$$w \sim \sum_{j=1}^{\infty} \epsilon^j w_j(t, t_1, t_2, y, y_1, \xi), \tag{3.20}$$

$$\Psi \sim \sum_{j=1}^{\infty} \epsilon^j \psi_j(t, t_1, t_2, y, y_1). \tag{3.21}$$

We note that Ψ, which describes the position of the propagating front, may grow linearly in time, while the other physical variables are bounded. We therefore assume in our expansions, that all terms are bounded except for those terms which are associated with the average position of the front. These terms are assumed to grow no faster than linearly in time.

Upon formal substitution of (3.17)-(3.21) into (3.11)-(3.16), we equate coefficients of like powers of ϵ and obtain a sequence of problems for the recursive determination of w_j and ψ_j $j = 1, 2, 3, \ldots$

$$L w_j = f_j, \tag{3.22}$$

$$[w_j] = \psi_j + p_j, \tag{3.23}$$

$$\left[\frac{\partial w_j}{\partial \xi}\right] = \psi_j + \frac{\partial \psi_j}{\partial t} + q_j, \tag{3.24}$$

$$\frac{\partial \psi_j}{\partial t} + Z_0 w_j \big|_{\xi=0^+} + r_j = 0 , \tag{3.25}$$

$$w_j \to 0 \text{ as } \xi \to -\infty, \tag{3.26}$$

$$| w_j | < \infty \text{ as } \xi \to +\infty, \tag{3.27}$$

$$| w_j | < \infty \text{ as } y \to \pm\infty. \tag{3.28}$$

The functions f_j, p_j, q_j and r_j for $j = 1, 2, 3$ are given by

$$f_1 = 0, \ p_1 = 0, \ q_1 = 0, \ r_1 = 0, \tag{3.29}$$

$$f_2 = -\frac{\partial w_1}{\partial t} + 2\frac{\partial^2 w_1}{\partial y \partial y_1}, \tag{3.30}$$

$$p_2 = -\frac{1}{2}\psi_1^2 - \psi_1 \frac{\partial \psi_1}{\partial t}, \tag{3.31}$$

$$q_2 = -\frac{1}{2}\psi_1^2 - 2\psi_1\frac{\partial\psi_1}{\partial t} + \frac{\partial\psi_1}{\partial t_1} + \psi_1\frac{\partial^2\psi_1}{\partial y^2} + \left(\frac{\partial\psi_1}{\partial y}\right)^2, \tag{3.32}$$

$$r_2 = Z_c\psi_1\frac{\partial w_1}{\partial\xi}\bigg|_{\xi=0^+} + \frac{1}{2}\left(\frac{\partial\psi_1}{\partial t}\right)^2 + \frac{1}{2}\left(\frac{\partial\psi_1}{\partial y}\right)^2 + \frac{\partial\psi_1}{\partial t_1}, \tag{3.33}$$

$$f_3 = -\frac{\partial w_2}{\partial t} + 2\frac{\partial^2 w_2}{\partial y\partial y_1} - \frac{\partial w_1}{\partial t_2} + \frac{\partial^2 w_1}{\partial y_1^2}, \tag{3.34}$$

$$p_3 = -\left(1 + \frac{\partial}{\partial t}\right)(\psi_1\psi_2) + \frac{1}{6}\left(1 + 2\frac{\partial}{\partial t} - \frac{\partial^2}{\partial y^2}\right)\psi_1^3 - \frac{1}{2}\frac{\partial\psi_1^2}{\partial t_1}, \tag{3.35}$$

$$q_3 = -\left(1 + 2\frac{\partial}{\partial t} - \frac{\partial^2}{\partial y^2}\right)(\psi_1\psi_2) +$$

$$+\frac{1}{6}\left(1 + 3\frac{\partial}{\partial t} - 2\frac{\partial^2}{\partial y^2} + \frac{\partial^2}{\partial t^2} - \frac{\partial^3}{\partial t\partial y^2}\right)\psi_1^3 + \tag{3.36}$$

$$+\frac{\partial^2\psi_1^2}{\partial y\partial y_1} - \frac{\partial\psi_1^2}{\partial t_1} + \frac{\partial\psi_2}{\partial t_1} + \frac{\partial\psi_1}{\partial t_2},$$

$$r_3 = \frac{\partial\psi_2}{\partial t_1} + \frac{\partial\psi_1}{\partial t_2} - \sigma\frac{\partial\psi_1}{\partial t}$$

$$+\frac{\partial\psi_1}{\partial t}\left(\frac{\partial\psi_2}{\partial t} + \frac{\partial\psi_1}{\partial t_1}\right) + \frac{\partial\psi_1}{\partial y}\left(\frac{\partial\psi_2}{\partial y} + \frac{\partial\psi_1}{\partial y_1}\right) + \tag{3.37}$$

$$+\frac{1}{3}\left(\frac{\partial\psi_1}{\partial t}\right)^3 + Z_c\left(\psi_1\frac{\partial w_2}{\partial\xi} + \psi_2\frac{\partial w_1}{\partial\xi} + \frac{1}{2}\psi_1^2\frac{\partial^2 w_1}{\partial\xi^2}\right)\big|_{\xi=0^+}.$$

We consider the case of an already established propagating front, rather than the problem of the evolution of arbitrary initial data to form the propagating front. Thus we assume that the transient period of formation has passed. There-fore, we consider the general long time solution of (3.22)-(3.28) for $j = 1$, which is given by

$$w_1 = (R_1 e_1 + S_1 e_2)g(\xi) + c.c. - H_1\hat{\Theta}', \tag{3.38}$$

$$\psi_1 = R_1 e_1 + S_1 e_2 + c.c. + H_1, \quad e_{1,2} \equiv \exp\{i(\omega_0 t \pm k_0 y)\}. \tag{3.39}$$

This solution describes periodic waves traveling in opposite directions along the front. The as yet undetermined complex amplitudes R_1 and S_1 of these waves are bounded functions of the slow variables and H_1 can grow linearly in time on the slow time scales. Solvability conditions which are described below will provide equations for these amplitudes.

Since the long time homogeneous problem (3.22)-(3.28) with $j = 1$ has nontrivial solutions, it is necessary that solvability conditions be satisfied in order for solutions to exist for $j \geq 2$. To derive the solvability conditions we introduce the inner product

$$\langle f, g \rangle = \frac{1}{PQ} \int_0^P dt \int_0^Q dy \left(\int_{-\infty}^0 + \int_0^\infty \right) f\bar{g} d\xi,$$

for functions f and g that are bounded continuous functions of ξ on $(-\infty, 0]$ and $[0, \infty)$, with a possible jump discontinuity at 0. We assume as well that the product $f\bar{g} \longrightarrow 0$ as ξ tends to $\pm\infty$ in such a way that the above integrals exist. Furthermore, since (3.38)-(3.39) are periodic in t and y, it is clear that the functions f_j, p_j, q_j, and r_j will also be periodic in t and y. Therefore we have confined consideration to functions f and g which are periodic in t and y, with periods $P = \frac{2\pi}{\omega_0}$ and $Q = \frac{2\pi}{k_0}$ respectively. We also define an inner product (ϕ, ψ) for any functions ϕ, ψ which are independent of ξ, as

$$(\phi, \psi) = \frac{1}{PQ} \int_0^P dt \int_0^Q \phi\bar{\psi} dy.$$

The adjoint problem to (3.22)-(3.28) with $j = 1$ is

$$\frac{\partial v}{\partial t} + \frac{\partial v}{\partial \xi} + \frac{\partial^2 v}{\partial \xi^2} + \frac{\partial^2 v}{\partial y^2} = 0 \tag{3.40}$$

subject to the jump conditions at $\xi = 0$

$$[v] = 0, \quad Z_c \frac{\partial v}{\partial t}\bigg|_{\xi=0} + Z_c \frac{\partial v}{\partial \xi}\bigg|_{\xi=0^-} + \left[\frac{\partial^2 v}{\partial t \partial \xi}\right] = 0, \tag{3.41}$$

and boundary conditions

$$|v| \rightarrow 0 \text{ as } \xi \rightarrow +\infty, \quad |v| < \infty \text{ as } \xi \rightarrow -\infty, \tag{3.42}$$

$$|v| < \infty \text{ as } y \rightarrow \pm\infty.$$

The long time adjoint solutions, which are P and Q periodic in t and y respectively, are

$$v_1 = \begin{cases} e_1 \exp\left(-\bar{\mu}_2 \xi\right), & \xi < 0 \\ e_1 \exp\left(-\bar{\mu}_1 \xi\right), & \xi > 0 \end{cases}, \tag{3.43}$$

$$v_2 = \begin{cases} e_2 \exp\left(-\bar{\mu}_2 \xi\right), & \xi < 0 \\ e_2 \exp\left(-\bar{\mu}_1 \xi\right), & \xi > 0 \end{cases}, \tag{3.44}$$

$$v_0 = \begin{cases} 1, & \xi < 0 \\ \exp\left(-\xi\right), & \xi > 0 \end{cases}, \tag{3.45}$$

where μ_1, μ_2 are given by

$$\mu_{1,2} = (1 \pm d)/2, \quad d = \sqrt{1 + 4i\omega_0 + 4k_0^2} = 2 + i\omega_0. \tag{3.46}$$

The solvability conditions can then be written in the form

$$\langle f_j, v \rangle + ((p_j - q_j), v|_{\xi=0}) + (p_j, \frac{\partial v}{\partial \xi}|_{\xi=0-}) - \frac{1}{Z_c}(r_j, [\frac{\partial v}{\partial \xi}]) = 0, \quad (3.47)$$

where the functions f_j, p_j, q_j, and r_j are those in (3.22)-(3.25), and v is given by (3.43)-(3.45) respectively.

Thus for $j = 2$, the three solvability conditions (3.47) imply that

$$\frac{\partial R_1}{\partial t_1} - \omega_0' \frac{\partial R_1}{\partial y_1} = 0, \quad (3.48)$$

$$\frac{\partial S_1}{\partial t_1} + \omega_0' \frac{\partial S_1}{\partial y_1} = 0, \quad (3.49)$$

$$\frac{\partial H_1}{\partial t_1} = -\frac{1}{4}(|R_1|^2 + |S_1|^2). \quad (3.50)$$

>From (3.48)-(3.49) it is clear that R_1 and S_1 represent waves propagating with group velocity $\omega_0' = \frac{d\omega}{dk}(k_0) = \sqrt{2}$ along y_1-axis, in opposite directions.

Integrating (3.50) we have

$$H_1 = H_1(t_1 = a) - \frac{1}{4}\int_a^{t_1} (|R_1|^2 + |S_1|^2)dt_1 \quad (3.51)$$

for some constant a. Differentiating (3.51) with respect to y_1 and using (3.48) and (3.49) we obtain

$$\frac{\partial H_1}{\partial y_1} = -\frac{1}{4\omega_0'}(|R_1|^2 - |S_1|^2) + h(y_1, t_2), \quad (3.52)$$

where

$$h(y_1, t_2) = \frac{\partial H_1}{\partial y_1}|_{t_1=a} + \frac{1}{4\omega_0'}(|R_1|^2 - |S_1|^2)|_{t_1=a}.$$

Introducing the characteristic coordinates

$$\eta_1 = y_1 + \omega_0't_1, \quad \eta_2 = y_1 - \omega_0't_1. \quad (3.53)$$

and using (3.53) in (3.48) and (3.49) we obtain

$$\frac{\partial R_1}{\partial \eta_2} = 0, \frac{\partial S_1}{\partial \eta_1} = 0. \quad (3.54)$$

We now seek solutions w_2 and ψ_2 in the form

$$w_2 = w_2^h + w_2^p, \quad (3.55)$$

$$\psi_2 = \psi_2^h + \psi_2^p, \quad (3.56)$$

where w_2^h, ψ_2^h are solution of the homogeneous problem, so that

$$w_2^h = (R_2 e_1 + S_2 e_2)g(\xi) + c.c. - H_2\hat{\Theta}', \tag{3.57}$$

$$\psi_2^h = R_2 e_1 + S_2 e_2 + c.c. + H_2, \tag{3.58}$$

where R_2, S_2 and H_2 are to be determined. Expanding the right–hand sides of (3.30)-(3.33) we find that w_2^p, ψ_2^p are given by

$$w_2^p = (R_1^2 e_1^2 + S_1^2 e_2^2)g_1(\xi) + R_1 S_1 e_1 e_2 g_2 + R_1 \overline{S}_1 e_1 \overline{e}_2 g_3 + \tag{3.59}$$

$$+ e_1 g_4 + e_2 g_5 + c.c. + g_6,$$

$$\psi_2^p = (R_1^2 e_1^2 + S_1^2 e_2^2)a_1 + a_2 R_1 S_1 e_1 e_2 + a_3 R_1 \overline{S}_1 e_1 \overline{e}_2 + \tag{3.60}$$

$$+ a_4 e_1 + a_5 e_2 + c.c. + a_6,$$

where

$$g_1(\xi) = \begin{cases} b_{11} \exp(\mu_7 \xi), & \xi < 0 \\ b_{12} \exp(\mu_8 \xi), & \xi > 0 \end{cases},$$

$$\mu_{7,8} = (1 \pm d_2)/2, \quad d_2 = \sqrt{1 + 8i\omega_0 + 16k_0^2},$$

$$g_2(\xi) = \begin{cases} b_9 \exp(\mu_5 \xi), & \xi < 0 \\ b_{10} \exp(\mu_6 \xi), & \xi > 0 \end{cases},$$

$$\mu_{5,6} = (1 \pm d_1)/2, \quad d_1 = \sqrt{1 + 8i\omega_0},$$

$$g_3(\xi) = \begin{cases} b_7 \exp(\mu_3 \xi), & \xi < 0 \\ b_8 \exp(\mu_4 \xi), & \xi > 0 \end{cases},$$

$$\mu_{3,4} = (1 \pm d_0)/2, \quad d_0 = \sqrt{1 + 16k_0^2},$$

$$g_4(\xi) = \begin{cases} b_5 \exp(\mu_1 \xi) + \frac{c_1}{2\mu_1 - 1}\left(\frac{\partial R_1}{\partial t_1} - 2ik_0\frac{\partial R_1}{\partial y_1}\right)\xi \exp(\mu_1 \xi), & \xi < 0 \\ b_6 \exp(\mu_2 \xi) + \frac{c_2}{2\mu_2 - 1}\left(\frac{\partial R_1}{\partial t_1} - 2ik_0\frac{\partial R_1}{\partial y_1}\right)\xi \exp(\mu_2 \xi), & \xi > 0 \end{cases},$$

$$g_5(\xi) = \begin{cases} b_3 \exp(\mu_1 \xi) + \frac{c_1}{2\mu_1 - 1}\left(\frac{\partial S_1}{\partial t_1} + 2ik_0\frac{\partial S_1}{\partial y_1}\right)\xi \exp(\mu_1 \xi), & \xi < 0 \\ b_4 \exp(\mu_2 \xi) + \frac{c_2}{2\mu_2 - 1}\left(\frac{\partial S_1}{\partial t_1} + 2ik_0\frac{\partial S_1}{\partial y_1}\right)\xi \exp(\mu_2 \xi), & \xi > 0 \end{cases},$$

$$g_6(\xi) = \begin{cases} b_1 \exp\xi - \frac{\partial H_1}{\partial t_1}\xi \exp\xi, & \xi < 0 \\ b_2, & \xi > 0 \end{cases}.$$

The quantities a_j, b_j, and d_j in the above formulas are given by

$$a_1 = -0.01 - 2.46i, \quad a_2 = -0.61 - 1.97i,$$

$$c_1 = -1 - i\omega_0/Z_c, \quad c_2 = -i\omega_0/Z_c,$$

$$a_3 = -0.39, \quad a_4 = a_5 = a_6 = 0,$$

$$d_0 = 2.236, \quad d_1 = 2.49 + 2.28i, \quad d_2 = 2.95 + 1.92i,$$

$$b_1 = |R_1|^2 + |S_1|^2 + H_1^2/2, \quad b_2 = 0,$$

$$b_3 = \frac{1}{4}\left(H_1 S_1(5 + 3.5i\omega_0) - \frac{\partial S_1}{\partial t_1}\right), \quad b_4 = \frac{1}{4}\left(H_1 S_1(1 - 0.5i\omega_0) - \frac{\partial S_1}{\partial t_1}\right),$$

$$b_5 = \frac{1}{4}\left(H_1 R_1(5 + 3.5i\omega_0) - \frac{\partial R_1}{\partial t_1}\right), \quad b_6 = \frac{1}{4}\left(H_1 R_1(1 - 0.5i\omega_0) - \frac{\partial R_1}{\partial t_1}\right),$$

$$b_7 = 1.45, \quad b_8 = 0.0625,$$

$$b_9 = 1.15 + 4.87i, \quad b_{10} = -0.46 + 0.08i,$$

$$b_{11} = 1.52 + 3.7i, \quad b_{12} = 1.21 - 0.17i.$$

Next, applying the solvability condition (3.47) to (3.22)-(3.25) with $j = 3$, using the expressions (3.50), (3.52) in the new coordinate system (3.53), we obtain

$$-2\omega_0'\frac{\partial R_2}{\partial \eta_2} = -\frac{\partial R_1}{\partial t_2} + \sigma\alpha R_1 + A\frac{\partial^2 R_1}{\partial \eta_1^2} + BR_1\,|R_1|^2 + cR_1\,|S_1|^2 - \quad (3.61)$$

$$-\frac{1}{2}iR_1 h(\frac{\eta_1 + \eta_2}{2}, t_2)|_{\eta_1 = \eta_2 + 2\omega_0' a},$$

$$2\omega_0'\frac{\partial S_2}{\partial \eta_1} = -\frac{\partial S_1}{\partial t_2} + \sigma\alpha S_1 + A\frac{\partial^2 S_1}{\partial \eta_2^2} + BS_1\,|S_1|^2 + cS_1\,|R_1|^2 + \quad (3.62)$$

$$+\frac{1}{2}iS_1 h(\frac{\eta_1 + \eta_2}{2}, t_2)|_{\eta_1 = \eta_2 + 2\omega_0' a},$$

where

$$A = \frac{8}{9} - \frac{5}{18}i\omega_0, \quad \alpha = 2 + i\omega_0,$$

$$B = 1.12 + 6.77i,$$

$$c = -1.82 - 0.56i.$$

Equations (3.61)-(3.62) are inhomogeneous versions of (3.54). Therefore, in order that solutions R_2 and S_2 exist, solvability conditions must be satisfied. Since the equations for R_2 and S_2 decouple we treat them separately. We first consider (3.61). The solvability condition is that

$$\lim_{L\to\infty}\frac{1}{2L}\int_{-L}^{L}\phi\overline{r_3}d\eta_2 = 0,$$

where r_3 denotes the right-hand side of (3.61) and ϕ is any element of the kernel of the adjoint of (3.54), that is any function of η_1 only. A similar solvability condition can be written for (3.62), with η_1 replaced by η_2. The solvability conditions yield (3.63)-(3.63) below.

We observe that the same equations can be obtained directly by integrating (3.61) and (3.62) on $[-L, L]$ with respect to η_2 and η_1 respectively, dividing by $2L$, taking the limit $L \to \infty$, and using the condition that R_2 and S_2 are bounded at infinity, to obtain

$$\frac{\partial R_1}{\partial t_2} = \left(\alpha\beta - \frac{i}{8\omega_0'} \widetilde{|R_1|^2} \right) R_1 +$$

$$+ A \frac{\partial^2 R_1}{\partial \eta_1^2} + B R_1 |R_1|^2 + C R_1 \widetilde{|S_1|^2},$$

$$\frac{\partial S_1}{\partial t_2} = \left(\alpha\beta - \frac{i}{8\omega_0'} \widetilde{|S_1|^2} \right) S_1 + \tag{3.63}$$

$$+ A \frac{\partial^2 S_1}{\partial \eta_2^2} + B S_1 |S_1|^2 + C S_1 \widetilde{|R_1|^2},$$

where

$$\lim_{\eta_1 \to +\infty} |R_1|^2 = \lim_{\eta_1 \to -\infty} |R_1|^2,$$

$$\lim_{\eta_2 \to +\infty} |S_1|^2 = \lim_{\eta_2 \to -\infty} |S_1|^2 .$$

We note that these conditions are satisfied for the plane wave solutions which are considered below.

Here,

$$C = c + \frac{i}{8\omega_0'},$$

and the averaged quantities are defined as

$$\tilde{f} = \lim_{L \to \infty} \frac{1}{2L} \int_{-L}^{L} f(\sigma, t_2) d\sigma.$$

When taking the average we used the fact that the average of $\partial H_1/\partial y_1|_{t_1=a}$ equals zero, and

$$\lim_{L \to \infty} \frac{1}{2L} \int_{-L+b}^{L+b} f(\eta) d\eta = \lim_{L \to \infty} \frac{1}{2L} \int_{-L}^{L} f(\eta) d\eta$$

for any constant b.

Equations (3.63) and (3.63) constitute a coupled set of complex Ginzburg-Landau equations. They differ from the usual complex Ginzburg-Landau equations by the appearance of the averaged terms, which may be interpreted as follows. The functions R_1 and S_1 (together with their complex conjugates) are the amplitudes of waves traveling in opposite directions on the t scale. On the t_1 scale they also describe waves traveling in opposite directions, each with speed ω_0'. On the t_2 scale the R_1 wave sees the S_1 wave going by very rapidly. Therefore, the effect on R_1 of the S_1 wave enters only through its average. The effect of the R_1 wave on S_1 is similar. In addition, the spatial variables in (3.63) – (3.63) are the characteristic variables.

Even for the simplest solutions, such as standing waves, the averages manifest themselves by providing different stability criteria than for complex Ginzburg–Landau equations without averages.

Equations similar to (3.63)–(3.63) were obtained by Knobloch and De Luca [17] who used symmetry arguments to derive the form of the equations, without determining the coefficients in terms of the parameters of a specific problem. Furthermore, they assumed that R_1 and S_1 were periodic. Since the basic solution in our problem is a uniformly propagating combustion wave rather than a homogeneous state, we have additional terms in our amplitude equations. Nevertheless, equations (3.63)– (3.63) can be reduced to those in [17] by an appropriate change of variables.

In [23] we derived plane wave solutions of the nonlocal amplitude equations and investigated their stability. The plane wave solutions correspond either to traveling waves, to standing waves or to quasiperiodic waves, depending on parameters. Note that plane wave solutions of the amplitude equations correspond to quasiperiodic solutions of the original model.

4. Computational Results

In this section we report on the results of computations which appear in [6, 7]. Unless otherwise stated for all of the computations reported here we fixed $N = 25$ and $\sigma = 0.6$ so that $Z = 5$. For these parameters, the uniformly propagating planar solution is unstable. The stability boundary is approximately $Z = 4.2$ and is very close to the analytically predicted value which is calculated for $\delta-$function kinetics rather than the distributed kinetics that we employ. The control parameter is R which is varied over approximately the interval $0 < R < 20$.

An overview of our results is shown in Figure 1 where we plot the mean axial flame speed V for the different solution branches that we have found. We compute V by first computing the z location of the front for each value of ψ, as described above. We next compute $\phi(t)$ by integrating the velocity ϕ_t in time. We then compute $x_f(\psi, t)$, the front location in the fixed frame. The mean axial flame speed V is then obtained by performing a linear least squares fit to $x_f(t)$ and taking the slope as V. We do not consider a solution to be equilibrated until we find that V does not vary with different choices of angles ψ employed in computing the average V, or if the run is continued longer in time. We note that all solution branches are described (and computed) with respect to a coordinate system moving with the flame.

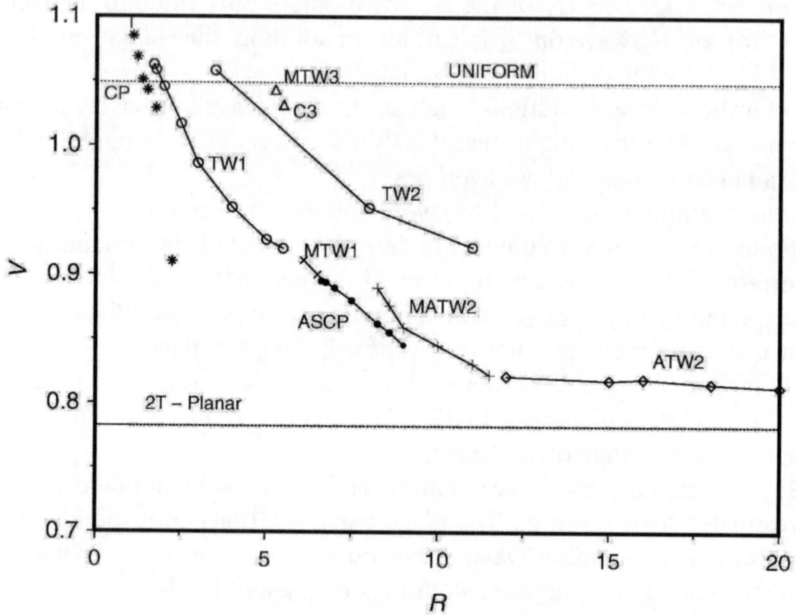

Figure 1. V as a function of R for various solution branches. Solid (dotted) lines indicate stable (unstable) solutions.

Figure 1 incorporates several solution branches. A common characteristic of these solution branches is the formation of hot spots. The branches are characterized by different dynamics exhibited by the spots. For small values of R the flame speed is close to that of the unstable, uniformly propagating mode. However, our results show that even for small values of R the primary conversion of reactant to product occurs via the spinning of the spots rather than through axial diffusion of heat. As R increases the spots become increasingly localized in space and exhibit increasing temperature. As a result large regions of the front are nearly uniform and thus prone to additional instabilities. Furthermore, as the spots become localized, the behavior at a fixed angle ψ becomes increasingly relaxational in time analogous to the behavior of the slowly propagating pulsating planar solution. Since for all solution branches the mean axial speed V is nonincreasing as R increases, the branches can be loosely thought of as providing a transition from behavior analogous to the higher speed uniform mode to behavior analogous to the lower speed planar pulsating mode as R increases. Since slowly propagating flames are more likely to be extinguished by heat losses, the modes at the left end of the branches may be more likely to survive heat loss effects than those at the right end. Due to the increased localization more of the front (in terms of ψ) attains the rather cool temperatures away from the spot, thus resulting in the lower mean speeds. This behavior is similar to that observed for planar modes of propagation, where the development of relaxation oscillations is accompanied by a decrease in the mean speed of the flame, e.g., [1, 21].

Broadly speaking, hot spot dynamics can be classified into two categories. Counterpropagating (CP) behavior in which pairs of spots propagate in opposite directions around the cylindrical surface and spinning modes in which one or more spots rotates around the cylinder in the same direction. As seen in Figure 1, the nonplanar solutions which appear first, i.e., for the smallest values of R, are CP solutions. For values of R below those indicated in the figure, only planar pulsating solutions are found. Spin behavior develops only for larger values of R. The CP modes indicated in Figure 1 correspond to a variety of behaviors. An extensive discussion of different CP modes of solid flames, computed for different parameter values than those in this chapter, is presented in [11]. Here, we briefly describe various types of CP behavior that we have found.

In Figure 2 we show a space time plot of a CP mode for $R = 1$. We note that there are collisions at $\psi = 0$ and $\psi = \pi$. The collision sites are determined by the initial conditions. This CP mode is periodic and symmetric in that the behavior at the two collision sites is the same, just $180°$ out of phase in time. In this mode, the hot spots enter and leave the collision essentially unchanged except for a phase shift, much like the behavior of solitons. Our computations indicate that the amplitude of the symmetric CP mode, i.e., the maximum temperature achieved at a collision, does not approach 0 as R approaches the transition point. Thus, the stable CP modes develop with finite amplitude. Our results also indicate that the mean speed V for these modes exceeds the speed for the unstable, uniformly propagating planar solution. Thus, near the transition point finite amplitude CP modes can propagate faster than the uniformly propagating mode. The mean propagation speed decreases as R increases.

As R increases there is a transition to CP modes which exhibit different behavior at each of the two collision sites. We illustrate such a solution in Figure 3 for $R = 1.2$, where we plot Θ_f as a function of t at the two collision sites $\psi = 0$ and $\psi = \pi$. Each spike in the time history of Θ_f corresponds to a collision. The solution is periodic, however, we note that had we chosen an angle different from one of the collision sites Θ_f would exhibit two spikes per period corresponding to the two counterpropagating hot spots. In contrast to the solution for $R = 1$ an asymmetry between the two collision sites has developed. Spots entering the collision at $\psi = 0$ are stronger than when they exit the collision and conversely for collisions at $\psi = \pi$. The collisions take on the character of apparent creation/annihilation sites as the asymmetry between

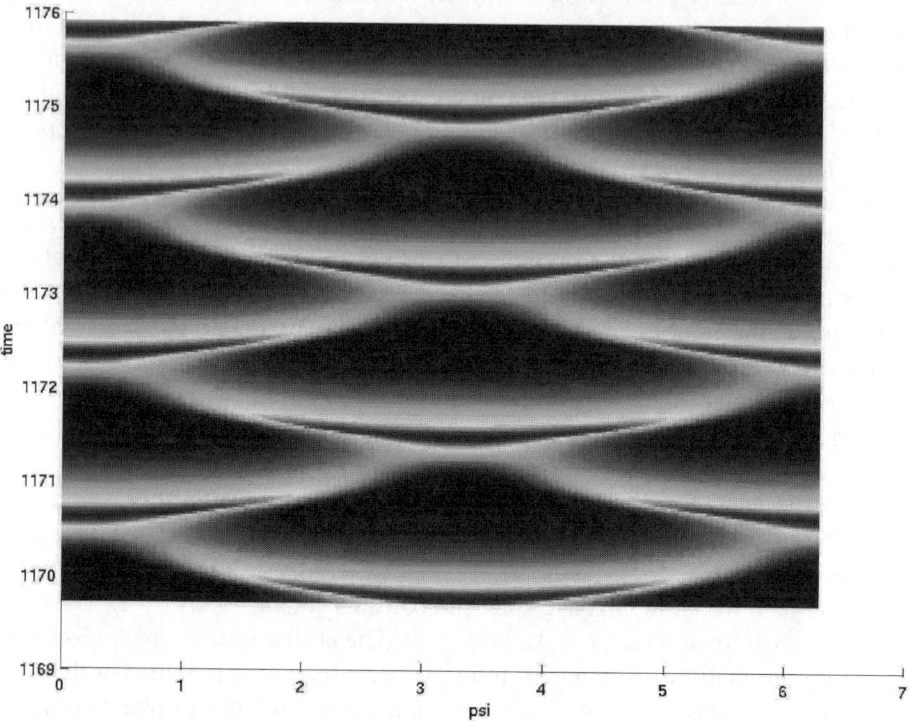

Figure 2. Θ_f as a function of ψ for a range of times for CP solution with $R = 1$. Time increases along the vertical axis.

the two sites increases. The development of nonlinear behavior for the CP solutions is accompanied by a reduction in the mean flame speed V.

The CP modes in Figures 2 and 3 are periodic in time. In Figure 4 we illustrate a CP mode for $R = 2.25$ in which the collisions do not occur in a periodic manner. We note that for this figure the collision sites, which are determined by initial conditions, are no longer at $\psi = 0$ and $\psi = \pi$. The problem is invariant under shifts in ψ, so that solutions which are shifted in ψ are also stable. The CP solution in Figure 4 exhibits apparent creation and annihilation in that at one collision site strong waves interact producing weak waves (apparent annihilation) while at the other collision site weak waves interact to produce strong waves (apparent creation). These and other CP modes were described in detail in [11] for different parameters. We have not traced the CP solutions further here, however, additional dynamical behavior, e.g., CP modes in which sites alternate between being an apparent creation site and an apparent annihilation site, was found in [11]. Such behavior is likely to occur for the parameters

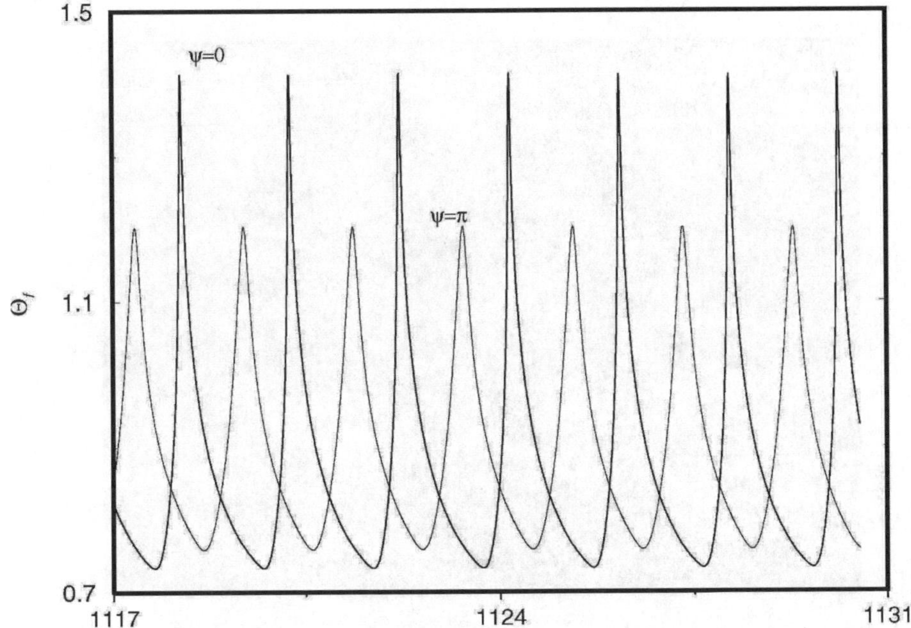

Figure 3. Θ_f as a function of t at the two collision sites $\psi = 0$ and $\psi = \pi$ for CP mode with $R= 1.2$.

employed in this chapter as well. In [11] we also showed that additional stable CP solutions existed, including solutions with multiple pairs of CP spots, e.g., one pair may execute the same dynamics as the others but with a time lag.

We next consider the spin mode solution branches shown in Figure 1. The first branch we focus on corresponds to 1-headed spin modes which are traveling waves (TW1). These modes require larger radii for stability than the CP modes. Stable CP modes may be able to exist for smaller values of R as they are, in a sense, reinforced by collisions after circuiting only half the cylinder. The TW1 modes also appear to enter with nonzero amplitude of the rotating hot spot. The smallest value of R for which we can find a stable TW1 mode is R=1.68. For this value of R the temperature of the hot spot is approximately 1.21. Unstable TW1 modes may well exist for smaller values of R. If we decrease R further solutions with TW1 initial data evolve to CP modes. We note that the mean axial speeds V for the TW1 modes near onset are below those of the CP modes near onset, probably due to the fact that R is larger.

We find that the TW1 modes are stable for $1.68 \leq R \leq 5.5$. Over this interval V decreases with R as indicated in Figure 1. In addition, the spinning spot becomes hotter and more localized as R increases. In Figure 5 we plot Θ_f as a function of the distance $R\psi$ around the cylinder at specific times for TW1

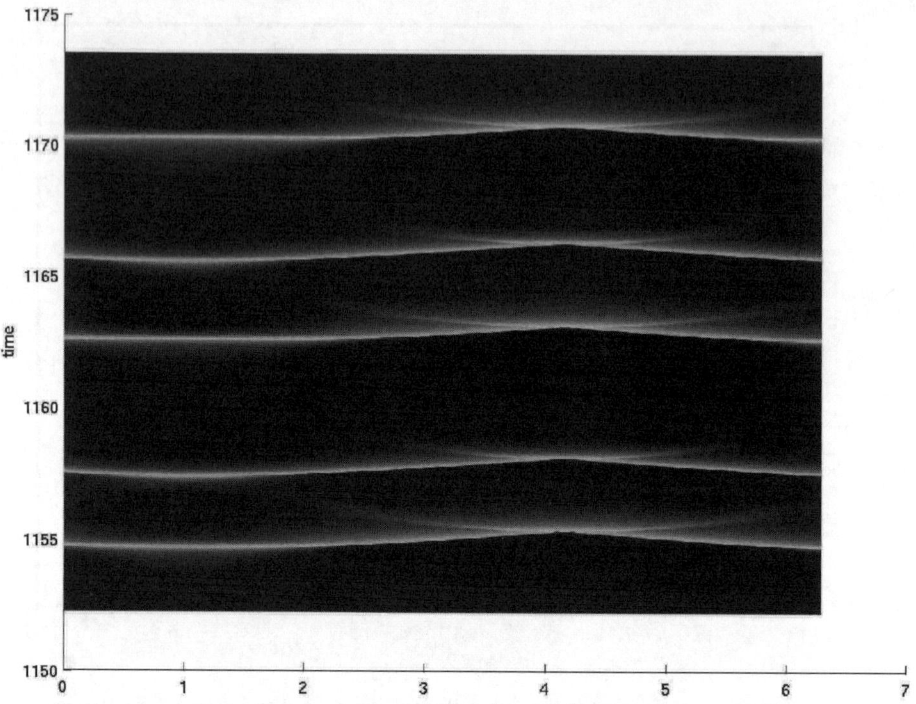

Figure 4. Θ_f as a function of ψ for a range of times for CP solution with $R = 2.25$. Time increases along the vertical axis.

modes with $R = 1.75$ and $R = 5.5$. The figure shows that as R increases the temperature on the front becomes nearly uniform over most of the front away from the hot spot.

For $R \geq 6.1$ we are unable to compute stable TW1 modes. Rather, we find modulated traveling waves consisting of a single hot spot which alternately speeds up and slows down and its intensity alternately increases and decreases as it rotates around the cylinder, thus describing a modulated traveling wave for a 1-headed spin (MTW1). A space time plot of such a mode is shown in Figure 6 for $R = 6.1$.

More complex behavior occurs for larger values of R. In Figure 7 we show a space time plot for $R = 7.5$. We note that there are periods of time where the trajectory of the spot (streak in the figures) is nearly horizontal. In order to analyze the nature of these solutions and relate them to the modes that occur for larger values of R we refer to Figure 8 which shows a space time plot of the front temperature over a small time interval around one of these events. A

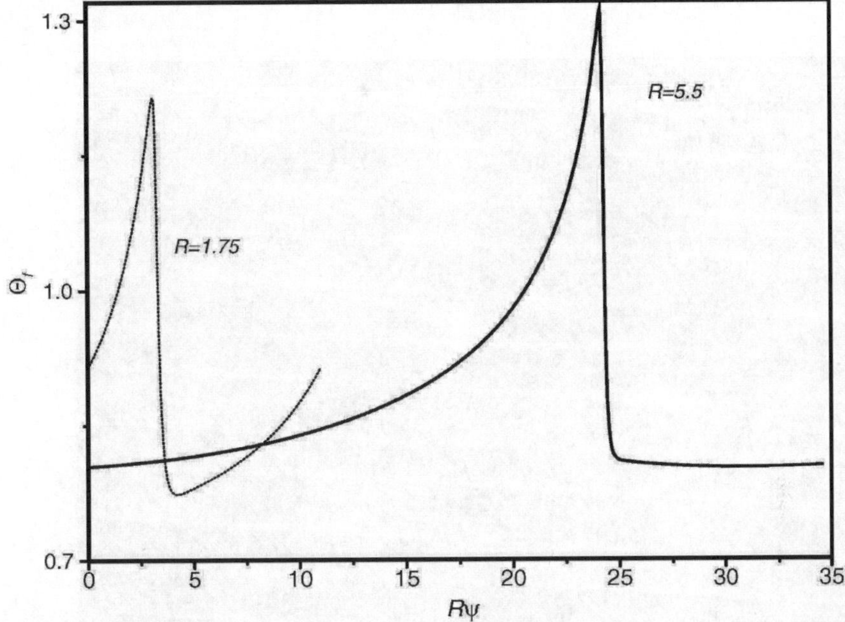

Figure 5. Θ_f as a function of distance $R\psi$ around the cylinder for TW1 solutions with $R = 1.75$ and $R = 5.5$.

small spot develops spontaneously ahead of the main propagating spot. The new spot splits into two counterpropagating spots, one of which collides with the original spot leading to their eventual mutual annihilation, while the other spot continues to rotate. Thus, for most of the time there is only one spot, but at times there can be either two or three spots on the front. We refer to these modes as Alternating Spin CP (ASCP) modes. As can be seen from Figure 7 these events occur periodically in time so that the mode is quasiperiodic. In Figure 9 we illustrate the CP event by showing Θ as a function of z and ψ at selected times around a CP event. In this figure the lateral direction is ψ so that propagation around the cylinder corresponds to lateral movement of a hot spot. The upper part of each figure corresponds to burned material (lighter shading) and the lower part (darker shading) is unburned material. The shading scale is the same for all frames in the figure. The first frame ($t = 948.8$) shows a single spot propagating counterclockwise (to the right in the figure). At $t = 950.1$ a new spot forms ahead of the main spot. At $t = 950.45$ the new spot splits into two daughter spots, one of which propagates to the right and the other to the left. At $t = 950.58$ the leftward propagating spot collides with the original spot, leading to the highest temperatures in the sequence of frames, while the rightward moving spot continues to propagate. At $t = 950.88$ the

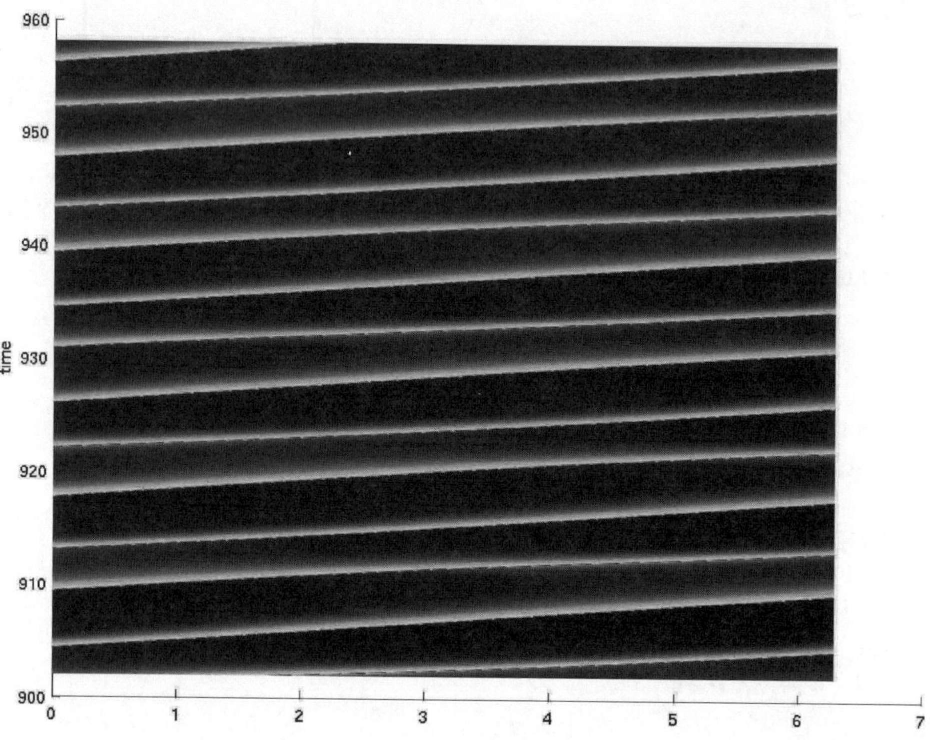

Figure 6. Θ_f as a function of ψ for a range of times for MTW1 solution with $R = 6.1$. Time increases along the vertical axis.

temperature at the collision site has decreased. Ultimately the two colliding spots completely annihilate each other. At $t = 957.8$ there is again a single spot as in the first frame. The large temperatures at the collision illustrated at $t = 950.58$ together with the subsequent annihilation can be explained by examining the degree of conversion $\eta = 1 - Y$. In Figure 10 we plot η at $t = 950.45$ and $t = 950.58$. In this plot light corresponds to complete burning ($\eta = 1$) and dark corresponds to unburned reactant ($\eta = 0$). For this problem the internal layer connecting $\eta = 0$ and $\eta = 1$ is an extremely thin strip. At $t = 950.48$ the new spot has split and one of the daughter spots is counterpropagating toward the original spot. It can be seen from Figure 10 that a pocket of unburned reactant penetrates the burned region between these two counterpropagating spots. At $t = 950.58$ when the two spots have collided, the pocket of unburned reactant has been essentially consumed in the high temperatures at the resulting collision. After this time the remnant of the collision decays due to thermal diffusion as there is no longer any unburned reactant to sustain

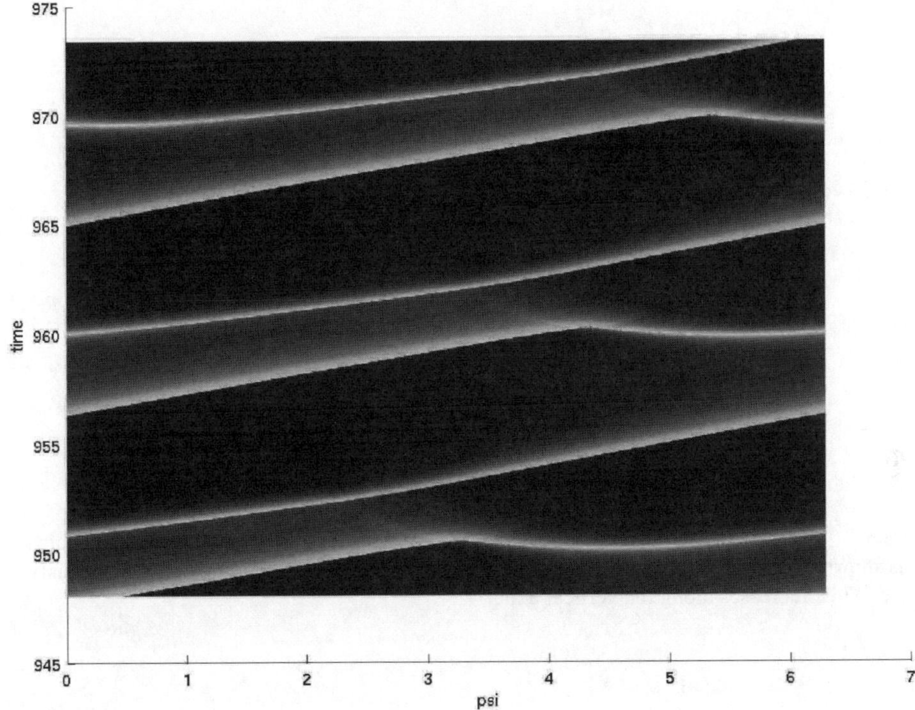

Figure 7. Θ_f as a function of ψ for a range of times for ASCP solution with $R = 7.5$. Time increases along the vertical axis.

the high temperature. The ASCP mode demonstrates that both spin and CP behavior can occur in a periodic fashion in the same solution. We note that the spontaneous creation of new spots ahead of rotating spots also occurs in gaseous combustion stabilized on a rotating burner ([10]).

We conjecture that the mechanism for ASCP behavior is the increasing localization of the spot as R increases. When R is sufficiently large, the portion of the front excluding the hot spot (topologically a circle in the moving coordinate) has a nearly uniform temperature (cf. Figure 5). Since the uniformly propagating planar mode is unstable additional spots are expected to form in the region away from the propagating spot. These spots can either exhibit spin or CP behavior. In the ASCP mode the new spot exhibits CP behavior.

We believe that the ASCP mode evolves continuously from the MTW1 branch which evolved from the TW1 branch. This mode, while exhibiting both spin and CP behavior, is essentially a one-headed spin except for brief intervals of time. As R increases V decreases and the instantaneous temperature

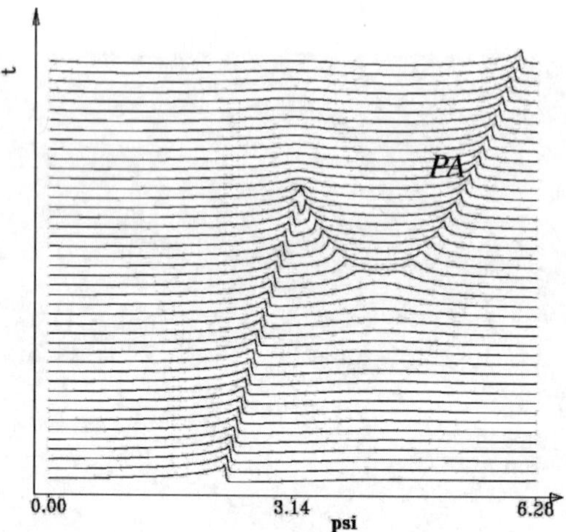

Figure 8. Θ_f as a function of ψ for a range of times for ASCP solution with $R = 7.5$. Solution is shown over a small interval of time around an event in which a new spot is spontaneously created. Time increases along the vertical axis.

at collision increases. We have computed ASCP modes from $R = 6.625$ up to $R = 9$. We have not attempted to compute this mode for larger values of R.

We next consider multi-headed spins. We first refer to the branch labeled TW2 in Figure 1. These solutions are replicates of analogous TW1 modes. Thus, for any value of R the TW2 solution consists of two replicates of the TW1 mode at $R/2$ and separated by π radians. We have tested the stability of these solutions by imposing perturbations in the initial conditions which do not correspond to replicated perturbations. (We note that replicated conditions correspond in Fourier space to even order modes. Thus, we impose perturbations which have odd Fourier modes.) We have found stable replicated TW2 modes for $3.5 \leq R \leq 11$, corresponding to TW1 modes for $1.75 \leq R \leq 5.5$. We note that for $R = 3.36$, corresponding to the TW1 at $R = 1.68$, the smallest value of R for which we found stable TW1 modes, we were unable to compute a TW2 solution. Rather, the perturbed replicated initial conditions evolved to a TW1 solution. Thus, our computations indicate that the replicated TW2 branch is stable, but not for all values of R for which a stable TW1 solution exists for $R/2$. The replicated TW2 solutions near $R = 3.5$ have the same large mean speed as the TW1 mode near $R = 1.75$, suggesting that these modes would be less prone to extinction and thus they may be more readily observed in experiments.

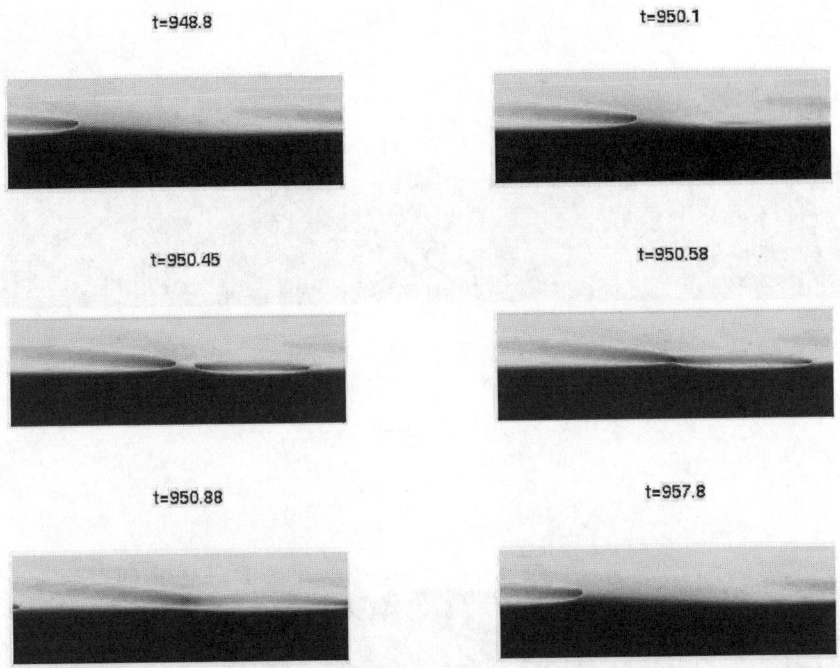

Figure 9. Θ as a function of ψ and z for selected times for ASCP solution with $R = 7.5$. Axial direction is vertical.

In the region of bistability between the TW1 and TW2 branches, the two spin modes are differentiated by lower axial flame speeds for the TW1 solutions, along with more localized, higher temperature spots than for the TW2 mode. The more localized spots result in a large region of the front being at a relatively low temperature thus possibly explaining the slower axial flame speed as compared to the TW2 solution. Ultimately, the localization is such that a new spot is spontaneously created and ASCP behavior ensues. For the TW2 mode the spots are at lower temperature and less localized since they are replicates of TW1 spots corresponding to lower values of R. Thus, the resulting front speed V is larger and closer to that of the uniform mode.

For $R > 11$ we are unable to compute stable TW2 modes. We note that the MTW1 branch described above appears to evolve continuously from the TW1 branch. We were unable to find any such branch evolving from the TW2 branch. When we tried to compute 2-headed spins for $R = 12$ using continuation, i.e., employing equilibrated TW2 data for R=11 as initial conditions, we found that the solution evolved to the mode shown in Figure 111, an asymmetric 2-headed spin, which we denote by ATW2, accompanied by a jump in V. This solution is a 2-headed traveling wave in that two spots rotate around the

Figure 10. η as a function of ψ and z for selected times for ASCP solution with $R = 7.5$. Axial direction is vertical.

cylinder at a uniform rate and without any change in shape. However, the two spots are not symmetrically placed, i.e., are not separated by $\psi = \pi$ and they are of unequal strength. The stronger (hotter) spot leads the weaker (cooler) spot. The spots rotate together as a bound state.

Bound states were also described in one dimensional computations of an excitable system ([29]). We have computed ATW2 modes up to $R = 20$. The mean axial speed V is nearly independent of R along the ATW2 branch as seen in Figure 1.

There are transitions from the ATW2 branch to non TW modes still exhibiting leading and trailing spot behavior at both its extremities, i.e., for smaller R and larger R. Upon increasing R beyond $R = 20$ we find slowly varying bound states in which a new spot is spontaneously created ahead of the leading spot, and subsequently ahead of the trailing spot in the bound state. The new spot, together with the leading (trailing) spot, exhibits episodes of ASCP behavior, Denote the strong leading spot in the bound state by S and the trailing weak spot by s. As the slowly varying bound state propagates, events occur in which a new spot forms ahead of S. Call this new spot SS. The new spot SS together with the spot S then exhibits ASCP behavior. That is, SS splits into

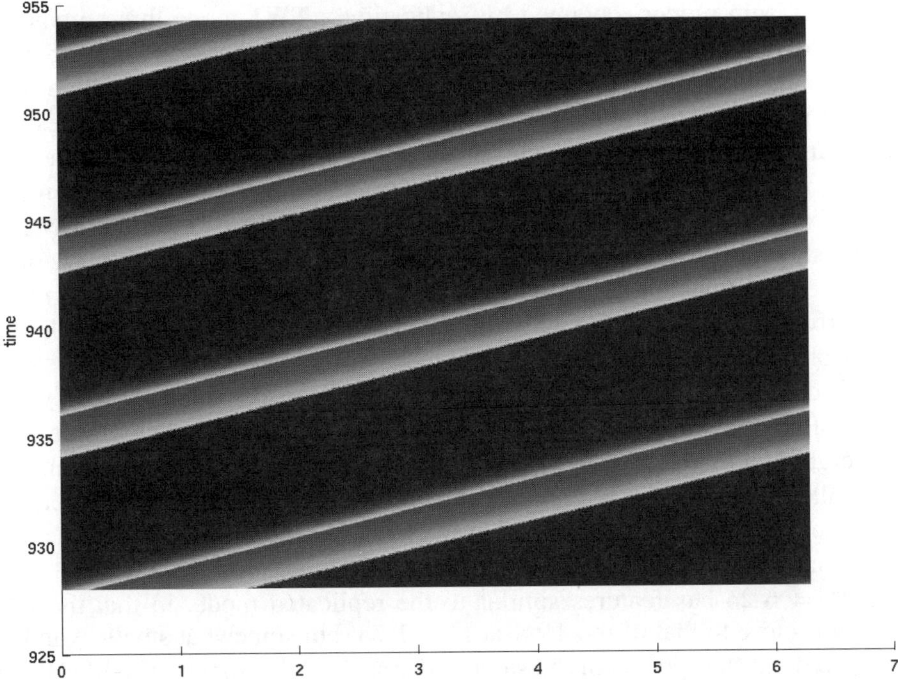

Figure 11. Θ_f as a function of ψ for a range of times for ATW2 solution with $R = 12$. Time increases along the vertical axis.

a pair of counterpropagating spots, one of which annihilates with S when they collide while the other continues to propagate. The effect is a speed up of the leading spot. Subsequently, similar behavior occurs for the trailing spot. We believe that the latter event is an attempt to maintain the bound state nature of the mode. We term this an Asymmetric Alternating Spin CounterPropagating (AASCP) mode. We note that the time interval between these events does not appear to be regular.

Upon decreasing R along the ATW2 branch we find a family of modulated ATW2 modes (MATW2), where the two bound spots alternately approach one another and then separate in an oscillatory manner. The solution is quasiperiodic, as the approaches and separations occur periodically. We note that this is an example of increasing spatiotemporal complexity accompanied by decreasing R. This is in contrast to the behavior of the 1-headed spins. We were unable to compute MATW2 modes for $R < 7.5$, where we found that MATW2 initial data evolved to a 1-headed spin solution.

Finally, we have investigated 3-headed spin modes. We find that replicated 3-headed spin modes, obtained by replicating a TW1 mode three times, are unstable. We note that for replicated 3-headed spins, replicated perturbations correspond to perturbations whose Fourier wave numbers are divisible by 3. This is in contrast to 2-headed spins, where replicated perturbations correspond to Fourier wave numbers which are divisible by 2. Since the TW1 mode is stable, replicated 2-headed spin modes are necessarily stable to all even $(2n)$ perturbations and potentially unstable only to odd $(2n + 1)$ perturbations. In contrast, replicated 3-headed spin modes are necessarily stable to all mode $(3n)$ perturbations and potentially unstable to mode $(3n + 1)$ and $(3n + 2)$ perturbations. This may be why replicated 3-headed spins are less stable than replicated 2-headed spins.

We do find 3-headed spins modes that are stable over a small range of values of R. For small values of R, these solutions are characterized by three spots undergoing a modulated rotation around the cylinder (MTW3), in which each spot alternately approaches and separates from its neighbors. These events do not occur periodically, in contrast to the MATW2 mode described above. We note that while the replicated TW3 mode is unstable, the stable MTW3 mode for $R = 5.25$ has features similar to the replicated mode, in that the front speed is close to that of the TW1 at $R = 1.75$ (but somewhat smaller) and the amplitude of the spots is of the same order as the TW1 mode. The MTW3 solutions have a small domain of attraction and unless very small steps are taken in R, MTW3 initial data evolves to either 1-headed or 2-headed spins. Upon increasing R, we find that the 3-headed spin modes exhibit apparently chaotic behavior. For certain, apparently random, times two spots come close together and nearly touch. One of the spots is then nearly extinguished and moves rapidly away from its neighbor. The spot then becomes hotter and continues its spinning behavior around the cylinder.

Upon increasing R further, we find it difficult to obtain equilibrated 3-headed spins. Starting with initial data corresponding to a 3-headed spin we find that during the transient, two spots actually collide, leading to the annihilation of one spot and the mode collapses to a steady state 2-headed spin. Thus, for increasing R, the 3-headed spins are either unstable or have a small domain of attraction.

References

[1] A. P. Aldushin, A. Bayliss, B. J. Matkowsky, *Dynamics of Layer Models of Solid Flame Propagation,* Physica D, **143** (2000), 109.

[2] A. P. Aldushin, B. A. Malomed, Ya. B. Zeldovich, *Phenomenological Theory of Spin Combustion,* Combustion and Flame, **42** (1981), 1.

[3] A. P. Aldushin, T. M. Martemyanova, A. G. Merzhanov, B. I. Khaikin, K. G. Shkadinsky, *Autooscillatory Propagation of Combustion Front in Heterogeneous Condensed Media,* Combustion Explosion and Shock Waves **9** (1973), 531.

[4] A. Bayliss, D. Gottlieb, B. J. Matkowsky, M. Minkoff, *An Adaptive Pseudo-Spectral Method for Reaction Diffusion Problems,* J. Comput. Phys. **81** (1989), 421.

[5] A. Bayliss, R. Kuske, B. J. Matkowsky, *A Two-Dimensional Adaptive Pseudo-Spectral Method,* J. Comput. Phys. **91** (1990), 174.

[6] A. Bayliss, B. J. Matkowsky, A. P. Aldushin, *Dynamics of Hot Spots in Solid Fuel Combustion.* Physica D, **166** (2002), 104.

[7] A. Bayliss, B.J. Matkowsky, A.P. Aldushin, *Solid Flame Waves,* Chapter 4 in Perspectives and Problems in Nonlinear Science: A Celebratory Volume in Honor of Larry Sirovich, Eds. E. Kaplan, J. Marsden. K. Sreenivasan, Springer Verlag, 2003

[8] A. Bayliss, B. J. Matkowsky, *Fronts, Relaxation Oscillations, and Period Doubling in Solid Fuel Combustion,* J. Comput. Phys. **71** (1987), 147.

[9] A. Bayliss, B. J. Matkowsky, *Two Routes to Chaos in Condensed Phase Combustion,* SIAM J. Appl. Math. **50** (1990), 437.

[10] A. Bayliss, B. J. Matkowsky, *Structure and Dynamics of Kink and Cellular Flames Stabilized on a Rotating Burner,* Physica D **99** (1996), 276.

[11] A. Bayliss, B. J. Matkowsky, *Interaction of Counterpropagating Hot Spots in Solid Fuel Combustion,* Physica D **128** (1999), 18.

[12] M. Booty, S.B. Margolis, B.J. Matkowsky, *Interaction of pulsating and spinning waves in condensed phase combustion,* SIAM J. Appl. Math. **46** (1986), 801.

[13] A. V. Dvoryankin, A. G. Strunina, A. G. Merzhanov, *Trends in the Spin Combustion of Thermites,* Combustion, Explosion and Shock Waves **18** (1982), 134.

[14] T. P. Ivleva, A. G. Merzhanov, K. G. Shkadinsky, *Mathematical Model of Spin Combustion,* Soviet Physics Doklady **23** (1978), 255.

[15] T. P. Ivleva, A. G. Merzhanov, K. G. Shkadinsky, *Principles of the Spin Mode of Combustion Front Propagation,* Combustion, Explosion and Shock Waves **16** (1980), 133.

[16] H.G. Kaper, G.K. Leaf, S.B. Margolis, B.J. Matkowsky, *On nonadiabatic condensed phase combustion,* Combustion Science and Technology **53** (1987), 289.

[17] E. Knobloch, J. De Luca, *Amplitudes equation for travelling wave convection,* Nonlinearity **3** (1990), 975.

[18] S. B. Margolis, *An Asymptotic Theory of Condensed Two-Phase Flame Propagation,* SIAM J. Appl. Math. **43** (1983), 351.

[19] S. B. Margolis, H. G. Kaper, G. K. Leaf, B. J. Matkowsky, *Bifurcation of Pulsating and Spinning Reaction Fronts in Condensed Two- Phase Combustion,* Comb. Sci. and Tech **43** (1985), 127.

[20] S.B. Margolis, B.J. Matkowsky, *New modes of quasi-periodic combustion near a degenerate Hopf bifurcation point,* SIAM J. Appl. Math. **48** (1988), 828. i

[21] B. J. Matkowsky, G. I. Sivashinsky, *Propagation of a Pulsating Reaction Front in Solid Fuel Combustion,* SIAM J. Appl. Math. **35** (1978), 465.

[22] B.J. Matkowsky, V.A. Volpert, *Spiral gasless condensed phase combustion,* SIAM J. Appl. Math. **54** (1994), 132.

[23] B.J. Matkowsky, V.A. Volpert, *Coupled nonlocal complex Ginzburg-Landau equations in gasless combustion,* Physica **D 54** (1992), 203.

[24] A. G. Merzhanov, *SHS Processes: Combustion Theory and Practice,* Arch. Combustionis **1** (1981), 23.

[25] A. G. Merzhanov, *Self-Propagating High-Temperature Synthesis: Twenty Years of Search and Findings,* in: Combustion and Plasma Synthesis of High-Temperature Materials, Z.A. Munir, J.B. Holt, Eds., VCH, (1990), 1.

[26] A. G. Merzhanov, A. K. Filonenko, I. P. Borovinskaya, *New Phenomena in Combustion of Condensed Waves,* Soviet Phys. Dokl. **208** (1973), 122.

[27] A. S. Mukasyan, S. C. Vadchenko, L. O. Khomenko, *Combustion Modes in the Titanium-Nitrogen System at Low Nitrogen Pressures,* Combustion and Flame **111** (1997), 65.

[28] Z. A. Munir, U. Anselmi-Tamburini, *Self-Propagating Exothermic Reactions: The Synthesis of High-Temperature Materials by Combustion,* Material Science Reports, A Review Journal **15** (1989), 277.

[29] M. Or-Guil, I. G. Kevrekidis, M. Bar, *Stable Bound States of Pulses in an Excitable System,* Physica D **135** (1999), 157.

[30] J.H. Park, A. Bayliss, B.J. Matkowsky, A.A. Nepomnyashchy, *Period doubling cascades on the route to extinction in the interfacial motion of a nonadiabatic solid flame,* submitted for publication (2005).

[31] C. Raymond, A. Bayliss, B. J. Matkowsky, V. Volpert, *Transitions to Chaos in Condensed Phase Combustion with Reactant Melting,* Int'l. J. Self-Propagating High-Temperature Synthesis, **10** (2001), 133.

[32] K. G. Shkadinsky, B. I. Khaikin, A. G. Merzhanov, *Propagation of a Pulsating Exothermic Reaction Front in the Condensed Phase,* Combustion, Explosion and Shock Waves **1** (1971), 15.

[33] G. I. Sivashinsky, *On Spinning Propagation of Combustion Waves,* SIAM J. Appl. Math. **40** (1981), 432.

[34] A. G. Strunina, A. V. Dvoryankin, A. G. Merzhanov, *Unstable Regimes of Thermite System Combustion,* Combustion, Explosion and Shock Waves **19** (1983), 158.

SELF-ORGANIZATION OF MICROTUBULES AND MOTORS

Igor S. Aranson
Argonne National Laboratory, 9700 South Cass Avenue, Argonne, Illinois, 60439
aronson@msd.anl.gov

Lev S. Tsimring
Institute for Nonlinear Science, University of California, San Diego,
La Jolla, CA 92093-0402
ltsimring@ucsd.edu

Abstract   Here we introduce a model for spatio-temporal self-organization of an ensemble of microtubules interacting via molecular motors. Starting from a generic stochastic model of inelastic polar rods with an anisotropic interaction kernel we derive a set of equations for the local rods concentration and orientation. At large enough mean density of rods and concentration of motors, the model describes orientational instability. We demonstrate that the orientational instability leads to the formation of vortices and (for large density and/or kernel anisotropy) asters seen in recent experiments. The corresponding phase diagram of vortex-asters transitions is in qualitative agreement with experiment.

Keywords: Microtubules, Molecular Motors, Self-Organization, Master Equation

Introduction

One of the central questions in biology concerns the formation of complex highly organized microscopic structures from initially disordered states. Complex ordered structures are established and maintained through a dynamic interplay between self-assembly and regulatory processes. Molecular motors play essential role in carrying out these developmental tasks. One of the most important functions of molecular motors is to organize a network of long stiff hollow filaments (microtubules) during cell division to form cytosceletons of daughter cells [1]. Recent in vitro experiments with purified motors and microtubules in thin microchambers [2–7] have shown a surprising variety of large-scale two-dimensional structures: *vortices*, in which filaments are oriented at some angle with respect to the radial direction, and *asters*, in which filaments

283

A. A. Golovin and A. A. Nepomnyashchy (eds.),
Self-Assembly, Pattern Formation and Growth Phenomena in Nano-Systems, 283–294.
© 2006 *Springer. Printed in the Netherlands.*

are oriented radially, see Fig. 1. The final structure depends on the relative
concentrations of the molecular components: at large enough concentration of
molecular motors and microtubules, the latter organize in asters and vortices
depending on the type and concentration of molecular motors, see Fig. 1. The
experiments also indicated that the final structure can be reached through different assembly pathways and that dynamic transitions from one state to the
other can be induced by varying the kinetic parameters.

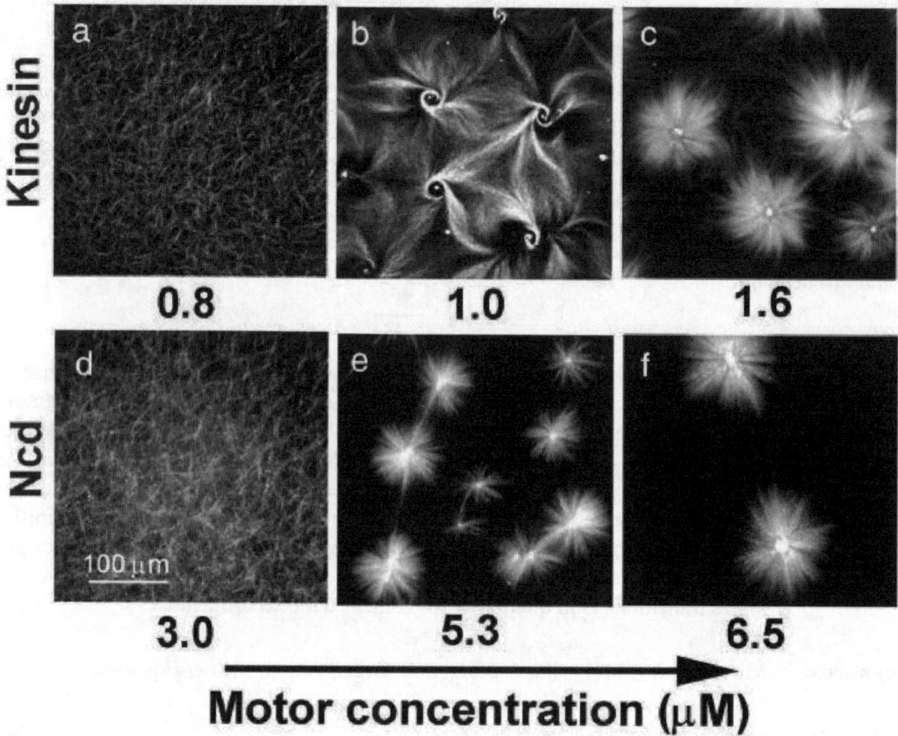

Figure 1.   Vortices and asters observed experimentally in microtubules/molecular motors mixtures for various concentration of motors and for two different types of motors (kinesin and
NCD), reprinted from [5].

After a molecular motor binds to a microtubule at a random position, it
marches along it in a fixed direction for some time until it unbinds. Since the
mass of a molecular motor is small in comparison with that of a microtubule,
the motion of molecular motors does not lead to appreciable displacement of
the microtubule. If a molecular motor binds to *two* microtubules, it can change
their mutual position and orientation significantly. In Ref. [5], the interaction of rod-like filaments via motor binding and motion has been studied, and

patterns resembling experimental ones were observed in small-scale molecular dynamics type simulations. In [8] a phenomenological model for the molecular motor density and the microtubule orientation has been proposed, which included transport of molecular motors along microtubules and alignment of microtubules mediated by molecular motors. Ref. [9, 10] generalized this model by including separate densities of free and bound molecular motors, as well as the density of microtubules. A transition from asters to vortices was found as the density of molecular motors was increased, in apparent disagreement with experimental evidence [7] that the asters give way to vortices with the *decrease* of the molecular motor concentration. A phenomenological flux-force relation for active gels was suggested in [11]. While vortex and aster solutions were obtained, an analysis of that model is difficult because of a large number of unknown parameters. In Ref. [12] a set of equations for microtubule density and orientation was derived by averaging conservation laws for microtubule probability distribution function. However, this model, contrary to experiment, does not exhibit orientation transition for the homogeneous microtubules distributions. Moreover, according to Ref. [12], even initially oriented microtubule states appear to decay for any concentration of the motors.

Here we derive a model for the collective spatio-temporal dynamics of microtubules starting with a master equation for interacting inelastic polar rods [13]. Our model differs from the transport equations [12] in that it maintains the detailed balance of rods with a certain orientation. The model exhibits an onset of orientational order for large enough density of microtubules and molecular motors, formation of vortices and then asters with the increase in the molecular motor concentration, in a qualitative agreement with experiment.

1. Maxwell Model and Orientational Instability

Molecular motors enter the model implicitly by specifying the interaction rules between two rods. Since the diffusivity of molecular motors is about 100 times larger than that of microtubules, as a first approximation we neglect spatial variations of the molecular motor density. While the varying concentration of molecular motors affects certain quantitative aspects [7], our analysis captures salient features of the phenomena and the collision rules are spatially homogeneous. All rods are assumed to be of equal length l and diameter $d \ll l$, and are characterized by their centers of mass, $\mathbf{r}$, and the orientation angles ϕ.

We consider the orientational dynamics only and ignore the spatial coordinates of interacting rods (an analog of the Maxwell model of binary collisions in kinetic theory of gases, see e.g. [15]). Since the motor residence time on microtubules (about 10 sec) is much smaller than the characteristic time of pattern formation (10 min or more), we model molecular motor – microtubule inelastic interaction as an instantaneous collision in which two rods change

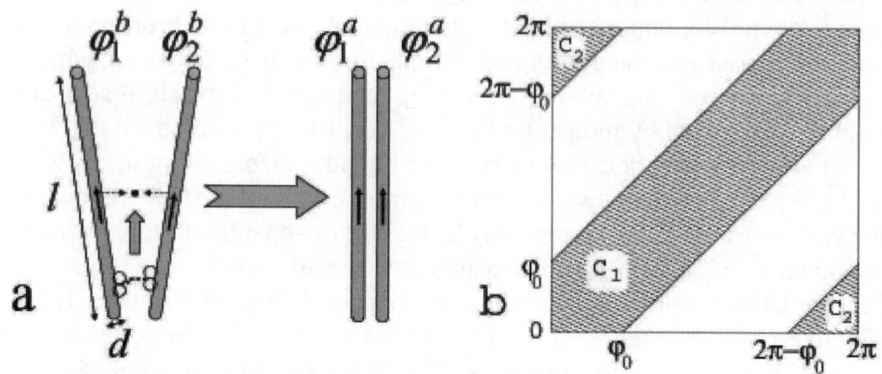

Figure 2.   a - sketch of motor-mediated two-rod interaction for $\gamma = 1/2$, b - integration regions $C_{1,2}$ for Eq.(1.2).

their orientations:

$$\begin{pmatrix} \phi_1^a \\ \phi_2^a \end{pmatrix} = \begin{pmatrix} \gamma & 1-\gamma \\ 1-\gamma & \gamma \end{pmatrix} \begin{pmatrix} \phi_1^b \\ \phi_2^b \end{pmatrix}, \tag{1.1}$$

where $\phi_{1,2}^b$ and $\phi_{1,2}^a$ are orientations before and after the collision, respectively, and γ characterizes the collision inelasticity. After the collision, the angle between two rods is reduced by the factor $2\gamma - 1$. Here, $\gamma = 0$ corresponds to a totally elastic collision (rods exchange their angles) and $\gamma = 1/2$ corresponds to a totally inelastic collision in which rods acquire identical orientations, $\phi_{1,2}^a = (\phi_1^b + \phi_2^b)/2$ (see Fig. 2,a). Here we assume that two rods interact only if the angle between them is less than ϕ_0, $|\phi_2^b - \phi_1^b| < \phi_0 < \pi$. Because of 2π-periodicity, two rods with mutual angle between $2\pi - \phi_0$ and 2π also interact. In this case we have to replace $\phi_1^{b,a} \to \phi_1^{b,a} + \pi$, $\phi_2^{b,a} \to \phi_2^{b,a} - \pi$ in Eq. (1.1). In the following we will only consider the case of totally inelastic rods ($\gamma = 1/2$) and $\phi_0 = \pi$; the generalization for arbitrary γ and ϕ_0 is straightforward. The probability $P(\phi)$ obeys the following master equation (compare with Ref. [15]):

$$\partial_t P(\phi) = D_r \partial_\phi^2 P(\phi) + g \int_{C_1} d\phi_1 d\phi_2 P(\phi_1) P(\phi_2) \tag{1.2}$$

$$\times [\delta(\phi - \phi_1/2 - \phi_2/2) - \delta(\phi - \phi_2)] + g \int_{C_2} d\phi_1 d\phi_2$$

$$\times P(\phi_1) P(\phi_2) [\delta(\phi - \phi_1/2 - \phi_2/2 - \pi) - \delta(\phi - \phi_2)],$$

where g is the "collision rate" proportional to the number of molecular motors, the diffusion term $\propto D_r$ describes the thermal fluctuations of the rod orien-

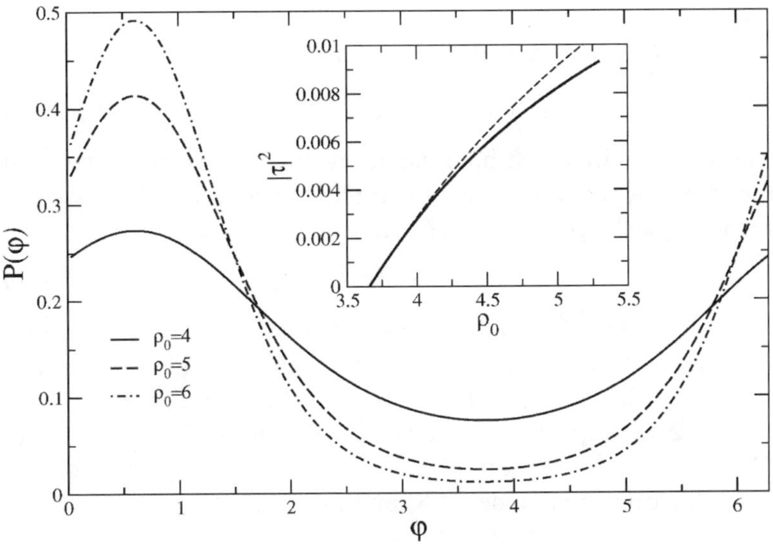

Figure 3. Stationary solutions $P(\phi)$ for different ρ. Inset: the stationary value of $|\tau|$ vs ρ obtained from the Maxwell model (1.3), dashed line - truncated model (1.8).

tation, and the integration domains C_1, C_2 are shown in Fig.2a. Changing variables $t \to D_r t$, $P \to gP/D_r$, $w = \phi_2 - \phi_1$, we arrive at

$$\partial_t P(\phi) = \partial_\phi^2 P(\phi) + \int_{-\pi}^{\pi} dw$$
$$\times \left[P(\phi + w/2)P(\phi - w/2) - P(\phi)P(\phi - w) \right]. \tag{1.3}$$

The rescaled number density $\rho = \int_0^{2\pi} P(\phi, t)d\phi$ now is proportional to the *density of rods multiplied by the density of motors*. Let us consider the Fourier harmonics:

$$P_k = \langle e^{-ik\phi} \rangle = \frac{1}{2\pi} \int_0^{2\pi} d\phi e^{-ik\phi} P(\phi, t). \tag{1.4}$$

The zeroth harmonic $P_0 = \rho/2\pi = const$, and the real and imaginary parts of P_1 represent the components $\tau_x = \langle \cos\phi \rangle, \tau_y = \langle \sin\phi \rangle$ of the average orientation vector $\boldsymbol{\tau}$, $\tau_x + i\tau_y = P_1^*$. Substituting Eq.(1.4) into Eq.(1.3) we obtain

$$\dot{P}_k + (k^2 + \rho)P_k = 2\pi \sum_m P_{k-m}P_m S[\pi k/2 - m\pi], \tag{1.5}$$

where $S(x) = \sin x/x$. Due to the angular diffusion term, the amplitudes P_k decay exponentially with $|k|$. Assuming $P_k = 0$ for $|k| > 2$ one obtains from Eq.(1.5)

$$\dot{P}_1 \ + \ P_1 = P_0 P_1 2(4 - \pi) - \frac{8}{3} P_2 P_1^* \tag{1.6}$$

$$\dot{P}_2 \ + \ 4P_2 = -P_0 P_2 2\pi + 2\pi P_1^2. \tag{1.7}$$

Since near the instability threshold the decay rate of P_2 is much larger than the growth rate of P_1, we can neglect the time derivative $\dot{P}_2$. Thus we obtain $P_2 = A P_1^2$ with $A = 2\pi(\rho + 4)^{-1}$, and arrive at:

$$\dot{\boldsymbol{\tau}} = \epsilon \boldsymbol{\tau} - A_0 |\boldsymbol{\tau}|^2 \boldsymbol{\tau}, \tag{1.8}$$

where $\epsilon = \rho(4\pi^{-1} - 1) - 1 \approx 0.273\rho - 1$ and $A_0 = 8A/3$. For large enough $\rho > \rho_c = \pi/(4 - \pi) \approx 3.662$, an ordering instability leads to spontaneous rod alignment. This instability saturates at the value determined by ρ. Near threshold, $A_0 \approx 2.18$. Fig. 3 shows stationary solutions $P(\phi)$ obtained from Eq. (1.3). As seen from the inset in Fig. 3, the corresponding values of $|\tau|$ are consistent with the truncated model (1.8) up to $\rho < 5.5$.

2.   Spatial Localization

In order to describe the *spatial localization* of interactions, we introduce the probability distribution $P(\mathbf{r}, \phi, \mathbf{t})$ to find a rod with orientation ϕ at location $\mathbf{r}$ at time t. By analogy with Eq. (1.2), the corresponding master equation for $P(\mathbf{r}, \phi, t)$ can be written as

$$\partial_t P(\mathbf{r}, \phi) = \partial_\phi^2 P(\mathbf{r}, \phi) + \partial_i D_{ij} \partial_j P(\mathbf{r}, \phi)$$

$$+ \ \int\int d\mathbf{r_1} d\mathbf{r_2} \int_{-\phi_0}^{\phi_0} dw \left[W(\mathbf{r}_1, \mathbf{r}_2, \phi + w/2, \phi - w/2) \right.$$

$$\times P(\mathbf{r}_1, \phi + w/2) P(\mathbf{r}_2, \phi - w/2) \delta\left(\frac{\mathbf{r}_1 + \mathbf{r}_2}{2} - \mathbf{r}\right)$$

$$- \ \left. W(\mathbf{r}_1, \mathbf{r}_2, \phi, \phi - w) P(\mathbf{r_2}, \phi) P(\mathbf{r}_1, \phi - w) \delta(\mathbf{r}_2 - \mathbf{r}) \right], \tag{2.9}$$

where we performed the same rescaling as in Eq.(1.3) and dropped the argument t for brevity. The first two terms on the r.h.s. of (2.9) describe angular and translational diffusion of rods with the diffusion tensor

$$D_{ij} = \frac{1}{D_r} \left(D_\| n_i n_j + D_\perp (\delta_{ij} - n_i n_j) \right), \tag{2.10}$$

where $\mathbf{n} = (\cos(\phi), \sin(\phi))$ and $D_r, D_\|, D_\perp$ are known from polymer physics [16],

$$D_\| = \frac{k_B T}{\xi_\|}, D_\perp = \frac{k_B T}{\xi_\perp}, D_r = \frac{4 k_B T}{\xi_r}. \tag{2.11}$$

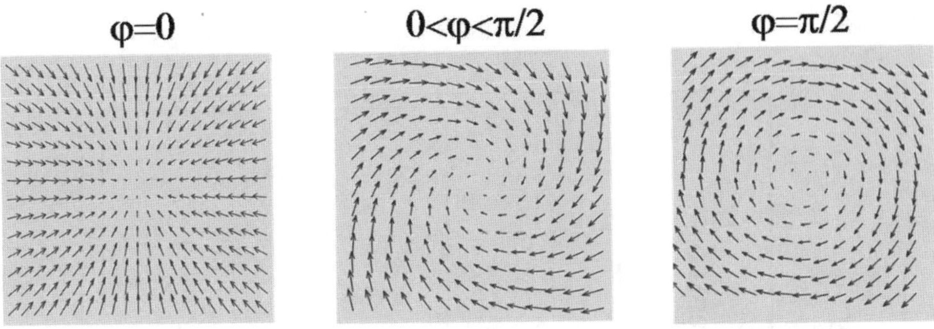

Figure 4. Schematics of orientation field τ for three different values of φ: aster ($\varphi = 0$); generic vortex ($0 < \varphi < \pi/2$) and ideal vortex ($\varphi = \pm\pi$), see Eq. (3.15).

Here, $\xi_\parallel, \xi_\perp, \xi_r$ are the corresponding drag coefficients. For rod-like molecules, $\xi_\parallel = 2\pi\eta_s l/\log(l/d); \xi_\perp = 2\xi_\parallel; \xi_r \approx \pi\eta_s l^3/3\log(l/d)$ where η_s is shear viscosity [16].

The last term in Eq.(2.9) describes molecular motor-mediated interaction of rods. We assume that after the interaction, the two rods share the same orientation and the same spatial location in the middle of their original locations. The interaction kernel W is localized in space, but in general does not have to be isotropic. On the symmetry grounds we assume the following form (we assume 2D geometry and neglect higher-order anisotropic corrections):

$$W = \frac{1}{b^2\pi} \exp\left[-\frac{(\mathbf{r}_1 - \mathbf{r}_2)^2}{b^2}\right] (1 + \beta(\mathbf{r}_1 - \mathbf{r}_2) \cdot (\mathbf{n}_1 - \mathbf{n}_2)), \qquad (2.12)$$

with $b \approx l = const$. This form implies that only nearby microtubules interact effectively due to molecular motors. The $O(\beta)$ anisotropic term describes the dependence of the coupling strength on the mutual orientation of microtubules: "diverging" polar rods (such as shown in Fig. 2,a) interact stronger than "converging" ones. This is the simplest term yielding non-trivial coupling between density and orientation. The parameters b and β depend on the type of the motor. Our calculations show that β is small for kinesin and large for NCD molecular motors due to the difference in the motor's dwelling time at the end of the microtubules, [5].

We perform Fourier expansion in ϕ and truncate the series at $|n| > 2$. Then, $2\pi P_0$ gives the local number density $\rho(\mathbf{r}, t)$, and $P_{\pm 1}$ the local orientation $\tau(\mathbf{r}, t)$. Omitting tedious but straightforward calculations (see [13] for details), rescaling space by l, and introducing the dimensionless parameters $B = b/l, H = \beta l B^2$, we arrive at

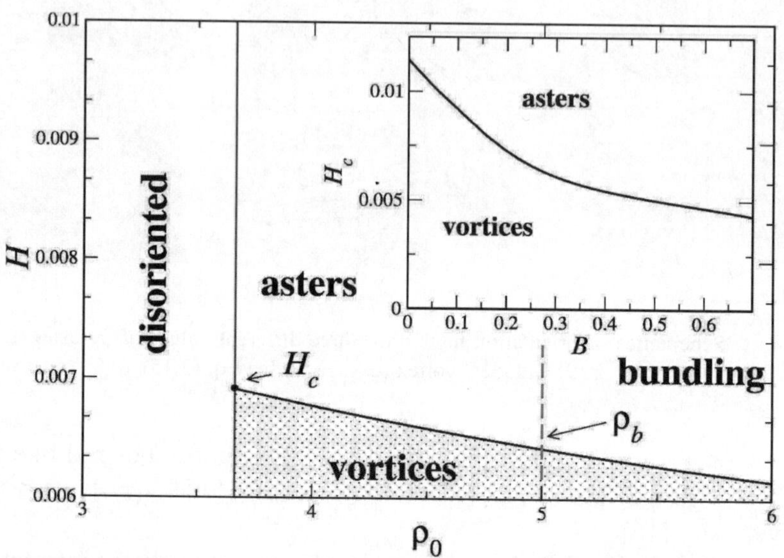

Figure 5.   Phase boundaries obtained form the linear stability analysis of aster solution for $B^2 = 0.05$, dashed line shows bundling instability limit $\rho_0 > \rho_b = 5$. Inset: Position of critical point H_c vs B at $\rho_0 = 4.5$.

$$\partial_t \rho = \nabla^2 \left[\frac{\rho}{32} - \frac{B^2 \rho^2}{16} \right] + \frac{\pi B^2 H}{16} \left[3\nabla \cdot \left(\tau \nabla^2 \rho - \rho \nabla^2 \tau \right) \right.$$

$$\left. + \; 2\partial_i \left(\partial_j \rho \partial_j \tau_i - \partial_i \rho \partial_j \tau_j \right) \right] - \frac{7\rho_0 B^4}{256} \nabla^4 \rho, \tag{2.13}$$

$$\partial_t \tau = \frac{5}{192} \nabla^2 \tau + \frac{1}{96} \nabla(\nabla \cdot \tau) + \epsilon \tau - A_0 |\tau|^2 \tau$$

$$+ \; H \left[\frac{\nabla \rho^2}{16\pi} - \left(\pi - \frac{8}{3} \right) \tau (\nabla \cdot \tau) - \frac{8}{3} (\tau \nabla) \tau \right]$$

$$+ \frac{B^2 \rho_0}{4\pi} \nabla^2 \tau. \tag{2.14}$$

The last two terms in Eqs. (2.13), (2.14) are linearized near the mean density, $\rho_0 = \langle \rho \rangle$. The last term in Eq.(2.13) regularizes the short-wave instability when the diffusion term changes sign for $\rho_0 > \rho_b = 1/4B^2$. This instability leads to strong density variations associated with the formation of microtubule bundles, see Fig. 5.

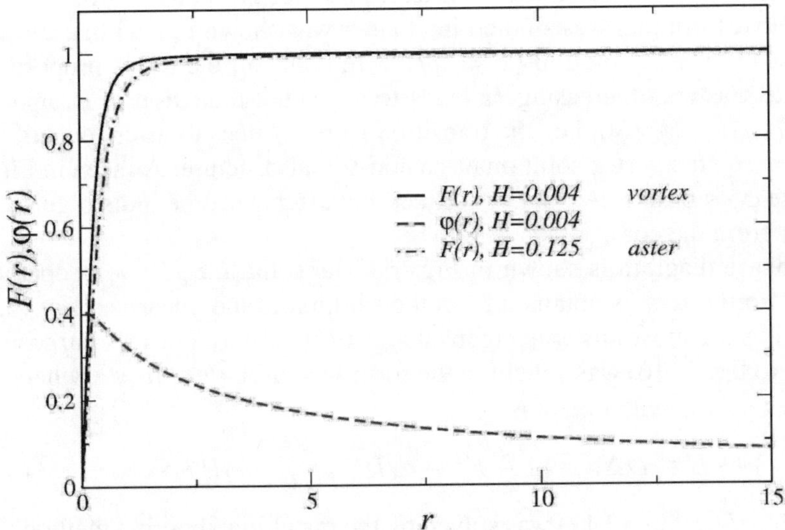

Figure 6.　　Stationary vortex and aster solutions $\tau_x + i\tau_y = F(r)\exp[i\theta + i\varphi(r)]$ to Eq. (3.16), for $\rho_0 = 4, B^2 = 0.05$.

3.　　Aster and Vortex Solutions

If $B^2 H \ll 1$, the density modulations are rather small, and Eq. (2.14) for orientation τ decouples from Eq. (2.13). It is convenient to rewrite Eq. (2.14) for complex variable $\psi = \tau_x + i\tau_y$ in polar coordinates r, θ:

$$\psi = F(r)\exp[i\theta + i\varphi(r)], \tag{3.15}$$

where the amplitude $F(r)$ and the phase $\varphi(r)$ are real functions. For the aster solution $\varphi(r) = 0$ and for the vortex $\varphi(r) \neq 0$, see Fig. 4. Asters and vortices can be examined in the framework of one-dimensional problem for $V = \sqrt{A_0}F(r)\exp[i\varphi(r)]$:

$$\partial_t V = D_1 \Delta_r V + D_2 \Delta_r V^* + \left(1 - |V|^2\right) V$$
$$-H\left(a_1 V \mathrm{Re}\nabla_r V + a_2 \partial_r V \mathrm{Re}V + \frac{a_2 V \mathrm{Im}V}{r}\right), \tag{3.16}$$

where $\Delta_r = \partial_r^2 + r^{-1}\partial_r - r^{-2}$, $\nabla_r = \partial_r + r^{-1}$, $D_1 = 1/32 + \rho_0 B^2/4\pi$, $D_2 = 1/192$, $a_1 = (\pi - 8/3)/\sqrt{A_0} \approx 0.321$, $a_2 = 8/3\sqrt{A_0} \approx 1.81$, and we rescaled time and space by $t \rightarrow t/\epsilon$ and $r \rightarrow r/\sqrt{\epsilon}$, respectively. The aster and vortex solutions for certain parameter values obtained by numerical integration of Eq. (3.16) are shown in Fig. 6. Vortices are observed only for small values of H and give way to asters for larger H. For $H = 0$, Eq.(3.16) reduces to a form that was studied in [17]. It was shown in [17] that the term $\Delta_r V^*$ favors vortex solution ($\varphi = \pi/2$). In contrast, the terms proportional to H select asters. Increasing H leads to a gradual reduction of φ, and at a finite $H_0(\rho_0)$ $\phi(r) = 0$, i.e. the transition from vortices to asters occurs. For $0 < H < H_0$, the vortex solution has a non-trivial structure. As seen in Fig. 6, the phase $\varphi \to 0$ for $r \to \infty$, i.e. vortices and asters become indistinguishable far away from the core.

The phase diagram is shown in Fig. 5. The solid line $H_0(\rho_0)$, separating vortices from asters, is obtained from the solution of the linearized Eq. (3.16) by tracking the most unstable eigenvalue λ of the aster. For this purpose the solution to Eq. (3.16) was sought in the form $V = F + iw \exp(\lambda t)$, where real w obeys $\hat{L} = \lambda w$ with operator

$$\hat{L} \equiv \bar{D}\Delta_r + \left(1 - F^2 - a_1 H \nabla_r F\right) - a_2 H F \nabla_r, \tag{3.17}$$

($\bar{D} = D_1 - D_2$). Eq. (3.17) was solved by the matching-shooting method. The dashed line corresponds to the orientation transition limit $\rho_0 = \rho_c$. The lines meet at the critical point $H_c = H_0(\rho_c)$ above which vortices are unstable for arbitrarily small $\epsilon > 0$. The phase diagram is consistent with experiments, see Ref. [5]: for low value of kernel anisotropy $H < H_c$ (possibly corresponding to kinesin motors) the increase of density ρ_0 first leads to the formation of vortices and then to asters. For $H > H_c$ (possibly corresponding to Ncd) only asters are observed.

For $H \neq 0$ well-separated vortices and asters exhibit exponentially weak interaction. For asters this follows from the fact that $\hat{L}$ is not a self-adjoint operator. The null-space of $\hat{L}^\dagger$ exponentially decays at large r, $w \sim \exp[-r/L_0]$ with the screening length in the original length units $L_0 = \bar{D}/a_2 H \sqrt{\epsilon}$ (see [18]). Thus, L_0 diverges for $H \to 0$, and at the threshold $\rho_0 \to \rho_c$.

We studied the full system (2.13),(2.14) numerically. The integration was performed in a two-dimensional square domain with periodic boundary conditions by a quasi-spectral method. For small H we observed vortices and for large H asters, in agreement with the above analysis. For the same integration time the number of vortices is typically smaller than the number of asters due to the fact that the screening length of asters is smaller. Therefore, vortices are more keen to annihilation then asters. As seen in Fig. 7, due to topological constraints, asters and vortices co-exist with a saddle-point type defect (total topological charge in a periodic domain must be zero). Asters have a unique orientation of the microtubules (here, towards the center). Asters with the opposite orientation of τ are unstable. In large domains asters form a disordered network of cells with a cell size of the order of L_0. The neighboring cells

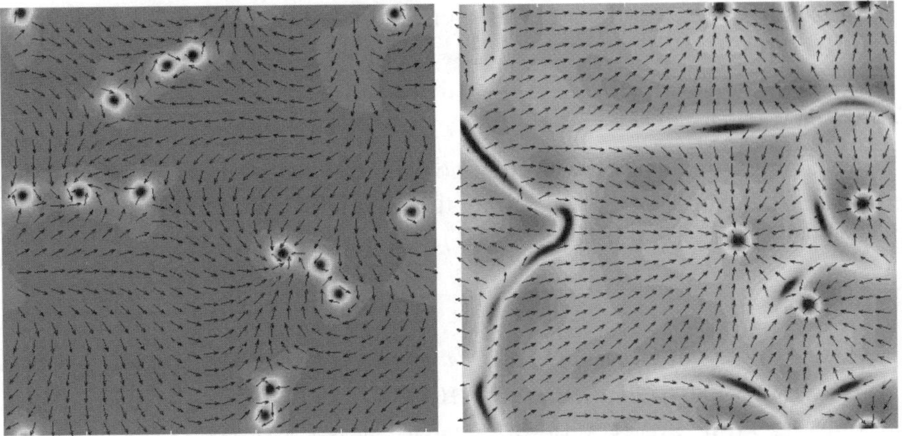

Figure 7. Orientation τ for vortices ($H = 0.006$, left) and asters ($H = 0.125$, right) obtained from Eqs. (2.13,2.14) from random initial conditions. Color code indicates the intensity of $|\tau|$ (red corresponds to maximum and blue to zero), $B^2 = 0.05$, $\rho_0 = 4$, domain of integration 80×80 units, time of integration 1000 units.

are separated by "shock lines" containing saddle-type defects. The pattern of asters resembles the "frozen" or spiral glass state [18, 19]. Starting from random initial conditions we observed initial merging and annihilation of asters. Eventually, annihilation slows down due to the exponential weakening of the aster interaction.

4.   Conclusion

We have derived continuous equations for the evolution of microtubule concentration and orientation. However, in this derivation we have neglected spatial non-uniformity of motors that can be an important factor in microtubule pattern formation [7]. The inclusion of the transport equations for the densities of free and bounded molecular motors in the model, in principle, is straightforward, but it can complicate the analysis (cf. [7–9]). We are planning to address this issue in our future work. We have found that an initially disordered system exhibits an ordering instability similar to the nematic phase transition in ordinary polymers at high density. The important difference is that here the ordering instability is mediated by molecular motors and can occur at arbitrarilly low densities of microtubules provided that the density and processivity of molecular motors are sufficiently high. At the nonlinear stage, the instability leads to the experimentally observed formation of asters and/or vortices.

Note that somewhat similar vortices were observed in a system of interacting granular rods [20, 17].

Acknowledgments

We thank Leo Kadanoff, Valerii Vinokur, and Jacques Prost for useful discussions. This work was supported by the U.S. DOE, grants W-31-109-ENG-38 (IA) and DE-FG02-04ER46135 (LT).

References

[1] Howard J. (2000), *Mechanics of Motor Proteins and the Cytosceleton*, Springer, New York.

[2] Takiguchi K. (1991), J.Biochem. (Tokyo) **109**, 250.

[3] Urrutia R., M. McNiven, J. Albanesi, D. Murphy, and B. Kachar, (1991), Proc. Natl. Acad. Sci. USA **88**, 6701.

[4] Nedelec F., T. Surrey, A.C. Maggs, and S. Leibler (1997), Nature (London) **389**, 305.

[5] Surrey T., F. Nedelec, S. Leibler, and E. Karsenti (2001), Science, **292**, 1167.

[6] Humphrey D., C. Duggan, D. Saha, D. Smith, and J. Kas (2002), Nature (London) **416**, 413.

[7] Nedelec F., T.Surrey, A.C.Maggs (2001), Phys. Rev. Lett. **86**, 3192.

[8] Lee H.Y. and M.Kardar (2001), Phys. Rev. E **64**, 056113.

[9] Kim J., Y. Park, B. Kahng, and H. Y. Lee (2003), J. Korean Phys. Soc., **42** 162.

[10] Sankararaman S., G.I. Menon, P.B.S. Kumar (2004), Phys. Rev. E **70**, 031905

[11] Kruse K., J.F. Joanny, F. Julicher, J. Prost, and K. Sekimoto, (2004), Phys. Rev. Lett. **92**, 078101.

[12] Liverpool T.B. and M.C. Marchetti (2003), Phys. Rev. Lett. **90**, 138102; Europhys. Lett. **69**, 846 (2005).

[13] Aranson I.S. and L.S. Tsimring (2005), Phys. Rev. E **71**, 050901

[14] Dogterom M. and S. Leibler (1993), Phys. Rev. Lett. **20**, 1347.

[15] Ben-Naim E. and P.L. Krapivsky (2000), Phys. Rev. E **61**, R5.

[16] Doi M. and S.F. Edwards (1988), *The Theory of Polymer Dynamics*, Clarendon Press, Oxford.

[17] Aranson I.S. and L.S.Tsimring (2003), Phys. Rev. E **67**, 021305.

[18] Aranson I.S. and L.Kramer (2002), Rev. Mod. Phys. **74**, 99.

[19] Brito C., I.S. Aranson and H. Chate (2003), Phys. Rev. Lett. **90**, 068301.

[20] Blair D. L., T. Neicu, and A. Kudrolli (2003), Phys. Rev. E **67**, 031303.

PHYSICS OF DNA

Maxim D. Frank-Kamenetskii
Center for Advanced Biotechnology and Department of Biomedical Engineering, Boston University, 36 Cummington St., Boston, MA 02215, USA

Abstract   The physical aspects of DNA structure and function are overviewed. Major DNA structures are described, which include: the canonical Watson-Crick double helix (B form), B', A, Z duplex forms, parallel-stranded DNA, triplexes and quadruplexes. Theoretical models, which are used to treat DNA, are considered with special emphasis on the elastic-rod model. DNA topology, supercoiling and their biological significance are extensively discussed. Recent developments in the understanding of molecular interactions responsible for the stability of the DNA double helix are presented.

Keywords: DNA; RNA; proteins; central dogma; DNA topology; knots; links; statistical mechanics; knot theory; Monte Carlo method; DNA supercoiling; base pairs; stacking interactions; triplexes; quadruplexes; DNA melting.

1.   Introduction

DNA plays a crucial role in all living organisms because it is the key molecule responsible for storage, duplication, and realization of genetic information. DNA is a heteropolymeric molecule consisting of residues (nucleotides) of four types, A, T, C and G. Fig. 1 shows the chemical structure of the DNA single strand and the complementary base pairs.

The genetic message is "written down" in the form of continuous text consisting of four letters (DNA nucleotides A, G, T and C). This continuous text, however, is subdivided, in its biological meaning, into sections. The most significant sections are genes, parts of DNA, which carry information about the sequence of amino acids in proteins.

The importance of the DNA molecule cannot be overestimated. It is therefore natural that the molecule has been attracting attention not only of biologists and physicians but also of chemists and physicists, even theorists (for an introduction into the field of DNA science, see Frank-Kamenetskii, 1997 [19]).

A. A. Golovin and A. A. Nepomnyashchy (eds.),
Self-Assembly, Pattern Formation and Growth Phenomena in Nano-Systems, 295–326.

2.   Major structures of DNA

In spite of the enormous versatility of living creatures and, accordingly, variability of genetic texts that DNA molecules in different organisms carry, they all have virtually identical physical, spatial structure: the double-helical B form discovered by Watson and Crick (1953) [89]. Sequences of the two strands of the double helix obey the complementarity principle. This principle is the most important law in the field of DNA, and probably, is the most important law of living nature. It declares that, in the double helix, A always opposes T and visa versa, whereas G always opposes C and visa versa (see Fig. 1b). In the cell, a copy of the gene is taken in the form of an RNA molecule, in the process known as transcription. This messenger RNA (mRNA) is utilized in the special molecular machines, ribosomes, to synthesize a protein molecule. This translation process is realized according to the genetic code. The unidirectional flow of information in the cell, DNA→RNA→ protein, constitutes the famous *central dogma* of molecular biology first formulated by Francis Crick (1916-2004). The RNA molecule is chemically very similar to DNA. The only significant difference is in the 2' position of sugar: RNA carries the OH group while in DNA the oxygen is absent. Another difference is that the role of T (thymine) in RNA is played by U (uracile). These two bases are different only by the methyl group.

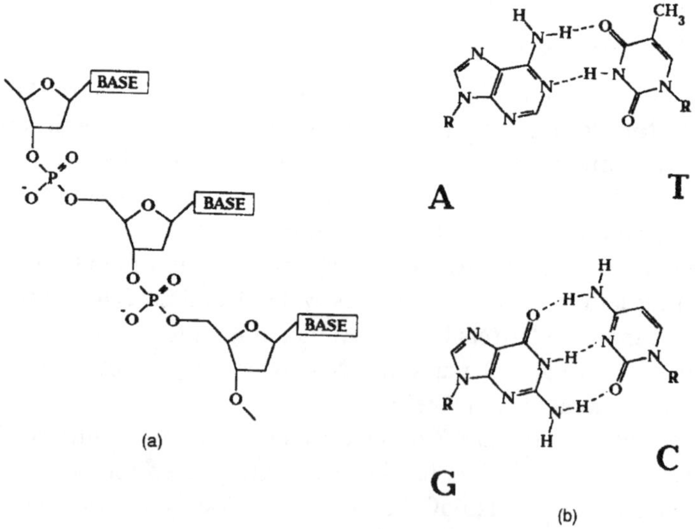

Figure 1.   (a) DNA single strand and (b) the Watson-Crick complementary base pairs.

2.1 B-DNA

B-DNA (Fig. 2a) consists of two helically twisted sugar-phosphate backbones stuffed with base pairs of two types, AT and GC. The helix is right-handed with 10 base pairs per turn. The base pairs are isomorphous: the distances between glycosidic bonds, which attach bases to sugar, are virtually identical for AT and GC pairs. Because of this isomorphism, the regular double helix is formed for an arbitrary sequence of nucleotides and the fact that DNA should form a double helix imposes no limitations on DNA texts. The surface of the double helix is by no means cylindrical. It has two very distinct grooves: the major groove and the minor groove. These grooves are extremely important for the functioning of DNA because, in the cell, numerous proteins recognize specific sites on DNA via binding with the grooves.

Each nucleotide has a direction and therefore the chemical direction is inherent in each of the DNA single strands. In the B-DNA double helix, the two strands have opposite directions. In B-DNA, base-pairs are planar and perpendicular to the axis of the double helix.

Under normal conditions in solution, often referred to as "physiological" (neutral pH, room temperature, about 200 mM NaCl), DNA adopts the B form. All available data indicate that the same is true for the totality of DNA within the cell. It does not exclude, however, the possibility that separate stretches of DNA carrying special nucleotide sequences would adopt other conformations.

2.2 B'-DNA

Up to now, only one such conformation is demonstrated, beyond any doubts, to exist under physiological conditions. When several A residues in one strand (and, accordingly, several T residues in the other DNA strand) occur, they adopt the B' form. In many respects, the B' form is similar to the classical B form but there are also significant differences. The main difference consists in the fact that base pairs in B'-DNA are not planar: They form a kind of propeller with a propeller twist of $20°$.

Stretches of A residues produce bends in the double helix (reviewed by Sinden, 1994 [66]). Such bends play a very important role in DNA functioning. Although the structural basis of these bends is not fully understood, the involvement of the B' form in the DNA bending is very probable. In spite of its importance, the B' form does not differ dramatically from the B conformation. Other helical conformations have been found in the course of DNA biophysical studies, which are significantly different from B-DNA.

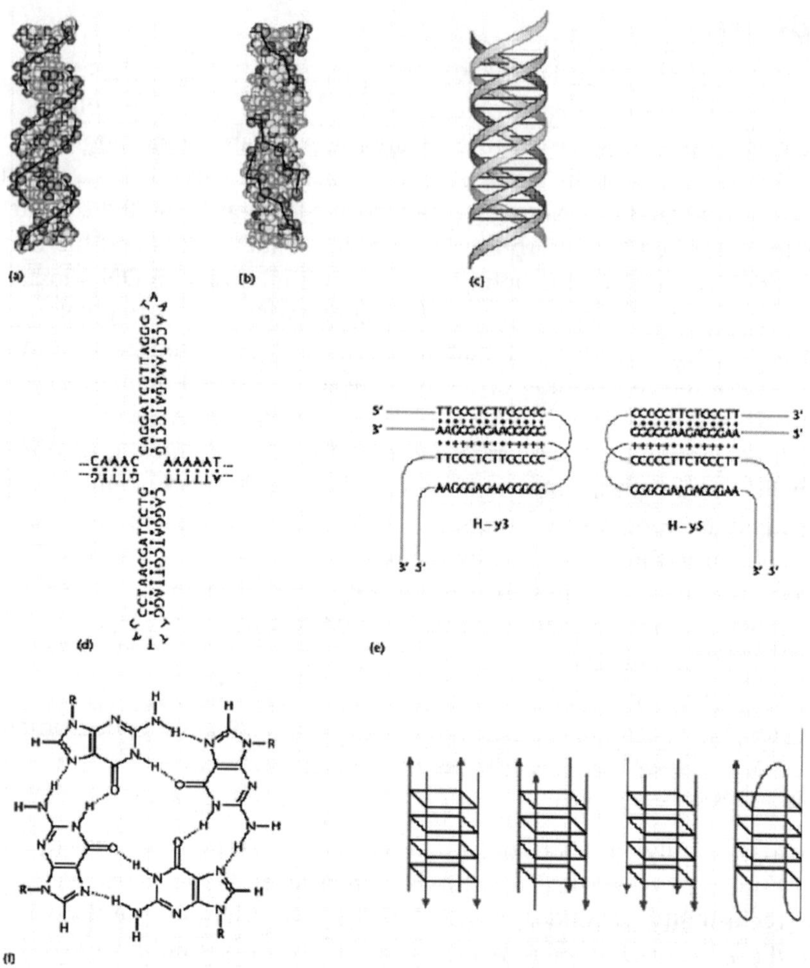

Figure 2. Various DNA structures. (a) B-DNA; (b) Z-DNA; (c) intermolecular triplex; (d) cruciform; (e) H-DNA (intramolecular triplex); (f) G-quadruplex and its various foldings.

2.3 A-DNA

Similarly to B-DNA, the A form can be adopted by an arbitrary sequence of nucleotides. Like in B-DNA, in A-DNA the two complementary strands are antiparallel and form right-handed helices. DNA undergoes transition from the B to A form under dehydration conditions (reviewed by Ivanov and Krylov, 1992 [34]). In A-DNA, the base pairs are planar but their planes make a considerable angle with the axis of the double helix. In doing so, the base pairs shift from the center of the duplex forming an empty channel in the center.

Although A-DNA plays a rather modest role in DNA functioning, the structure is very important because it is the predominant structure for duplex RNA. Right now we witness a new revolution in molecular biology due to the discovery of crucial role of small duplex RNA molecules in regulation of gene activity (see below).

2.4 Z-DNA

Z-DNA (Fig. 2b) presents the most striking example of how different from the B form the DNA double helix can be (Wang et. al., 1979 [87]). Although in Z-DNA the complementary strands are antiparallel like in B-DNA, unlike in B-DNA, they form left-handed, rather than right-handed, helices. There are many other dramatic differences between Z- and B-DNA (reviewed by Dickerson, 1992 [15]).

Not any sequence can adopt the Z form. To adopt the Z form, the regular alternation of purines (A or G) and pyrimidines (T or C) along one strand is strongly preferred. However, even this is not enough for Z-DNA to be formed under physiological conditions. Nevertheless, Z-DNA can be adopted by DNA stretches in cell due to DNA supercoiling (see Section 4.3.1). The biological significance of Z-DNA, however, remains to be elucidated.

2.5 ps-DNA

The complementary strands in a DNA duplex can be parallel. Such parallel-stranded (ps) DNA is formed most readily if both strands carry only adenines and thymines and their sequence excludes formation of the ordinary antiparallel duplex (reviewed by Rippe and Jovin, 1992 [58]). If these requirements are met, the parallel duplex is formed under quite normal conditions. It is right-handed, but the AT pairs are not the usual, Watson-Crick ones, but rather the so-called reverse Watson-Crick.

Some other sequences also can adopt parallel duplexes. For instance, at acidic conditions two strands carrying only C residues form parallel duplex consisting of protonated CC+ base pairs (see Section 2.7).

2.6 Triplexes

If DNA carries a homopurine-homopyrimidine tract, a homopyrimidine oligonucleotide can bind to this tract lying in the major groove (Fig. 2c) and forming Hoogsteen pairs with DNA bases (Moser and Dervan, 1987 [53]; Lyamichev et al., 1988 [45]; reviewed by Frank- Kamenetskii and Mirkin, 1995 [23]; Soyfer and Potaman, 1996 [69]). The canonical base-triads thus formed are shown in Fig. 3. In recent years, the variety of sequences, which have been found to be capable to form triplexes, has been significantly enlarged (reviewed by Frank- Kamenetskii and Mirkin, 1995 [23]; Soyfer and Potaman,

1996 [69]). In addition to intermolecular triplexes, intramolecular triplexes or H-DNA (Fig. 2e) can be formed under certain conditions (see Section 4.4.3).

2.7 Quadruplexes

Of all nucleotides, guanines are the most versatile in forming different structures. They may form GG pairs but the most stable structure, which is formed in the presence of monovalent cations (especially potassium), is G4 quadruplex (see Fig. 2f). G- quadruplexes may exist in a variety of modifications: all-parallel, all-antiparallel and others (reviewed by Sinden, 1994 [66]). As a result, G-quadruplexes are easily formed both inter- and intramolecularly, again with a variety of modifications.

A totally unusual quadruplex structure was discovered by Gehring et al. (1993) [28]. It contains two hemiprotonated parallel-stranded duplexes consisting of CC+ pairs. The two parallel-stranded duplexes are associated in a mutually antiparallel manner so that CC+ base pairs from one duplex are "layered" by CC+ pairs from the other duplex, thus alternating along the structure (Gehring et. al. (1993) [28]; Chen et. al. (1994) [10]).

3.   DNA functioning

According the central dogma, there are two fundamental processes, in which DNA participates: replication and transcription. They are both far from trivial to be executed because the DNA chemical structure imposes serious constrains on how DNA can function. A major limitation consists in the fact that DNA and RNA chains can be extended only from one of two ends, the so-called 3'-end, which carries the 3'-OH group. Another limitation in case of replication is less fundamental: DNA polymerases have not acquired in the course of evolution the ability to start DNA synthesis without the presence of the 3'-OH group. In other words, in addition to the template and the precursors of DNA (dNTPs) the DNA polymerase always requires a primer to start synthesis of the new DNA chain. It is not the case for transcription: RNA polymerase can start RNA synthesis on DNA template without any primer.

3.1 Replication

Due to all these constrains, the DNA replication becomes a complicated process (see schematics in Fig. 4). The process begins when a special type of RNA polymerase called primase synthesizes short RNA primers, which then extended by DNA polymerase. On one of the two strands, where primer extension proceeds in the 5'→3' direction, the synthesis is straightforward (it is called the leading strand, see Fig. 4). But how can primer be extended on the opposite, so-called lagging, strand as a template? It is done in a very non-elegant way: short pieces of DNA are synthesizes in the 5'→3' direction, i.e.

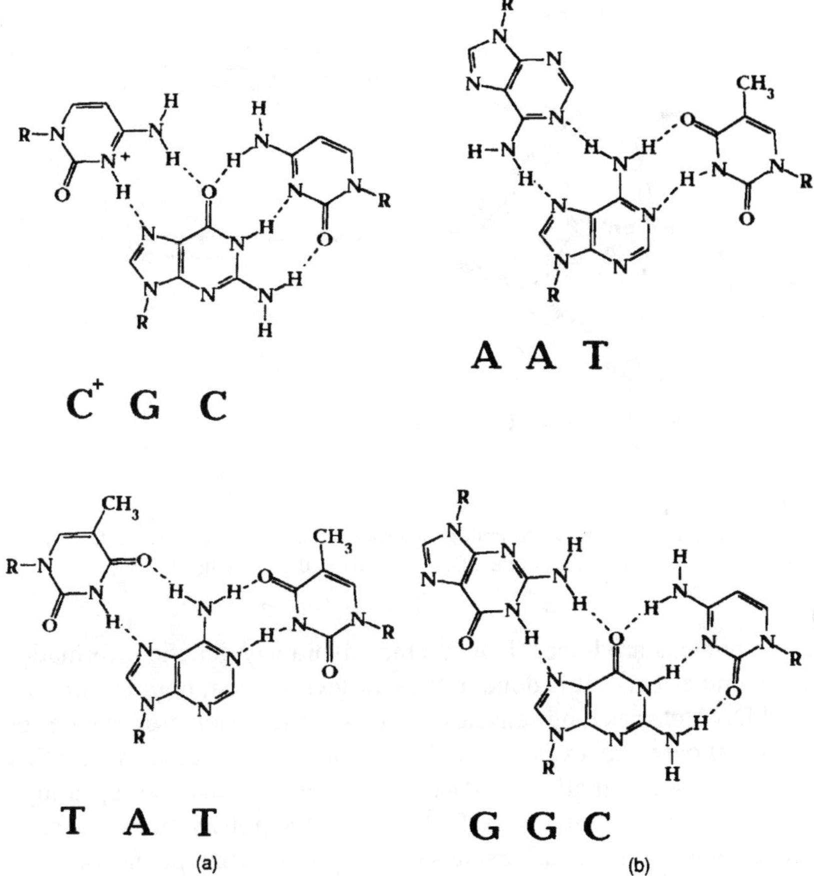

Figure 3. The structure of various base-triades.

in the direction opposite to the direction of the replication fork movement. A complicated task of removing RNA primers and repairing the gaps is required to finalize replication. But, again because of the constrains, the full copy of original DNA cannot be obtained: RNA primer at the 5'-end of the newly synthesized chain cannot be replaced with DNA since no 3'-end becomes available after the primer is removed.

Therefore, each cycle of replication ends up with duplex DNA carrying a 3' overhang at one of two termini of the newly synthesized molecule. How does living nature deal with this problem of ends? Interestingly, prokaryotes (bacteria) and eukaryotes (higher organisms) deal with the problem very differently. Prokaryotes rid out of ends entirely keeping DNA in the form of circular molecules. This allows them to avoid the end problem. DNA molecules in the

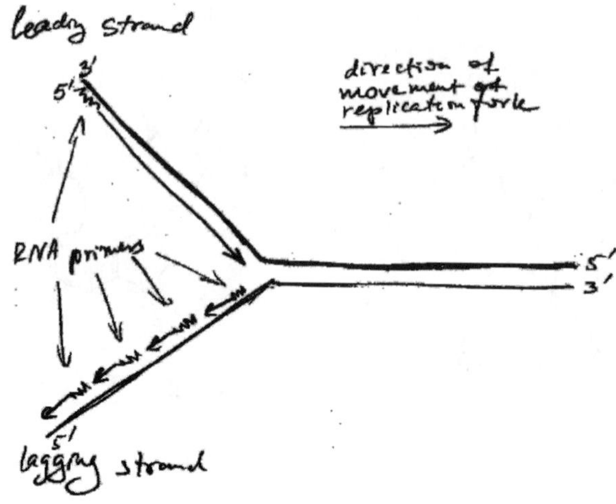

Figure 4. Schematics of the DNA replication process in the cell (see the text for explanation). DNA strands are shown in straight lines, DNA strands in the wavy lines.

eukaryote nucleus are linear. How do they avoid any genetic information loss due to the end effects? It is done in the two levels. First, both termini of chromosomal DNA carries long senseless repeats, called telomeres, which serve as buffers and allow many cycles of replication to occur before any genetic information may be lost. In all vertebrates, including humans, the repeating motif in one of two DNA strands is: 5'TTAGGG3'. Obviously, however, sooner or later the telomere repeats are exhausted. A special enzyme has been discovered, which extends the telomeric repeats (Chan and Blackburn, 2004 [9]). The enzyme uses the 3'-overhanging telomeric repeats at the ends of chromosomes as primers but where does it find the template? Astonishingly, the enzyme itself carries the template in the form for a RNA molecule, which has several repeats of the 5'CCCUAA5' sequence serving as a template for telomere extension. Interestingly, telomerase is inactive in somatic cells and is active only in gametes. Therefore, during the individual development telomeres must be shortened. This is exactly what is observed for humans and some researchers believe that this chromosomal DNA shortening during the life span may be one of the mechanisms of aging (see Epel et al., 2004 [17]).

3.2 Transcription and gene expression

The DNA molecule is a depository of information about the entire organism and its development, and virtually all cells of a given organism carry the same set of DNA molecules (the whole genome). However, at each particular moment and in each particular tissue (and even in the individual cell) some genes are switched on and some are switched off. This regulation of gene expression is essentially what the biology is all about. As we understand now, there are two levels of regulation of gene expression. The first one is universal for all living creatures and is a part of the central dogma. It consists in activation or repression of initiation of mRNA copying of particular gene by RNA polymerase. Although this part of central dogma has been subjected to a very serious revision with respect to eukaryotes after discovery of the exon-intron organization of eukaryotic genomes and the process of splicing, in the very core the principle of regulation remains unchanged. The second mechanism, which is specific only for eukaryotes, had been totally overlooked before five years ago. It is the so-called RNA interference (RNAi) pathway, whose discovery constitutes no less than a new revolution in molecular biology, which is well under way right now (see, e g., Novina and Sharp, 2004 [54]; Matzke and Bircher (2005) [50]). This is only a negative regulation consisting in the specific degradation of particular mRNA molecules on their way between the nucleus, where they are synthesized via the transcription process, and cytoplasm, where they are ultimately used in the translation process (protein syntheses on ribosomes). The specificity of the degradation process is realized via synthesis of short (about 20-bp-long) duplex RNA molecules, in which the sequence of one of the strands is identical to a part of the mRNA molecule destined to be degraded. A special complex of proteins, which included nucleases, secures the recognition of mRNA and its subsequent cleavage.

4.   Global DNA conformation

4.1. Elastic rod model of DNA

DNA behaves as an almost ideal polymer chain. No other polymer molecule is closer to the ideal polymer chain than the DNA double helix. Due to unusually high bending rigidity of DNA, the ratio of its persistence length, a, to its diameter, d, is very high. This leads to very small, sometimes negligible, excluded volume effects under a variety of ambient conditions, not only at the θ-point, like with ordinary polymers (see, e.g., Grosberg and Khokhlov, 1994 [30]). This unusual rigidity stems from the fact that DNA consists of two, rather than one, polymer chains. A common mechanism of polymer flexibility, due to rotation around single bonds, is excluded for the double helix. It exhibits bending flexibility only due to accumulation of small changes of angles between adjacent base pairs. As a result, the DNA double helix is best modeled as an elastic rod (see Fig. 5a). Within first approximation, one can neglect the sequence dependence of the DNA bending and torsional rigidities and treat DNA as a homogeneous and isotropic elastic rod. This model proved to be a remarkably good first approximation to treat global DNA macromolecular properties.

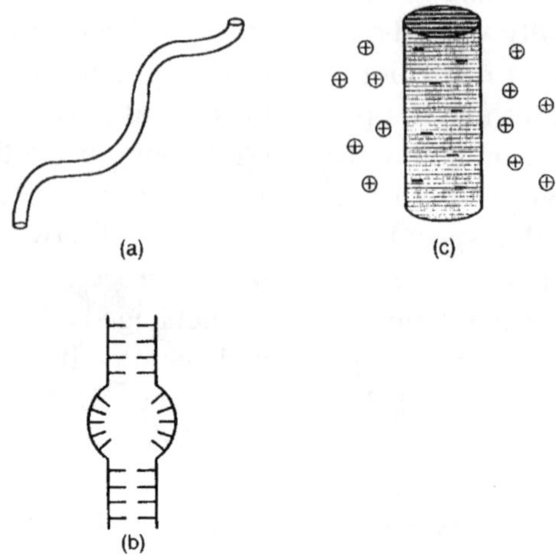

Figure 5.   Theoretical models of DNA: (a) elastic-rod model; (b) helix-coil model; (c) poly-electrolyte model.

Within the framework of this model, the DNA chain is characterized by three parameters: the bending rigidity, measured in terms of the persistence length, a, or the Kuhn statistical length ($b = 2a$); the torsional rigidity C; the DNA effective diameter, d. Numerous properties of linear and circular DNA molecules can be quantitatively understood in terms of the elastic rod model and the same set, under given ambient conditions, of the above three parameters.

Ambient conditions, especially the concentration of counter-ions in solution, may significantly affect some DNA parameters. This is the case for the DNA effective diameter. Because DNA is a highly charged polyion, the excluded volume effects strongly depend on the screening of Coulomb interaction between DNA segments approaching each other. As a result, the DNA effective diameter significantly exceeds its geometrical diameter of 2nm at a low concentration of counter-ions in solution. By contrast, DNA bending and torsional rigidities are ionic-strength-independent within a wide range of ambient conditions.

For the theoretical treatment of statistical-mechanical properties of DNA within the elastic-rod model, a Metropolis-Monte-Carlo-type approach was elaborated by Frank-Kamenetskii et al. (1985) [22]. In this approach, the DNA chain is modeled as a series of straight segments so that each Kuhn length contains k such segments. The total elastic energy is the sum of terms, each of which corresponds to a pair of adjacent straight segments and quadratically depends on the angle between them (see Frank-Kamenetskii et al., 1985 [22]; Vologodskii and Frank-Kamenetskii, 1992 [81] for details). The final results are obtained, within the framework of the model, as asymptotic ones for the large k values. Fortunately, all characteristics we studied leveled off very quickly with increasing k so that $k = 10$ proved to be a quite sufficient value to get very reliable quantitative asymptotic results (see Fig. 6). Asymptotically, this model corresponds to the elastic-rod model of the polymer chain (it is also often referred as the worm-like model; see, e.g., Grosberg and Khokhlov, 1994 [30]).

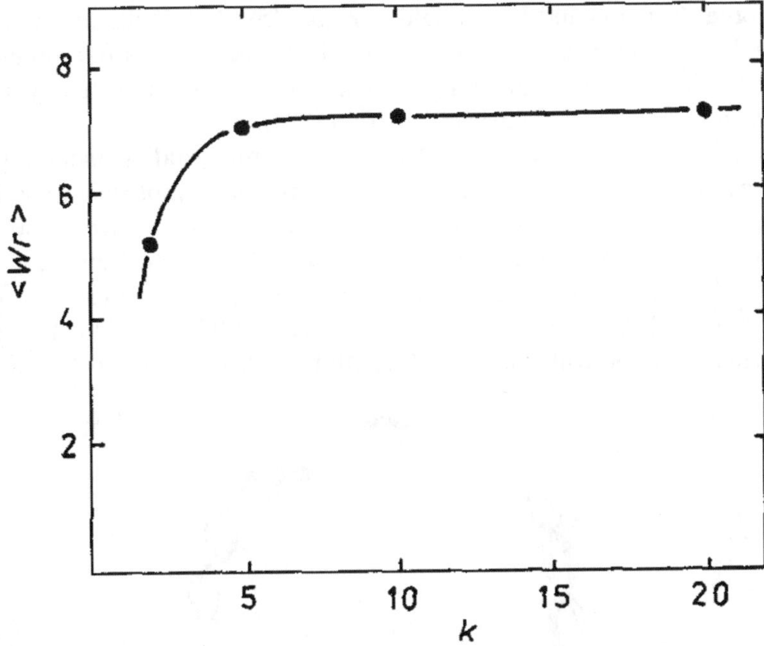

Figure 6. Typical results of Metropolis-Monte Carlo calculations on the dependence on the number of straight segments per Kuhn length, k, of a mean quantity (the mean writhing number, Wr, see Section 4.3.2.1, in the particular case) for closed polymer chain. The data are from Vologodskii and Frank-Kamenetskii (1992) [81].

4.2. Linear DNA

DNA is a unique object for experimental studies of a virtually ideal macromolecular coil. In addition to the already mentioned fact of an exceptionally high a/d ratio, DNA samples are strictly monodisperse and the length of the molecule can be varied in a very wide range: From below one persistence length up to hundreds of persistence lengths. Smith et al. (1992) [67] performed remarkable measurements of strain/extension relationship on single DNA molecules. Bustamante et al. (1994; 2003) [7, 8] showed that experimental data agree with theoretical predictions obtained within the framework of the elastic-rod model. After the DNA molecule was fully extended, further increase of force led to a sharp transition to what was first interpreted as a new more extended DNA conformation, in which the average distance between adjacent base pairs was 1.6 times larger than in the normal B-DNA (Smith et al., 1996 [68]; Strick et al., 1996 [71]). However, further experimental studies have strongly indicated that this extended conformation actually results from separation of DNA strands, i.e. the transition is just DNA melting under the external force (Rouzina and Bloomfield, 2001a,b [59, 60]; Williams et al., 2002 [92]; Williams and Rouzina, 2002 [91]). Most recently, this interpretation of the DNA behavior under stretching force has been fully supported by molecular dynamic simulations (Harris et al., 2005 [32]).

Normally, linear DNA is in the B-form. Numerous studies, including accurate measurements of force-extension curves in single-molecule experiments (see Bustamante et al. (2003) [8] and references therein) have made it possible to determine an accurate value of the DNA persistence length, a, which proved to be very close to 50 nm (Hagerman (1988) [31]; Taylor and Hagerman (1990) [72]; Vologodskaia and Vologodskii (2002) [75], Du et al. (2005) [16]). Therefore, the Kuhn statistical length for DNA is equal to 100 nm.

Figure 7.   In closed circular DNA, two complementary strands form linkage of a high order.

4.3. DNA topology

It was unexpectedly found in 1963 that DNA exists in certain viruses in a closed circular (cc) form. In this state, the two single strands of which the DNA consists are each closed on them- selves. Fig. 7 schematically illustrates ccDNA. One can see that the two complementary strands in ccDNA proved to be linked. They form a high-order linkage (of the order of N/γ_0, where N is the number of pairs in the DNA and γ_0 in the number of base pairs per turn of the double helix). Initially, the discovery of circular DNA was not seen to be very significant, since this form of DNA was regarded as exotic. However, over the years, the cc form of DNA was discovered in an even greater number of organisms. Currently, it is generally acknowledged that precisely this form of DNA is typical of prokaryotic DNAs, and also of the cytoplasm DNAs of animals. Also most virus DNAs pass through a stage of the cc form in the course of infection of cells. As we discussed in Section 3, the reason for such abundance of circular DNAs lies in the fact that circularization is one of two basic mechanisms developed in the course of evolution to deal with fundamental constraints imposed on DNA replication by DNA chemical structure.

The discovery of ccDNA has led to the formulation of fundamentally new problems, since it turned out that many of the physical properties of the closed circular form differ radically from those of the linear form. The difference between the properties of these two forms of DNA is not at all due to the existence of end effects in the one case but not in the other.

There are two levels of DNA topology. First, ccDNA as a whole can be unknotted (form the trivial knot, or unknot) or form knots of different types (see Fig. 8). Secondly, two complementary strands in DNA are linked with each other topologically (Fig.7).

4.3.1. Knots. The firstproblem that arises in theoretical analysis of ring polymer chains, including ccDNA, is formulated in the following way. Let a ring molecule be formed by fortuitous closure of a linear molecule consisting of n segments. What is the probability of forming a knotted chain, i.e., a nontrivial knot? This problem has been clearly formulated by Max Delbriick and solved by our group (Vologodskii et al., 1974 [76]; Frank-Kamenetskii et al., 1975 [21]).

4.3.1.1. Statistical mechanics of knots. To solve the problem of statistical mechanics of knots, one needs, first of all, a knot invariant. Indeed a closed chain can be unknotted or can form knots of different types. The very beginning of the table of knots is shown in Fig.8. However, an analytical expression for the knot in variant is unknown. Therefore, we had to use a computer and analgebraic invariant elaborated in the topological theory of knots. We found that the most convenient in variant was the Alexander polynomial (reviewed by Frank-Kamenetskii and Vologodskii, 1981 [24] and Vologodskii and Frank-Kamenetskii, 1992 [81]).

The next problem consisted of generating closed polymer chains. In our first calculations, we simulated DNA as a freely-joint polymer chain. Several methods exist to generate exclusively closed chains for this model (Frank-

Figure 8.   Knots.

Kamenetskii and Vologodskii, 1981 [24]; Vologodskii and Frank-Kamenetskii, 1992 [81]). Using these methods and teaching the computer to calculate the Alexander polynomials and therefore to distinguish the knots of different type, we could calculate the knotting probability.

Analogous calculations have been performed later by other researchers (reviewed by Frank-Kamenetskii and Vologodskii, 1981 [24]; see also more recent papers by Deruchi and Tsurusaki, 1993a, b, 1994 [12–14]). The data on

the relationship between the probability of knot formation and the number of
Kuhn lengths in the chain are shown in Fig. 9. Remarkably, the data are well
approximated by a simple equation:

$$P(n) = 1 - \exp(-\kappa n),$$

where $\kappa = 3 \cdot 10^{-3}$.

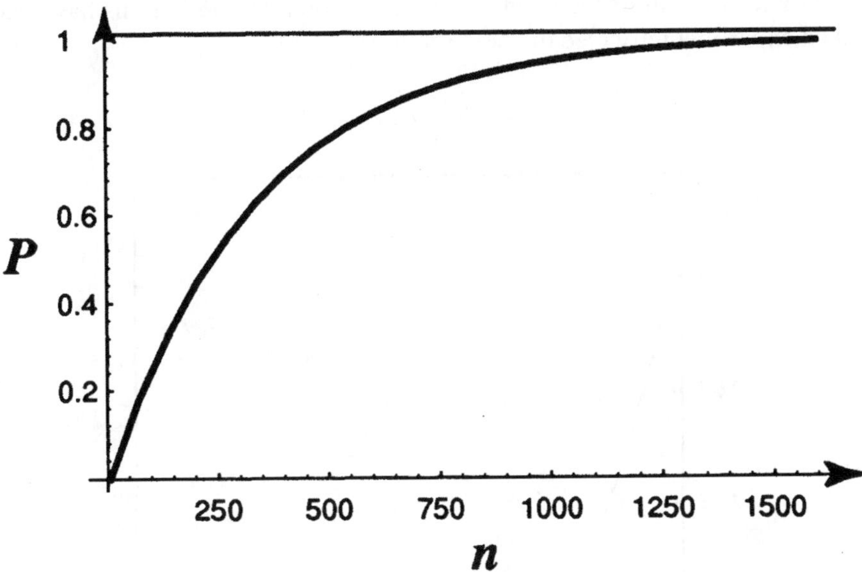

Figure 9.   Probability of knot formation, P, as a function of the number n of Kuhn statistical
lenghts for an infinitely thin polymer chain.

The above calculations were performed under the assumption that the polymer chain under consideration has zero diameter. In the very early stage
of our study of knots we already realized that the excluded volume effects
should significantly decrease the knotting probability (Vologodskii et al., 1974
[76], Frank-Kamenetskii et al., 1975 [21]). However, the knotting probability
proved to be even more sensitive to the excluded volume effects than we orig-
inally anticipated so that these effects could not be neglected even in the case
of DNA.

We arrived at this conclusion using the Metropolis-Monte Carlo approach to
calculate DNA topological characteristics within the framework of the elastic-
rod model (Frank- Kamenetskii et al., 1985 [22]).

This approach made it possible to simulate the behavior of DNA molecules
allowing for excluded volume effects (Klenin et al., 1988 [36]). So we arrived
at quantitative predictions about the dependence of knotting probability on the

DNA effective diameter, d. Fig.10 shows the results. One can see a dramatic dependence of the P value on d. Even in case of DNA geometric diameter, which corresponds to $d = 0.02$ in Fig. 10, the knotting probability is already significantly lower than for $d = 0$. However, in reality the effective diameter of DNA noticeably exceeds its geometric value due to the excluded volume effects, which are determined by the screened electrostatic interactions between highly charged DNA segments. Therefore, the d value can be varied by changing the ionic strength of the solution. These theoretical predictions have been checked experimentally (see the next section).

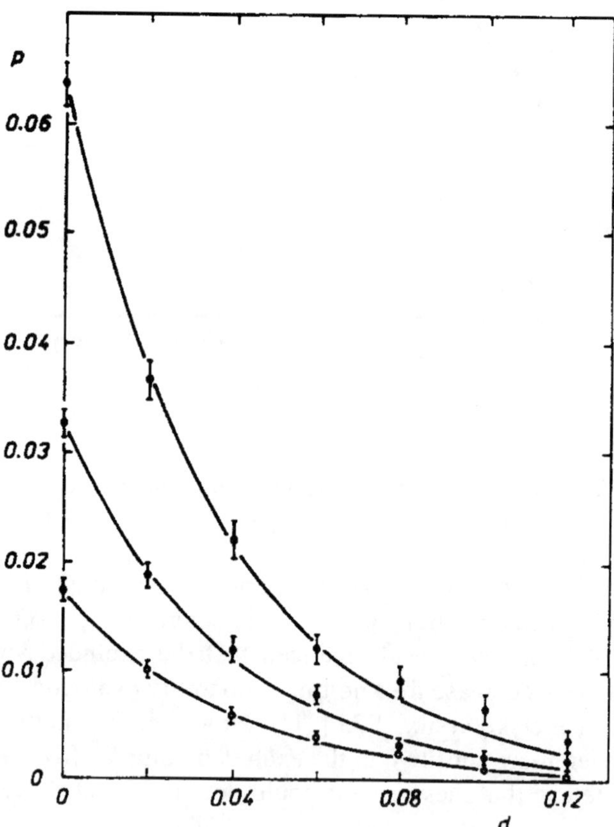

Figure 10.   Dependence of the equilibrium fraction of knotted molecules on DNA effective diameter, d, for closed DNA containing 14 Kuhn lengths (lower curve), 20 Kuhn lengths (middle curve) and 30 Kuhn lengths (upper curve). The data are from Klenin et al. (1988) [36]. The diameter is measured in Kuhn lengths; so, to obtain the d value in nanometers one has to multiply the figures on the abscissa by the factor of 100.

4.3.1.2. Knotted DNAs. As mathematical objects, knots and links have been studied already for more than hundred years The question of possible existence of such topological states in molecules has been raised at least since the late-1940s (see Frisch, 1993 [26]). It has acquired special interest since the disery of closed circular DNA molecules. The calculations of the probability of knot formation upon closing a polymer chain (see Section 4.3.1.1) have posed the problem of the possible existence of knotted DNAs. The results indicated that the equilibrium fraction of knotted DNAs must be appreciable for circular DNAs containing more than about 10^4 base pairs (30 Kuhn lengths). In most cases, DNA molecules have even greater length, and a hypothesis has been put forward of the existence in the cell of special mechanisms that prevent the formation of knotted DNAs (Frank-Kamenetskii et al., 1975 [21]). In fact, in the course of replication of a knotted chain (at least for some types of knots) the daughter strands cannot separate. That is, the replication of knotted DNAs involves serious problems.

Knotted molecules were first detected in preparations of single-stranded circular DNAs after they had been treated under special conditions with a type I topoisomerase (Liu et al., 1976 [41]). This was the first case when a knotted molecule was observed. However, the problem of knotting of normal, double-stranded DNAs continued to be very intriguing. It turned out that there is a special subclass of topoisomerases called type II topoisomerases, which are capable of untying and tying knots in ccDNAs. Moreover, these enzymes catalyze the formation of catenanes from pairs or from a larger number of molecules of ccDNA. Here entire networks are formed, similarly to those observed in vivo in kinetoplasts.

In contrast to type I topoisomerases, type II topoisomerases break, and then rejoin both strands of DNA molecules. It has been shown that the enzyme "draws" a segment of the same or of another molecule lying nearby through the "gap" that is formed in the intermediate state between the ends that arise through breakage. Thus, the type II topoisomerases catalyze the process of mutual penetration of segments of the double helix through one another. This process has been elaborated in details by Wang and his collaborators in their remarkable studies of the enzyme by various methods including X-ray crystallography (Berger et al., 1996 [5]; Wang, 1996 [86]).

Almost 20 years after theoretical estimations of the probability of DNA knotting were first published (Vologodskii et al., 1974 [76]; Frank-Kamenetskii et al., 1975 [21]), quantitative experimental data have been reported (Rybenkov et al., 1993 [61]; Shaw and Wang, 1993 [65]) which fully agreed with the theory. In these experiments, the equilibrium fraction of knotted DNA molecules at various ionic conditions was quantitatively measured while molecules carrying "cohesive" ends randomly closed, in the absence of any proteins. Comparing the fraction with theoretical predictions of Klenin et al. (1988) [36, the value of DNA effective diameter, d, was determined as a function of salt concentration. The obtained dependence proved to be in complete quantitative agreement with theoretical predictions of Stigter (1977) [70], which were based on the polyelectrolyte model of DNA schematically shown in Fig 5c.

While studying knotting and catenation of DNA molecules under the action of type II topoisomerases, Cozzarelly, Vologodskii and their co-workers made an astonishing discovery. They found that these enzymes are actually molecular motors, which directionally simplify DNA topology with respect to knots and links (Rybenkov et al., 1997 [62]). But how could that be? Are not topoisomerases much smaller than DNA molecules they untie? How such global characteristic as topology can be "sensed" by proteins? These intriguing questions are currently under intense theoretical and experimental investigation (Vologodskii et al., 2001 [83]).

4.3.2. Torus links and ribbons. From the schematics in Fig. 7 it is clear that the two complementary strands of DNA form a link, in the topological sense One can present a table of links similar to the table of knots in Fig.8 (see Frank-Kamenetskii and Vologodskii, 1981 [24]). However, because the two complementary strands of DNA are attached to each other forming the duoble helix, the links which DNA can form, belong to a subclass of all possible links. Namely, they form a class of the so-called torus links because the two strands could be put into a torus. For torus links, the well-known Gauss integral, which defines the linking number value, Lk, is a strict topological invariant (see Frank-Kamenetskii and Vologodskii, 1981 [24]).

There is another viewpoint on the torus links. The two strands in this case could be treated as the edges of a ribbon. Therefore, the topological theory of torus links is actually the theory of ribbons.

4.3.2.1. DNA supercoiling. The application of the topological ideas to studying the properties of ccDNA was started by Fuller (1971) [27] when he applied the results of the ribbon theory to analyzing the properties of these molecules. According to this theory (White, 1969 [90]; a simple derivation can be found in Frank-Kamenetskii and Vologodskii, 1981 [24]), besides the topological characteristic of a ribbon, the Lk value, two differential-geometric characteristics play an important role, the twist, Tw, of the ribbon, and its writhing, Wr. All three characteristics are interrelated by the condition:

$$Lk = Tw + Wr. \tag{4.1}$$

The ccDNA is generally not characterized by the total quantity Lk, but by the number of excess turns (the number of supercoils τ):

$$\tau = Lk - N/\gamma_0. \tag{4.2}$$

The number of base pairs per turn of the double helix, γ_0, is rigorously fixed under given ambient conditions. However, upon changing the ambient conditions (temperature, composition of solvent, etc.), it can vary. Therefore, the number of supercoils τ, in contrast to the Lk value, is a topological invariant of DNA only under fixed ambient conditions.

Very valuable information on the energy and conformation characteristics of ccDNA has arisen from experiments in which the value of Lk could vary, and the equilibrium distribution of the cc molecules over the Lk value was studied. The most convenient way to vary Lk is to employ special enzymes we have already mentioned above, the topoisomerases. The studies under discussion employed type I topoisomerases, which alter the topological state of ccDNA by breaking and rejoining only one of the strands of the double helix. The mechanism of action of these enzymes has been elaborated in great details (Wang, 1996 [86]). These enzymes relax the distribution of the molecules over the Lk value to its equilibrium state. The very sensitive gel- electrophoresis method was used to analyze the distribution of the ccDNA molecules over the Lk value. This method can easily separate two molecules of ccDNA that differ in Lk just by one.

Naturally, the maximum of the equilibrium distribution always corresponds to $\tau = 0$ because this minimizes the elastic energy. Note that, although the quantity τ can only adopt discrete values that differ by no less than unity, it is not required to be an integer. Therefore, as a rule, molecules having $\tau = 0$ do not appear in a preparation. A distribution, in which the molecules having positive and negative values of τ are separated, is obtained when the electrophoresis is performed under conditions differing from those under which the reaction with the topoisomerase is conducted. The change in the conditions means that we must substitute some other value γ_0' instead of γ_0 in Eq. (4.2) without changing the Lk value. This means that the entire distribution is shifted by the amount of $\delta\tau = N[(1/\gamma_0) - (1/\gamma_0')]$. Then the molecules that had the τ value in the original distribution will have the values $\tau' = \tau + \delta\tau$ in the new distribution. If the value $\delta\tau$ is large enough, all of the topoisomers are well separated.

Experiments have shown that the obtained distribution is always normal (Depew and Wang, 1975 [11]; Pulleyblank et al., 1975 [57]; Horowitz and Wang, 1984 [33]). The variance, $\langle \tau^2 \rangle$, of this normal distribution was measured for different DNAs. These experiments have played a very important role in studying the physical properties of ccDNA. They made it possible to determine the free energy of supercoiling, which is directly connected to the variance:

$$G = k_B T \tau^2 / 2\langle \tau^2 \rangle = 1100 k_B T N^{-1} \tau^2, \qquad (4.3)$$

where k_B is the Boltzmann constant and T is the absolute temperature.

4.3.2.2. Theoretical understanding of DNA supercoiling. Quantitative explanation and prediction of a variety of DNA topological characteristics, most notably the data on the equilibrium knotting probability and on the equilibrium distribution of ccDNA over topoisomers, demonstrated a remarkable success of the DNA elastic-rod model. The model also proved to be extremely successful in theoretical treatment of the phenomenon of DNA supercoiling.

In its traditional form, the Monte Carlo approach does not permit simulating highly or even moderately supercoiled molecules because the probability of their occurrence due to thermal motion is negligible. We have extended our Metropolis-Monte Carlo calculations (Frank- Kamenetskii et al., 1985 [22]) to make it possible to generate supercoiled DNA molecules with arbitrary supercoiling (Klenin et al., 1991 [37]; Vologodskii et al., 1992 [82]). The computational procedure is described at length by Vologodskii and Frank-Kamenetskii (1992) [81].

An ensemble of chains generated by the approach is used to calculate different averaged characteristics of supercoiled molecules and enables one to obtain theoretical images of supercoiled molecules. Fig. 11 presents examples of such images. Our theoretical predictions about the shape of supercoiled DNA molecules agree with most available experimental data.

Marko and Siggia (1994, 1995) [47, 48] developed an approximate analytical theory describing the structures of supercoiled DNA molecules. This theory provides insight into the role of entropic effects in the shapes of supercoiled DNA molecules of the type shown in Fig. 11.

4.4. Breakdown of the elastic-rod model: DNA unusual structures induced by supercoiling

With increasing negative supercoiling, the elastic-rod model breaks down. This happens when the elastic energy stored in the form of bending and torsional deformations exceeds the energy necessary for local formation of unusual DNA structures. These unusual structures release superhelical stress thus decreasing the total energy of the molecule. The competition between different unusual structures for the total pool of the superhelical energy dramatically depends on the presence of special sequence motifs, which favor various unusual structures. Before these unusual structures (cruciforms, Z-DNA, H-DNA) were discovered, the main reason for breakdown of the double helix was believed to be the local melting (separation of DNA complementary strands, see Section 5). Anshelevich et al. (1979) [1] and Vologodskii et al. (1979) [77] were the first to include DNA melting and cruciform formation into comprehensive statistical mechanical treatment of supercoiled DNA. As other unusual structures emerged and their energy parameter became available, the treatment has been modified accordingly (Vologodskii and Frank-Kamenetskii,

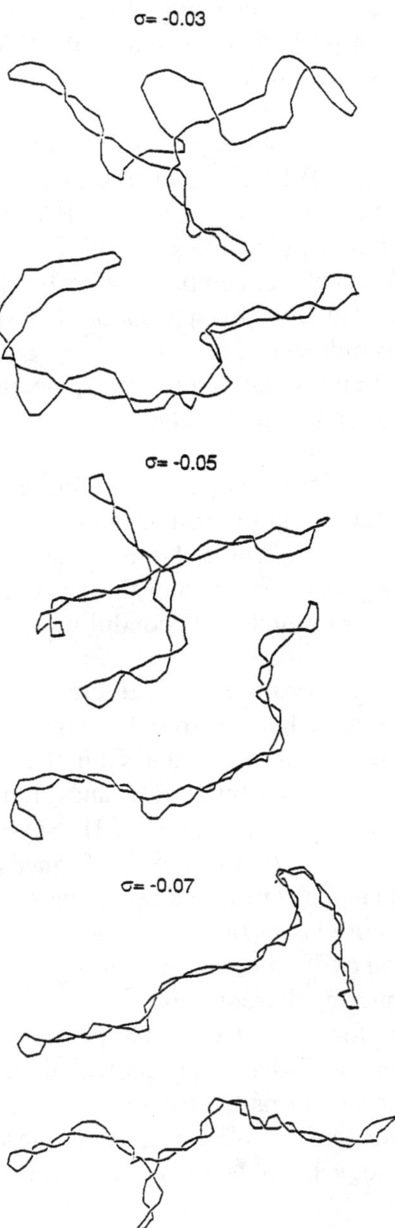

Figure 11. Results of computer simulations of supercoiled DNA molecules for different values of superhelical density $\sigma = \tau\gamma_0/N$. The data are from Klenin et al. (1991) [37].

1982 [79]; Frank-Kamenetskii and Vologodskii, 1984 [25]; Vologodskii and Frank-Kamenetskii, 1984 [80]; Anshelevich et al., 1988 [3]). These unusual structures are briefly described below.

4.4.1. Z-DNA. Negative supercoiling mostly favors formation of left-handed Z-DNA (see Section 2.4) because, in this case, the maximal release of superhelical stress per base pair adopting a non-B-DNA structure is achieved. As a result, although under physiological ambient conditions the Z form is energetically very unfavorable as compared with B-DNA, it is easily adopted in negatively supercoiled DNA by appropriate DNA sequences (with alternating purines and pyrimidines).

Linear DNA with the appropriate sequence adopts the Z conformation at a very high salt concentration (about 3M NaCl).

4.4.2. Cruciforms. Another structure readily formed under negative supercoiling is cruciform, which requires palindromic regions (see Fig. 2d). To form a cruciform, a palindromic region should be larger than a certain minimum. For example, six-base-pair-long palindromes recognized by restriction enzymes do not form cruciforms under any conditions.

4.4.3. H-DNA. H-DNA forms a special class of unusual structures, which are adopted under superhelical stress by sequences carrying purines (A and G) in one strand and pyrimidines (T and C) in the other, i.e. homopurine-homopyrimi sequences (reviewed by Mirkin and Frank-Kamenetskii, 1994 [52]; Frank-Kamenetskii and Mirkin, 1995 [23]; Soyfer and Potaman, 1996 [69]). The major element of H-DNA is a triplex formed by one half of the insert adopting theb H form and by one of the two strands of the second half of the insert (Fig. 2e).Two major classes of triplexes are known-pyrimidine-purine-pyrimidine (PyPuPy) and pyrimidine-purine-purine (PyPuPu). Fig.3 shows the canonical base-triads entering these triplexes.

Always two isomeric forms of H-DNA are possible, which are designated as H-y3, H-y5, H-r3 and H-r5, depending on which kind of triplex is formed and which half of the insert forms the triplex (see Fig. 2e). H-DNA may be considered as an intramolecular triplex (it is often referred to in this way). Its formation under physiological ambient conditions occurs only under superhelical stress.

The discovery of H-DNA (Lyamichev et al., 1985, 1986 [43, 44]; Mirkin et al., 1987 [51]; reviewed by Mirkin and Frank-Kamenetskii, 1994 [52]; Frank-Kamenetskii and Mirkin, 1995 [23]; Soyfer and Potaman, 1996 [69]) stimulated studies of intermolecular triplexes, which may be formed between homopurine-homopyrimidine regions of duplex DNA and corresponding pyrimidine or purine oligonucleotides (see Section 2.6).

5.   The DNA stability

Soon after the discovery of the double helix by Watson and Crick (1953) [89], the phenomenon of DNA melting was demonstrated experimentally. It was shown that when the DNA solution is heated, the complementary strands separate: instead of the regular double helix, two single-stranded DNA coils emerge (Marmur and Doty, 1962 [49]). This phenomenon is also called the helix-coil transition. The DNA melting may be monitored by various techniques. Two most popular methods are UV-spectrometry and microcalorimetry (reviewed by Breslauer et al., 1992 [6]). Instead of exhibiting a phase transition, DNA melts gradually, in a wide temperature range (Fig. 12b). DNAs from different organisms differ in their melting profiles.

5.1.   Helix-coil model

In attempts to understand the phenomenon of DNA melting, a simplified theoretical model was elaborated (see Fig. 5b), which treated DNA as a one-dimensional array of interacting spins. Each spin corresponded to a DNA base pair. Spin up corresponded to the helical state while spin down corresponded to the melted (open) state of the base pair. Two features made the problem much more difficult and much more interesting than the one-dimensional Ising model well known in the solid state physics. First, because open regions in DNA presented closed polymer chains, a long-range interaction between spins emerged. Secondly, because two base pairs in DNA (AT and GC) have different stability, DNA had to be modeled as a linear array of spins under the influence of disorder external magnetic field. Although irregular, the sequence is fixed so that the external field is quenched.

5.1.1. Theoretical development. In statistical-mechanical terms, the second feature of DNA helix-coil model (the linear memory due to the fixed sequence of DNA base pairs) means that one cannot average the partition function over different sequences of AT and GC pairs even if one assumes that the sequence itself is totally random. In reality, of course, the sequence is not random because it carries the genetic information. However, at early stages of treatment of the DNA melting phenomenon, long before the first real DNA sequences became available, the sequence was assumed to be random in theoretical studies. This made it possible to apply not only numerical but also analytical tools to treat the problem. Among others, important contributions of Vedenov et al. (1971) [74], Azbel (1972) [4] and Lifshitz (1973) [40] are worth mentioning.

As to the numerical solution, the challenge was to reduce the problem of direct computation of the partition function for a chain consisting of a very large number (N) of base pairs ("spins"), which required exponentially large computer time, to a procedure, which required polynomial time N^α with α as

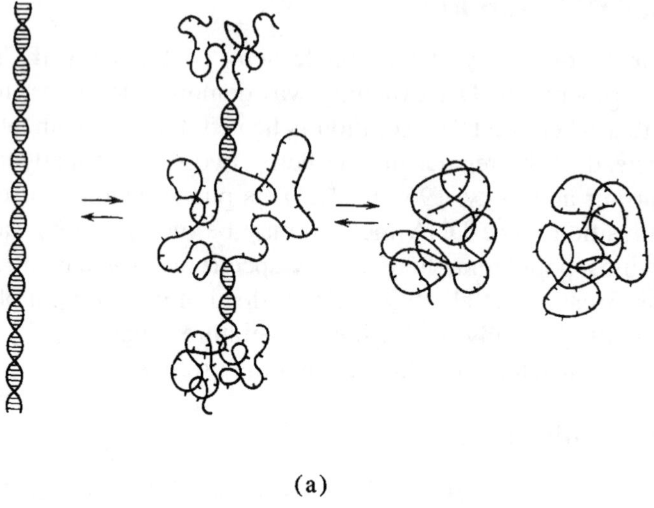

(a)

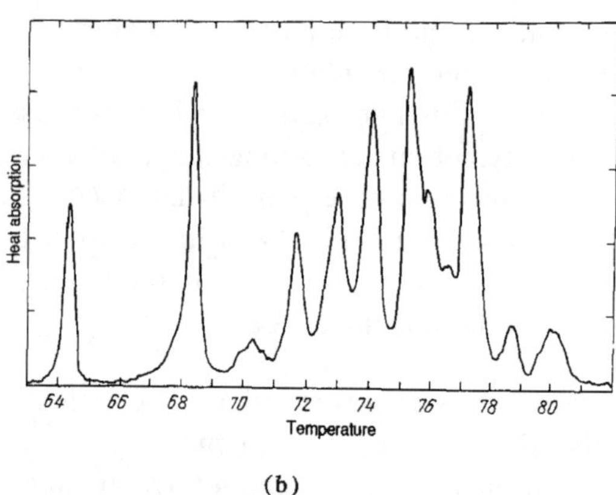

(b)

Figure 12.   Melting of DNA. (a) The helix-coil transition of a DNA molecule (intramolecular melting), (b) Typical DNA melting profile. This curve is also often called the differential melting curve. The curve was obtained for DNA which has the code name of ColEl and contains about 6500 nucleotide pairs.

small as possible. Several rigorous algorithms were proposed (Vedenov et al., 1967, 1971 [74, 73]; Poland, 1974 [55]).

However, an efficient way of solving the problem, which allowed for both the above features of the DNA helix-coil model, was not available until Fixman and Friere (1977) [18] proposed their algorithm. In so doing they heavily relied on the Poland (1974) [55] algorithm and some of our results (Frank-

Kamenetskii and Frank-Kamenetskii, 1969 [20]; Lukashin et al, 1976 [42]). Theoretical development of the helix-coil model has been extensively reviewed by Vedenov et al. (1971) [74], Wada et al. (1980) [85], Wada and Suyama (1986) [84], and Wartell and Benight (1985) [88].

It is worth mentioning that the helix-coil model without long-range interactions found applications far beyond the area of DNA biophysics. Among other applications, the model has been extensively used to study of helix-coil transition in polypeptides and most recently it was used by Selinger and Selinger (1996) [64] to explain experimental data on chiral order in random copolymers consisting of two enantiomers.

5.1.2. Comparison with experiment. When the very first full DNA sequence appeared in 1977 (of bacteriophage $\Phi X174$), DNA bio-physicists were well equipped to compare quantitatively experimental DNA melting profiles with theoretical predictions. It was first done by Lyubchenko et al. (1978) [46]. Essentially, it was the beginning of the end of the theme of DNA melting in DNA biophysics because theoretical prediction correlated with experiment sufficiently well. Even more direct comparison was done by Kalambet et al. (1985) [35] using electron-microscopy visualization of the melted regions in DNA with the known sequence on different stages of the melting process. Such comparisons and similar studies (reviewed by Wartell and Benight, 1985 [88]; Wada and Suyama, 1986 [84]) left no doubts that we correctly understood in quantitative terms major features of the phenomenon of DNA melting.

5.1.3. Heterogeneous stacking. A theme that dominated the field after the first demonstration of a success of the theory in achieving quantitative explanation of experimental data for DNAs with known sequences, was the so-called heterogeneous stacking. In the original helix-coil model, the external field could acquire only two values, corresponding to AT and GC pairs. This meant that interaction between all possible combinations of near neighbors along the DNA chain was assumed to be the same. Of possible 16 types of nearest neighbors, or stacks, only 10 are different because of the complementarity rule. It was quite natural to attribute some remaining differences between theory and experiment to the fact that these 10 parameters are different, i.e., to the effect of heterogeneous stacking. However, the very fact that the original model, which ignored the difference, worked well, indicated that the deviations from the mean interaction energy between adjacent base pairs were small as compared with the energy itself. In other words, these data indicated that the heterogeneous stacking was a small parameter.

In the first paper where heterogeneous stacking was allowed for, Gotoh and Tagashira (1981) [29] overlooked the fact that of 10 parameters of heterogeneous stacking only 8 of their combinations (invariants) actually determine the behavior of long DNA chains. When they adjusted all 10 parameters of heterogeneous stacking by comparing theory with experiment, a great confusion occurred because, unexpectedly, the effect of heterogeneous stacking proved to be very large. Vologodskii et al. (1984) [78] dispelled the confusion adjusting 8 invariants, not 10 parameters, by comparing theory with experiment. As a result, a reasonable set of relatively small parameters of heterogeneous stacking emerged (Vologodskii et al., 1984 [78]). Still, some discrepancy remained between the heterogeneous stacking parameters obtained from data on large DNA fragments (about 10^3 bp) and short duplexes (about 20 bp). This controversy has been resolved by SantaLucia (1998) [63] who arrived at a very solid "unified" set of DNA melting parameters.

5.2.   Slow relaxational processes

The remarkable success of statistical-mechanical theory in explaining the phenomenon of DNA melting overshadowed some significant limitation of the approach. For a long time experimental observations of hysteresis phenomena in DNA melting were largely ignored. However, when comparison of theory and experiment reached a high precision, kinetic effects in DNA melting could not be ignored any longer.

A comprehensive analysis of slow relaxation processes in DNA melting was performed by Anshelevich et al. (1984) [2]. The hysteresis phenomena in DNA melting are a direct consequence of the fact that very long regions are melted out cooperatively in the course of the process. The characteristic time of strands separation for a helical region consisting of m base pairs may be roughly estimated as (see Anshelevich et al., 1984 [2], for more accurate expressions):

$$\tau = (\tau_0/m)\, s^m \qquad (5.4)$$

where $\tau_0 \sim 10^6 s$ and s is the stability constant for a base pair. Although s is very close to unity within the melting range, because m is several hundreds the s^m value may be extremely large. Hence, very large τ values and a significant contribution of kinetic effects.

Subsequent thorough experimental studies completely confirmed all major theoretical predictions (Kozyavkin et al., 1984,1986 [38, 39]; Wada and Suyama, 1986 [84]).

5.3.   Stacking versus base pairing

Despite an impressive progress in detailed physical understanding of the DNA melting phenomenon, the central question on the nature of forces determining the stability of the double helix remained unanswered until very recently. Indeed, there are two radically different interactions within the double helix: stacking between adjacent base pairs and pairing between complementary bases. Melting experiments cannot answer the question of their respective contribution because the melting free energy includes sum of these two contributions. Therefore, independent determination of at least one of the two components, stacking or base paring, is required to separate the two contributions into the melting free energy. Such independent determination of the stacking parameters has recently been performed in my group (Protozanova et al. 2004 [56]).

It was found that a duplex DNA fragment carrying a single-strand break (nick) assumes two sates: closed conformation in which stacking between base pairs flanking the nick is preserved and open or kinked conformation, in which stacking is fully disrupted (see Fig. 13). The relative occupancy of these two states, i. e. $N_{\text{open}}/N_{\text{closed}}$, can be determined in gel electrophoresis mobility experiments. This ratio is shown to be governed by the stacking free energy parameter for the pairs flanking the nick:

$$N_{\text{open}}/N_{\text{closed}} = \exp(-\Delta G^{st}/RT). \qquad (5.5)$$

As a result, stacking parameters for all possible contacts have been determined (Protozanova et al. 2004 [56]). Comparing the obtained values of with melting parameters for various contacts, the conclusion was made that the DNA double helix is predominantly stabilized by stacking interactions. For AT pairs the base-pairing contribution is virtually absent while GC pairs stabilize the duplex state but their contribution is pretty small.

This conclusion is not surprising. Indeed, it has long been realized that

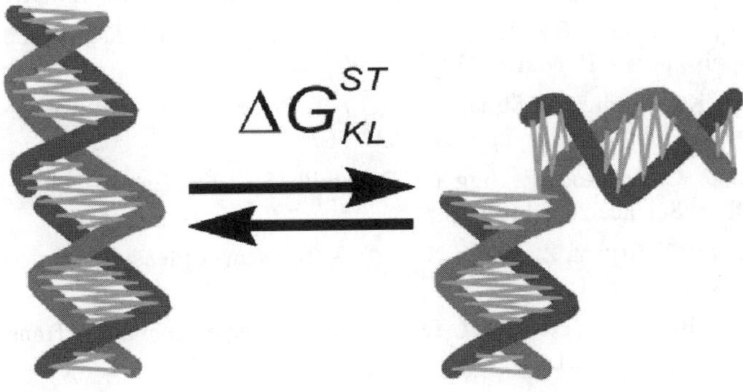

Figure 13.   Stacking-unstacking in the nick site of DNA carrying a single-strand break.

hydrogen bonding within complementary base pairs may virtually compensate at opening by hydrogen bonding between base pairs and water. Of course, if hydrogen bonds are not formed in the closed state, the close state becomes highly unfavorable. Therefore, complementary recognition is based not on favorability of complementary base pairs but rather on unfavorability of non-complementary (mismatched) pairs.

6.   Conclusion

DNA physics is a very well developed field, which forms a solid background for understanding the basic principles of DNA functioning in the cell and for numerous biotechnology applications of DNA. Indeed, such major tools of contemporary biotechnology as polymerase chain reaction (PCR), DNA chips and microarrays heavily rely on our knowledge and understanding of DNA physics. DNA physics also plays a decisive role in a rapid progress in design nanostructures and nanodevises based on DNA.

References

[1] Anshelevich, V.V., Vologodskii, A.V., Lukashin, A.V., Frank-Kamenetskii, M.D., 1979. Statistical-mechanical treatment of violations of the double helix in supercoiled DNA. Biopolymers 18: 2733-2744.

[2] Anshelevich, V.V., Vologodskii, A.V., Lukashin, A.V., Frank-Kamenetskii, M.D., 1984. Slow relaxational processes in the melting of linear biopolymers. A theory and its application to nucleic acids. Biopolymers 23: 39-58.

[3] Anshelevich, V.V., Vologodskii, A.V., Frank-Kamenetskii, M.D., 1988. A theoretical study of formation of DNA noncanonical structures under negative superhelical stress. J. Biomol. Struct. Dyn. 6: 247-259.

[4] Azbel', M.Y., 1972. The inverse problem for DNA. JETP Lett. 16: 128-131.

[5] Berger, J.M., Gamblin, S.J., Harrison, S.C., Wang, J.C., 1996. Structure and mechanism of DNA topoisomerase II. Nature 379: 225-232.

[6] Breslauer, K.J., Freire, E., Straume, M., 1992. Calorimetry: a tool for DNA and ligand-DNA studies. Methods Enzymol. 211: 533-567.

[7] Bustamante, C., Marko, J.F., Siggia, E.D., Smith, S., 1994. Entropic elasticity of lambda-phage DNA. Science 265: 1599-1600.

[8] Bustamante, C., Bryant, Z., Smith, S.B. 2003. Ten years of tension: single-molecule DNA mechanics. Nature 421: 423-427.

[9] Chan, S.R., Blackburn, E.H. 2004. Telomeres and telomerase. Philos. Trans. R. Soc. Lond. B Biol Sci. 359: 109-121.

[10] Chen, L., Cai, L., Zhang, X., Rich, A., 1994. Crystal structure of a four-stranded intercalated DNA: d(C4). Biochemistry 33: 13540-13546.

[11] Depew, R.E., Wang, J.C., 1975. Conformational fluctuations of DNA helix. Proc. Natl. Acad. Sci. USA, 72: 4275-4279.

[12] Deruchi, T., Tsurusaki, K., 1993a. A new algorithm for numerical calculation of link invariants. Phys. Lett. A 174: 29-37.

[13] Deruchi, T., Tsurusaki, K., 1993b. Topology of closed random polygons. J. Phys. Soc. Japan 62: 1411-1414.

[14] Deruchi, T., Tsurusaki, K., 1994. A statistical study of random knotting using the Vassiliev invariants. J. Knot Theory and Its Ramifications 3: 321-353.

[15] Dickerson, R.E., 1992. DNA structure from A to Z. Methods Enzymol. 211: 67-111.

[16] Du, Q., M. Vologodskaia, H. Kuhn, M. Frank-Kamenetskii and A. Vologodskii, 2005. Gapped DNA and cyclization of short DNA fragments, Biophys. J. 88: 4137-4145.

[17] Epel, E.S., Blackburn, E.H., Lin, J., Dhabhar, F.S., Adler, NE, Morrow, J.D., Cawthon, R.M. 2004. Accelerated telomere shortening in response to life stress. Proc. Natl. Acad. Sci. USA 101: 17312-17315.

[18] Fixman, M., Freire, J.J., 1977. Theory of DNA melting curves. Biopolymers 16: 2693-2704.

[19] Frank-Kamenetskii, M.D., 1997. *Unraveling DNA. The Most Important Molecule of Life.* Addison Wesley, Reading, MA.

[20] Frank-Kamenetskii, M.D., Frank-Kamenetskii, A.D., 1969. Theory of helix-coil transition for the case of double stranded DNA. Molek. Biol. 3: 375-382.

[21] Frank-Kamenetskii, M.D., Lukashin, A.V., Vologodskii, A.V., 1975. Statistical mechanics and topology of polymer chains. Nature 258: 398-399.

[22] Frank-Kamenetskii, M.D., Lukashin, A.V., Anshelevich, V.V., Vologodskii, A.V., 1985. Topsional and bending rigidity of the double helix from data on small DNA rings. J. Biomol. Struct. Dyn. 2: 1005-1012.

[23] Frank-Kamenetskii, M.D., Mirkin, S.M., 1995. Triplex DNA structures. Ann. Rev. Biochem. 64: 65-95.

[24] Frank-Kamenetskii, M.D., Vologodskii, A.V., 1981. Topological aspects of polymer physics: theory and its biophysical applications. Sov. Phys. Usp. 24: 679-696.

[25] Frank-Kamenetskii, M.D., Vologodskii, A.V., 1984. Thermodynamics of the B-Z transition in superhelical DNA. Nature 307: 481-482.

[26] Frisch, H.L., 1993. Macromolecular topology. Metastable isomers from pseudo interpenetrating polymer networks. New J. Chem. 17: 697-701.

[27] Fuller, F.B., 1971. The writhing number of a space curve. Proc. Natl. Acad. Sci. USA 68: 815-819.

[28] Gehring, K., Leroy, J.L.. Gueron, M., 1993. A tetrameric DNA structure with protonated cytosine-cytosine base pairs. Nature 363: 561-565.

[29] Gotoh, O., Tagashira, Y., 1981. Stabilities of nearest-neighbor doublets in double-helical DNA determined by fitting calculated melting profiles to observed profiles. Biopolymers 20: 1033-1042.

[30] Grosberg, A.I., Khokhlov, A.R., 1994. *Statistical Physics of Macromolecules*. AIP Press, New York.

[31] Hagerman, P.J., 1988. Flexibility of DNA, Ann. Rev. Biophys. Biophys. Chem. 17: 265-286.

[32] Harris, S.A., Sands, Z.A., and Laughton, C.A., 2005. Molecular dynamics Simulations of duplex stretching reveal the importance of entropy in determining the biomechanical properties of DNA. Biophys. J. 88: 1684-1691.

[33] Horowitz, D.S., Wang, J.C., 1984. Torsional rigidity of DNA and length dependence of the free energy of DNA supercoiling. J. Mol. Biol. 173: 75-91.

[34] Ivanov, V.I., Krylov, D.Y., 1992. A-DNA in solution as studied by diverse approaches. Methods Enzymol. 211: 111-127.

[35] Kalambet, Y.A., Borovik, A.S., Lyamichev, V.I., Lyubchenko, Y.L., 1985. Electron microscopy of the melting of sequenced DNA. Biopolymers 24: 359-377.

[36] Klenin, K.V., Vologodskii, A.V., Anshelevich, V.V., Dykhne, A.M., Frank-Kamenetskii, M.D., 1988. Effect of excluded volume on topological properties of circular DNA. J. Biomol. Struct. Dyn. 5: 1173-1185.

[37] Klenin, K.V., Vologodskii, A.V., Anshelevich, V.V., Dykhne, A.M., Frank-Kamenetskii, M.D., 1991. Computer stimulation of DNA supercoiling. J. Mol. Biol. 217: 413-419.

[38] Kozyavkin, S.A., Lyubchenko, Y.L., 1984. The nonequilibrium character of DNA melting: effects of heating rate on the fine structure of melting curves. Nucl. Acids Res. 12: 4339-4349.

[39] Kozyavkin, S.A.. Naritsin, D.B., Lyubchenko, Y.L., 1986. The kinetics of DNA helix-coil subtransitions. J. Biomol. Struct. Dyn. 3: 689-704.

[40] Lifshitz, I.M., 1973. On the statistical thermodynamics of fusion of long heteropolymer chains. Sov. Phys.-JETP 65: 1100-1110.

[41] Liu, L., Depew, R.E., Wang, J.C., 1976. Knotted single-stranded DNA rings: a novel topological isomer of circular single-stranded DNA formed by treatment with Escherichia coli w protein. J. Mol. Biol. 106: 439-452.

[42] Lukashin, A.V., Vologodskii, A.V., Frank-Kamenetskii, M.D., 1976. Comparison of different theoretical descriptions of helix-coil transition in DNA. Biopolymers 15: 1841-1844.

[43] Lyamichev, V.I., Mirkin, S.M., Frank-Kamenetskii, M.D., 1985. A pH-dependent structural transition in the homopurine-homopyrimidine tract in superhelical DNA. J. Biomol. Struct. Dyn. 3: 327-338.

[44] Lyamichev, V.I., Mirkin, S.M., Frank-Kamenetskii, M.D., 1986. Structures of homopurine-homopyrimidine tract in superhelical DNA. J. Biomol. Struct. Dyn. 3: 667-669.

[45] Lyamichev, V.I., Mirkin, S.M., Frank-Kamenetskii, M.D., Cantor, C.R., 1988. A stable complex between homopyrimidine oligomers and the homologous regions of duplex DNA. Nucl. Acids Res. 16: 2165-2178.

[46] Lyubchenko, Y.L., Vologodskii, A.V., Frank-Kamenetskii, M.D., 1978. Direct comparison of theoretical and experimental melting profiles for ϕX174 DNA. Nature 271: 28-31.

[47] Marko, J.F., Siggia, E.D., 1994. Fluctuations and supercoiling of DNA. Science 265: 506-508.

[48] Marko, J.F., Siggia, E.D., 1995. Statistical mechanics of supercoiled DNA. Phys. Rev. E 52: 2912-2938.

[49] Marmur, J., Doty, P., 1962. Determination of the base composition of deoxyribonucleic acid from its thermal denaturation temperature. J. Mol. Biol. 5: 109-118.

[50] Matzke, M.A., Birchler, J.A. 2005. RNAi-mediated pathways in the nucleus. Nat Rev Genet. 6: 24-35.

[51] Mirkin, S.M., Lyamichev, V.I., Drushlyak, K.N., Dobrynin, V.N., Filippov, S.A., Frank-Kamenetskii, M.D., 1987. DNA H form requires a homopurine-homopyrimidine mirror repeat. Nature 330: 495-497.

[52] Mirkin, S.M., Frank-Kamenetskii, M.D., 1994. H-DNA and related structures. Ann. Rev. Biophys. Biomol. Struct. 23: 541-576.

[53] Moser, H.E., Dervan, P.B., 1987. Sequence-specific cleavage of double helical DNA by triple helix formation. Science 238: 645-650.

[54] Novina, C.D., Sharp, P.A. 2004. The RNAi revolution. Nature 430: 161-164.

[55] Poland, D., 1974. Recursion relation generation of probability profiles for specific-sequence macromolecules with long-range correlations. Biopolymers 13: 1859-1871.

[56] Protozanova E., Yakovchuk, P., Frank-Kamenetskii, M.D. 2004. Stacked-unstacked equilibrium at the Nick site of DNA. J.Mol.Boil. 342: 775-785.

[57] Pulleyblank, D.E., Shure, D.E., Tang, D., Vinograd, J., Vosberg, H.-P. 1975. Action of nicking-closing enzyme on supercoiled and nonsupercoiled closed circular DNA: formation of a Boltzmann distribution of topological isomers. Proc. Natl. Acad. Sci. USA 72: 4280-4284.

[58] Rippe, K., Jovin, T.M., 1992. Parallel-stranded Duplex DNA. Methods Enzymol. 211: 199-220.

[59] Rouzina, I., Bloomfield, V.A. 2001a. Force-induced melting of the DNA double helix 1. Thermodynamic analysis. Biophys J. 80: 882-293.

[60] Rouzina, I., Bloomfield, V.A. 2001b. Force-induced melting of the DNA double helix 2. Effect of solution conditions. Biophys J. 80: 894-900.

[61] Rybenkov, V.V., Cozzarelli, N.R., Vologodskii, A.V., 1993. Probability of DNA knotting and the effective diameter of the DNA double helix. Proc. Natl. Acad. Sci. USA 90: 5307-5311.

[62] Rybenkov, V.V., Ullsperger C, Vologodskii, A.V., Cozzarelli, N.R. 1997. Simplication of DNA topology below equilibrium values by type II topoisomerases. Science 277: 690-693.

[63] SantaLucia, J. 1998. A unified view of polymer, dumbbell, and oligonucleotide DNA nearest-neighbor thermodynamics. Proc. Natl. Acad. Sci. USA. 95: 1460-1465.

[64] Selinger, J.V., Selinger, R.L.B., 1996. Theory of chiral order in random copolymers. Phys. Rev. Lett. 76: 58-61.

[65] Shaw, S.Y., Wang, J.C., 1993. Knotting of a DNA chain during ring closure, Science 260: 533-536.

[66] Sinden, R.R., 1994. *DNA Structure and Function*. Academic Press, San Diego.

[67] Smith, S.B., Finzi, L., Bustamante, C., 1992. Direct mechanical measurements of the elasticity of single DNA molecules by using magnetic beads. Science 258: 1122-1126.

[68] Smith, S.B., Cui, Y., Bustamante, C., 1996. Overstretching B-DNA: the elastic response of individual double-stranded and single-stranded DNA molecules. Science 271: 795-789.

[69] Soyfer, V.N., Potaman, V.V., 1996. *Triple-helical Nucleic Acids*. Springer, New York.

[70] Stigter, D., 1977. Interactions of highly charged colloidal cylinders with applications to double-stranded DNA. Biopolymers 16: 1435-1448.

[71] Strick, T.R., Allemand, J.F., Bensimon, D., Bensimon, A., Croquette, V., 1996. The elasticity of a single supercoiled DNA molecule. Science 271: 1835-1837.

[72] Taylor, W.H., Hagerman, P.J., 1990. Application of the method of phage T4 DNA ligase-catalyzed ring-closure to the study of DNA structure. II. NaCl-dependence of DNA flexibility and helical repeat. J. Mol. Biol. 212: 363-376.

[73] Vedenov, A.A., Dykhne, A.M., Frank-Kamenetskii, A.D., Frank-Kamenetskii, M.D., 1967. A contribution to the theory of helix-coil transition in DNA. Molek. Biol. 1: 313-319.

[74] Vedenov, A.A., Dykhne, A.M., Frank-Kamenetskii, M.D., 1971. The helix-coil transition in DNA. Sov. Phys. Usp. 14: 715-736.

[75] Vologodskaia, M. and A. Vologodskii. 2002. Contribution of the intrinsic curvature to measured DNA persistence length. J. Mol. Biol. 317: 205-213.

[76] Vologodskii, A.V., Lukashin, A.V., Frank-Kamenetskii, M.D., Anshelevich, V.V., 1974. The knot problem in statistical mechanics of polymer chains. Sov. Phys. JETP 39: 1059-1063.

[77] Vologodskii, A.V., Lukashin, A.V., Anshelevich, V.V., Frank-Kamenetskii, M.D., 1979b. Fluctuations in superhelical DNA. Nucleic Acids Res. 6: 967-982.

[78] Vologodskii, A.V., Amirikyan, B.R., Lyubchenko, Y.L., Frank-Kamenetskii, M.D., 1984. Allowance for heterogeneous stacking in the DNA helix-coil transition theory. J. Biomol. Struct. Dyn. 2: 131-148.

[79] Vologodskii, A.V., Frank-Kamenetskii, M.D., 1982. Theoretical study of cruciform states in superhelical DNA. FEES Lett. 143: 257-260.

[80] Vologodskii, A.V., Frank-Kamenetskii, 1984. Left-handed Z form in superhelical DNA: a theoretical study. J. Biomol. Struct. Dyn. 1: 1325-1333.

[81] Vologodskii, A.V., Frank-Kamenetskii, M.D., 1992. Modeling DNA supercoiling. Methods Enzymol. 211: 467-480.

[82] Vologodskii, A.V., Levene, S.D., Klenin, K.V., Frank-Kamenetskii, M.D., Cozzarelli, N.R., 1992. Conformational and thermodynamic properties of supercoiled DNA. J. Mol. Biol. 227: 1224-1243.

[83] Vologodskii, A.V., Zhang, W., Rybenkov, V.V., Podtelezhnikov, A.A., Subramanian, D., Griffith, J.D., Cozzarelli, N.R. 2001. Mechanism of topology simplification by type II DNA topoisomerases. Proc. Natl. Acad. Sci. USA. 98: 3045-3049.

[84] Wada, A., Suyama, A., 1986. Local stability of DNA and RNA secondary structure and its relation to biological functions. Prog. Biophys. Mol. Biol. 47: 113-157.

[85] Wada, A., Yabuki, S., Husimi, Y., 1980. Fine structure in the thermal denaturation of DNA: high temperature-resolution spectrophotometric studies. CRC Crit. Rev. Biochem. 9: 87-144.

[86] Wang, J.C., 1996. DNA topoisomerase. Annu. Rev. Biochem. 65: 635-692.

[87] Wang, A.H.-J., Quigley, G.J., Kolpak, F.K., Crawford, J.L., van Boom, J.H., van der Marel, G., Rich, A., 1979. Molecular structure of a left-handed double helical DNA fragment at atomic resolution. Nature 282: 680-685.

[88] Wartell, R.M., Benight, A.S., 1985. Thermal denaturation of DNA molecules: a comparison of theory with experiment. Phys. Rep. 126: 67-107.

[89] Watson, J.D., Crick, F.H.C. 1953. Molecular structure of nucleic acids. Nature 171: 737-738.

[90] White, J.H., 1969. Self-linking and the Gauss integral in higher dimensions. Am. J. Math. 91: 693-728.

[91] Williams, M.C., Rouzina, I. 2002. Force spectroscopy of single DNA and RNA molecules. Curr. Opin. Struct. Biol. 12: 330-336.

[92] Williams, M.C., Rouzina, I., Bloomfield, V.A. 2002. Thermodynamics of DNA interactions from single molecule stretching experiments. Acc. Chem. Res. 35: 159-166.

Topic Index